Nediljko Budisa

Engineering the Genetic Code

Related Titles

K. H. Nierhaus, D. N. Wilson (eds.)

Protein Synthesis and Ribosome Structure

Translating the Genome

597 pages with approx. 150 figures and approx. 40 tables
2004
Hardcover
ISBN 3-527-30638-2

S. Brakmann, A. Schwienhorst (eds.)

Evolutionary Methods in Biotechnology

Clever Tricks for Directed Evolution

227 pages with 115 figures and 11 tables
2004
Hardcover
ISBN 3-527-30799-0

N. Sewald, H.-D. Jakubke

Peptides: Chemistry and Biology

597 pages with 163 figures and 33 tables
2002
Hardcover
ISBN 3-527-30405-3

S. Brakmann, K. Johnsson (eds.)

Directed Molecular Evolution of Proteins

or How to Improve Enzymes for Biocatalysis

368 pages with 121 figures and 15 tables
2002
Hardcover
ISBN 3-527-30423-1

G. Walsh

Proteins

Biochemistry and Biotechnology

560 pages
2002
Softcover
ISBN 0-471-89907-0

Nediljko Budisa

Engineering the Genetic Code

Expanding the Amino Acid Repertoire for the Design of Novel Proteins

WILEY-VCH Verlag GmbH & Co. KGaA

Nediljko Budisa
Max-Planck-Institute of Biochemistry
Division of Bioorganic Chemistry
Am Klopferspitz 18A
82152 Martinsried
Germany

Library of Congress Card No.:
applied for
British Library Cataloguing-in-Publication Data
A catalogue record for this book is available from the British Library.

Bibliographic information published by Die Deutsche Bibliothek
Die Deutsche Bibliothek lists this publication in the Deutsche Nationalbibliografie; detailed bibliographic data is available in the Internet at ⟨http://dnb.ddb.de⟩.

Printed in the Federal Republic of Germany
Printed on acid-free paper

Typesetting Asco Typesetters, Hong Kong
Printing betz-druck GmbH, Darmstadt
Binding J. Schäffer GmbH i. G., Grünstadt

ISBN-13: 978-3-527-31243-6
ISBN-10: 3-527-31243-9

This book is dedicated to my scientific parents:

Prof. Robert Huber

Prof. Luis Moroder

Prof. Dieter Oesterhelt

Engineering the Genetic Code. Nediljko Budisa

ISBN: 3-527-31243-9

Foreword

„Vielleicht ist die physikalisch und chemisch orientierte Biologie im Augenblick die zukunftsreichste aller Realwissenschaften."

Carl Friedrich Freiherr von Weizsäcker
From the lecture: „Wie wird die Wissenschaft die Welt verändern?", Munich 1969.

The emerging field of chemical biology is based on blending chemistry with molecular biology, a process that is largely stimulated by the most recent advances in genetic code engineering. While common methods of protein engineering and site-directed mutagenesis allow to identify contributions of particular amino acids to the structure and/or function of proteins via permutation of the twenty canonical amino acids, expansion of the genetic code offers the perspectives of enriching this alphabet of building blocks with noncanonical α-amino acids of unique biophysical and chemical properties. In fact, chemistry should allow the access to a nearly infinite array of diverse structural elements for the design of novel proteins or for incorporation of new chemical functionalities that can be exploited at will for bioorthogonal posttranslational protein tagging and transformations to study and to enrich the chemistry of living cells. At the present stage of this forefront research a comprehensive treatise on the state of the art was indeed required, and the book *Engineering of the Genetic Code*, authored by our previous student and now colleague Dr. Nediljko Budisa who decisively contributed to advances in the field, elegantly fulfils this need. Since our own reseach addresses structure and function of biomolecules, we are fascinated by the new perspectives offered and we witness the fast development of this challenging field of research. We trust that the critical treatment and novel concepts described in Budisa's book *Engineering of the Genetic Code* will provide stimulus and guidance for scientists, researchers, teachers and students to learn about and apply this fascinating research area and its innovative tools for their questions at the interface of chemistry and biology.

Robert Huber
Luis Moroder
Dieter Oesterhelt

Engineering the Genetic Code. Nediljko Budisa

ISBN: 3-527-31243-9

Contents

Engineering the Genetic Code. Nediljko Budisa

ISBN: 3-527-31243-9

Preface

This book began as a review article in Angewandte Chemie ("Prolegomena to future experimental efforts on genetic code engineering by expanding its amino acid repertoire") based on my Habilitation Thesis at Technical University in Munich. The idea to write a book on code engineering crystallised in my mind much earlier. Finally, it was possible to transform my ideas into a book after intensive and constructive discussions with Dr. Peter Gölitz and Dr. Frank Weinreich and their strong encouragement and support.

The *Engineering of the genetic code* is not intended to be traditional text for undergraduates since code engineering unifies many research fields, each of them now covered with at least several good introductory books and monographs. The main aim of this book is to provide a balanced treatment of the major areas of organic and biological chemistry as well as molecular and theoretical biology (Chapters 1–4) important to understand the basics, principles and potentials of code engineering (Chapters 5–7). Medical researchers and biologists, such as those engaged in molecular biology, do need to know much more organic chemistry than is presented especially in first and last chapters. On the other hand, chemists and physicists need to know much more molecular biology than found in this book. Philosophers, theologians and social scientists interested in possible general impacts of code engineering on society are required to be familiar at least with basics of chemistry, biology and physics. Therefore, I presume they all will consult other sources where these topics are covered more extensively.

The majority of the book chapters (Chapters 1–4) have only a limited number of references and many of them are reviews and book chapters. Although this approach might offend some, it should be kept in mind that it is almost impossible to give proper priority among a vast number of primary sources in well established fields such as chemical modifications (Chapter 1), protein translation mechanisms (Chapter 3) and theoretical biology dealing with genetic code (Chapter 4). The same principle applies for the historical background of the field (Chapter 2). In the last three chapters and especially in the central one, that is Chapter 5, I tried my best to provide almost all references that appeared since the birth of this young research field (i.e. in the last fifteen years). For a last chapter dealing with applicative potentials of code engineering (Chapter 7), I may well imagine that a major criticism from some of my colleagues working in the field might be that I have

Engineering the Genetic Code. Nediljko Budisa

ISBN: 3-527-31243-9

left out some of their pet topics. I have tried very hard to describe everything that prospective code engineers, biochemists, biophysicists and life scientists to my opinion really need to know. But I might be wrong. Like many other book authors, I am well aware that job of book-writing makes one prone to commit (mostly unintentionally) many sins, sins of commission and sins of omission. I have taken particular care to ensure that the text is free of errors. This is difficult in a rapidly changing field, such as code engineering. Therefore I am eager to receive all suggestions for improvement of any kind and promise to take them seriously. Of course, all mistakes throughout the text should be assigned to me.

This book is also a result of my personal scientific development and I consider it as a work in progress. It was possible to write it only because I had luck to be at right place at right time and surrounded with right people. By right time I mean a lucky event that beginning of my Ph D work coincided with the birth of code engineering. By right place I mean those places where a precious institution of freedom is cultivated, i.e. where heads and leaders allow sometimes members of their groups to step beyond the strict boundaries of their charge. Such heads and leaders are Prof. Robert Huber, Prof. Luis Moroder and Prof. Dieter Oesterhelt to whom I would like to express my deep gratitude and thank. They generously provided me not only with an excellent infrastructure and technical support, but also outstanding help, support and advices, while their unyielding faith in me and willingness to give me freedom in pursuing my work allowed me to follow my visions. My technical assistants Traudl Wenger, Petra Birle and Tatjana Krywcun are also among the most important people at Max-Planck Institute. With their positive attitudes they provided friendly environment and created the circumstances that made works on writing of this book much easier for me. I apologise to my students for the fact that I had less time as usual for them during book writing and thank them for their patience, understanding and support. I am especially in debt to my student Prajna Paramita Pal who worked hard to ease my "Croatian English" and colleague Dr. Markus Seifert for his generous help in preparation of some figures in this book. And the last, but not the least expression of heartfelt gratitude goes to my beloved wife Monika Franke and my children Lukas and Andreas for their patience, tolerance and strong support. Finally, I would like to thank to Dr. Frank Weinreich and his co-workers from Wiley-VCH for their assistance, patience and support during the preparation of the manuscript and for the preparation of the finished book.

Martinsried, September 2005 *Nediljko Budisa*

1
Introduction

The important thing in science is not so much to obtain new facts as to discover new ways of thinking about them.

[W. L. Bragg]

1.1
Classical Approaches to Protein Modification

Chemical modification of solvent-accessible reactive side-chains has a long history in protein science and a number of group-specific modifying agents are well known. The N-terminal α-amino groups in protein sequences and ε-amino groups of lysine side-chains are common targets of chemical modifications. Chemical conjugations still play an important and general role in protein and cellular chemistry along with biotechnology. For example, human Annexin-V modified by conjugation of its lysine side-chains with the fluorescent dye fluoroisothiocyanate is commonly used as a marker to study cell apoptosis [1]. The modification of proteins and peptides with polyethylene glycol was, and continues to be, one of the most frequent chemical modifications used in the manufacture of biopharmaceuticals [2]. Moreover, a widely used protein modification method called "crystal soaking", where a protein crystal is "soaked" in a solution with certain heavy metals, is an essential tool for protein X-ray crystallography ("structural genomics") [3]. Chemical modifications were also extensively used for the introduction of different labels (fluorescence markers, spin labels, etc.) and crosslinking with various photolinkers, fluorophores and cages in order to study protein topology or protein–protein interactions [4].

In the era of classical enzymology, chemical modifications (acylations, amidations, reductive alkylations, cage reagents, etc.) of functional groups (lysine, cysteine, tyrosine, histidine, methionine, arginine and tryptophan side-chains) represented the only available approach for studies of structure–function relationships. At that time, a typical application of these approaches was to identify the residue involved in catalysis or binding (e.g. carboxyl groups of aspartic or glutamic acids, imidazole groups of histidine side-chains, hydroxy groups of tyrosine, serine and

Engineering the Genetic Code. Nediljko Budisa

ISBN: 3-527-31243-9

threonine, etc.) by a substance with specific modifying capacity. Cysteine side-chains were always an attractive target for site-specific modification in proteins. This is largely driven by the relative ease of specific modification of cysteine in proteins without concomitant modification of other nucleophilic sites such as lysine and histidine. As a result, a large number of reagents are available for the modification of cysteine. For example, Bender and Koschland used this approach to chemically modify ("chemical mutation") the active site of serine to cysteine in subtilisin, which resulted in the loss of the peptidase activity [5, 6]. These early protein structure–function studies also resulted in the mapping of the receptor-binding regions in insulin. The chemical modification of the N-terminal glycine of chain A provided a model to help understand the prerequisites of productive binding. However, it was not possible to resolve the influence of the steric bulk imposed upon modification from the effects of the positive charge neutralization at the N-terminal amino group [7].

The lack of specificity and unpredictability of the reaction accompanied by a distinct reduction or even loss of activity of the protein under study are, among others, the major reasons why protein modifications are no longer popular today. Site-directed mutagenesis offers a much more elegant and precise tool to address such issues. For example, in the absence of a precise three-dimensional structure, alanine (or glycine)-scanning mutagenesis [8] is suitable for systematic substitutions of all residues in (or around) the putative active center of the protein under study and the subsequent identification of the "essential" functional group.

1.2 Peptide Synthesis, Semisynthesis and Chemistry of Total Protein Synthesis

The "peptide theory" put forward in 1902 by Emil Fisher and Franz Hofmeister [9] correctly postulated that proteins are made up of α-amino acids that are linked head-to-tail by amide bonds. At that time chemists were mainly interested in the total chemical synthesis of protein by using the techniques of classical organic chemistry. Emil Fisher succeeded in synthesizing an 18-residue peptide composed of glycine and L-leucine in which amino acids were combined to yield small peptides that could be coupled together to produce larger peptides [10]. The amino acids could then be linked via the peptide (amide) in a stepwise manner [11]. Alternatively, in convergent fragment condensation, small peptides can be coupled together to give a larger peptide. Fragment synthesis of insulin, the first protein molecule to have been sequenced, was accomplished by three groups working independently at approximately the same time. The most recent achievement of solution synthesis was the preparation of a linear 238-residue protein, a precursor of green fluorescent protein (GFP) from *Aequorea victoria* [12]. Convergent condensation in combination with peptide bond synthesis in a stepwise manner was successfully applied in the solid phase, resulting in the first total synthesis of the enzyme ribonuclease S [13, 14]. In this method, the C-terminal residue of the peptide which is covalently anchored to an insoluble support is used to assemble the remaining amino acids in a stepwise manner and, finally, to cleave the synthesized product from the solid phase. These initial results have been followed by a series of successful syntheses of a variety of enzymes.

These methods require a high degree of chemical sophistication, advanced synthetic methods and skillful experimentation. These are some of the chief reasons why they are still not widespread among the community of synthetic chemists. These approaches have their own drawbacks, such as the poor solubility of the protected peptide intermediates formed by solution synthesis along with the accumulation of byproducts that block reactions in a stepwise solid-phase synthesis. For all these reasons, the total chemical synthesis of a homogeneous protein longer than 100 residues still presents a formidable challenge.

The strategy of convergent assembly (i.e. condensation) of synthetic and natural peptide fragments is termed "protein semisynthesis". The basic requirements for semisynthesis are: (i) the synthetic donor peptide has to be protected and activated, and (ii) an acceptor protein fragment that has to be prepared by enzymatic or chemical fragmentation of the parent protein should be available and properly protected. Offord and Rose pioneered the use of hydrazone- and oxime-forming reactions for chemically ligating such fragments [15]. Although these chemistries are selective, they were in practice often hampered by the insolubility of the large protected peptide building blocks. Most recently, the rediscovery of the Staudinger ligation has represented an additional breakthrough in this field (see Fig. 1.1) [16]. From a purely chemical perspective, it is an excellent tool for protein/peptide ligation that allows different protein/peptide fragments to be coupled at any desired position, without the requirement for a particular sequence composition [17].

In order to chemically create a native amide bond between interacting fragments, Kent and coworkers successfully performed a thiol–thioester exchange between unprotected fragments in aqueous solution – a technique they called "native chemical ligation" (NCL) [18]. The N-terminal fragment contains a C-terminal electrophilic α-thioester which can be conjugated via a thiol–thioester exchange reaction to the N-terminal thiol-harboring fragment. Such developments are based on the early discovery by Wieland and co-workers [19] that thioesters of amino acids and peptides form peptide bonds with N-terminal cysteine in neutral aqueous solution via a spontaneous $S \rightarrow N$-acyl shift, along with the procedures of Blake [20] and Yamashiro [21] for thioester preparation by solid-phase synthesis. See Fig. 1.1.

Some of the most striking examples that demonstrate the unique capacity of NCL are total protein synthesis of the D-chiral form of HIV-1 protease (100 residues) [28] and the preparation of the posttranslationally modified artificial variant of erythropoietin [29] (polymer modified; 166 residues). The D-enantiomer of HIV-1 protease is completely active and exhibits reciprocal chiral specificity just as much as the L-enantiomer, i.e. it is capable of cutting only the D-peptide substrates. Its three-dimensional structure is in all respects the mirror image of the "natural" L-protease. Indeed, these examples dramatically illustrate the considerable potential of NCL as a complementary approach for protein engineering based on templated ribosome-mediated protein synthesis.

Since peptide synthesis is generally limited by the size of polypeptide-chain (around 10 kDa) and genetically encoded protein modification (*vide infra*) still suf-

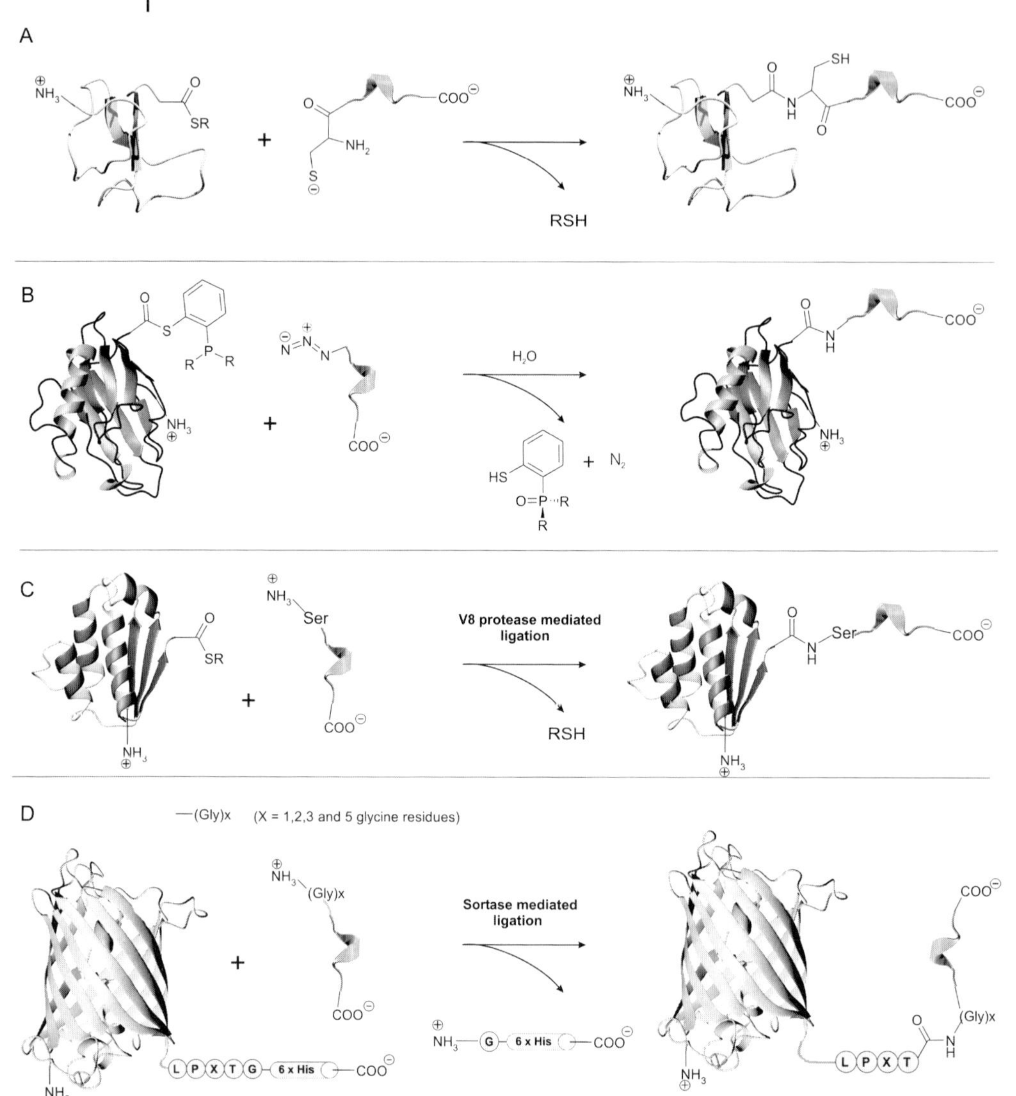

Fig. 1.1. Basic approaches for chemoselective ligations [22, 23]. (A) NCL between unprotected peptide fragments that contain the requisite reactive groups (N-terminal Cys and C-terminal thioester). (B) Staudinger ligation forms an amide bond from an azide and a specifically fictionalized phosphine. This method allows independent amino acid sequences to be coupled at any desired Xxx–Yyy bond and, in this respect, is potentially almost universal [24]. (C) EEL [25] uses specific thioesters (generated by "classical" intein-mediated approaches) as a substrate mimetic for a *Staphylococcus aureus* V8 serine protease in order to bypass the requirement for the cysteine at the ligation site. (D) Sortase from *S. aureus*, a membrane-anchored transpeptidase, cleaves any polypeptide provided with a C-terminal sorting signal (LPXTG motif). This strategy when applied to tagged green fluorescent protein (Nt-GFP-LPXTG-6His-Ct) enables its successful conjugation with various donor molecules containing two or more N-terminal glycines [26] (D- and L-peptides, nonpeptide fragments and even other GFPs). (From Budisa [27]; © Wiley 2004.)

fers from limited scope, NCL has no current rival in the various types of experimental designs that include sequential isotopic labeling, preparation of circular proteins, insertion of the non-native polypeptide fragments or nonpeptide molecules at predefined sites. The most commonly used ligation techniques for chemical transformation of proteins are thioester ligation, Staudinger ligation, oxime or hydrazone ligation and disulfide ligation.

1.3 Chemoselective Ligations Combined with Biochemical Methods

In parallel with these purely chemical approaches, a number of biochemical methods have also entered this arena. For example, protein splicing can now be coupled with native chemical ligation and to give rise to expressed protein ligation (EPL) [30]. This technique takes advantage of recombinant DNA technology to generate protein fragments of any size via ribosomal synthesis. In this way large proteins become accessible for chemoligation. An interesting alternative to these chemoligation procedures is enzyme-catalyzed condensation [31] which was demonstrated as early as 1938 by Max Bergman [32]. The studies of Kaiser and coworkers [33] on subtilisin in the 1960s and 1970s provided a solid base for the success of Wells and associates [34] in the engineering of an active site for this enzyme. They generated an enzyme ("subtiligase") capable of efficiently catalyzing the ligation of peptide fragments. Subtiligase exhibits a largely reduced proteolytic activity and is functionally active as an acyl transferase. This property was exploited for enzymatic condensation of six peptide fragments of ribonuclease A (each 12–30 residues, one of them containing the noncanonical amino acid 4-fluorohistidine) [35].

Recent research in this field yielded expressed enzymatic ligation (EEL) [25], which combines the advantages of EPL and the substrate mimetic strategy of protease-mediated ligation. However, genome and proteome mapping among different biological kingdoms might offer attractive tools for such purposes. For example, sortase-catalyzed proteolysis entered the arena of enzyme-mediated native protein ligation very recently. Sortases are bacterial (*Staphylococcus aureus*) enzymes that are responsible for the covalent attachment of specific proteins to the cell wall of Gram-positive bacteria in a two-step transpeptidation reaction either *in vivo* or *in vitro* [36]. This strategy has now been "borrowed" from nature and was shown to be suitable for protein–peptide and protein–protein ligations [26]. It is reasonable to expect that the recruitment of chemical strategies that living organisms have optimized and developed during their evolution and their application in chemistry in the future will be crucial for the development of novel technologies.

1.4 Methods and Approaches of Classical Protein Engineering

The use of the polymerase chain reaction (PCR) – originally developed by Kary Mullis for efficient multiplication of specific DNA sequences [37], their sequencing and cloning – has revolutionized the possibilities for protein engineering. The PCR greatly simplified experimental procedures to tailor new genes *in vitro* via fragment deletion or insertion or nucleotide substitutions. Indeed, it is only with the approach of oligonucleotide-mediated site-directed substitutions of particular amino acids in a target sequence [38] that the term "protein engineering" entered the vocabulary of protein science [39]. The design and identification of proteins with novel functions and properties was dramatically powered by methods that mimic Darwinian evolutionary processes, i.e. natural evolution produces a large number of variants by mutation and subsequently selects the "fittest" among them. Routine molecular biology methods of mutation/recombination and screening/selection in the test tube allow for rapid and direct isolation of biomolecules based on their functional properties. This collection of methods has been termed directed evolution [40] and provides a powerful tool for the development of biocatalysts with novel properties without requiring an understanding of their complicated structure–function relationships, or knowledge of enzyme structures or catalytic mechanisms.

However, the major limitation of these methods and approaches (usually neglected in the current literature) is that the changes introduced are limited to the repertoire of the canonical 20 amino acids. The above-mentioned methods, in combination with the experimental extension of the amino acid repertoire of the genetic code through its engineering, will open a new era for designing not only protein structure and function, but also the design of novel cell types. Thus, traditional ("classical") methods for protein engineering and design can be supplemented or even fully replaced by these novel approaches (Fig. 1.2).

1.5 Genetically Encoded Protein Modifications – Reprogramming Protein Translation

The ability and capacity of living cells to synthesize functional proteins is unrivaled. From the synthetic chemist's point of view, the basic features which demonstrate the power and versatility of ribosome-mediated protein synthesis over synthetic approaches are the structural homogeneity of the synthesized polypeptide and the possibility for the precise control of the (stereo)chemical composition of the desired sequence. On the other hand, the same mechanisms that ensure high fidelity of such templated protein synthesis limit the diversity of the amino acid basic building blocks in this process. Therefore, breaking these limits either *in vitro* or *in vivo* (eubacteria and eukaryotic cells) should offer the possibility to add novel amino acids into the existing repertoire of the genetic code. This would be possible only if the protein translational apparatus is reprogrammed, and subsequently the scope and utility of the protein engineering is greatly expanded. This means that genetically encoded (i.e. templated) protein modifications in combination with genetics, physiology and metabolic manipulation of living cells should have great advantages over classical chemical modification, peptide synthesis, chemoligation and even routinely used site-directed mutagenesis. This would make

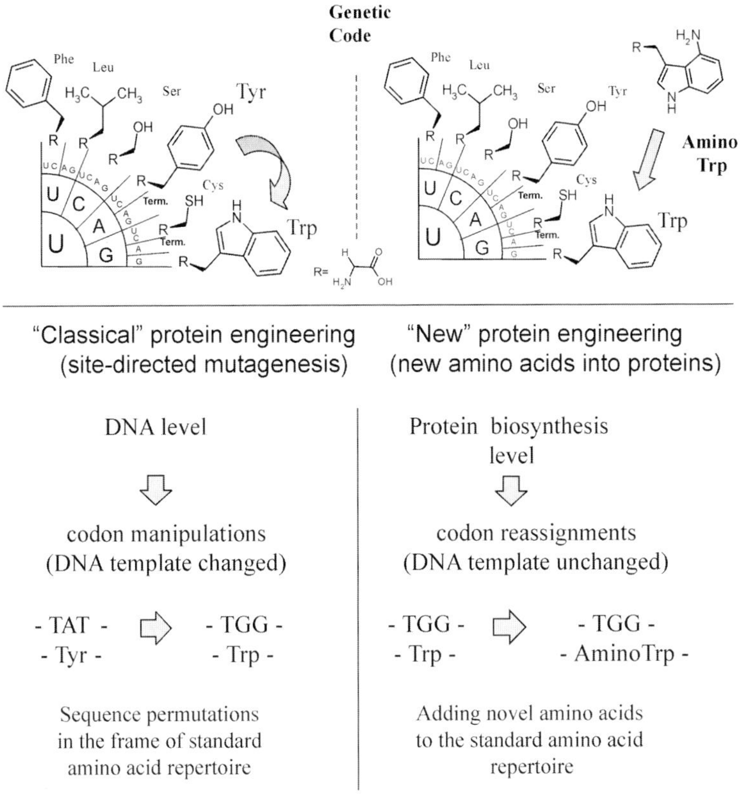

Genetic Code

Amino Trp

"Classical" protein engineering (site-directed mutagenesis) | "New" protein engineering (new amino acids into proteins)

DNA level

codon manipulations (DNA template changed)

- TAT - ⇨ - TGG -
- Tyr - ⇨ - Trp -

Sequence permutations in the frame of standard amino acid repertoire

Protein biosynthesis level

codon reassignments (DNA template unchanged)

- TGG - ⇨ - TGG -
- Trp - ⇨ - AminoTrp -

Adding novel amino acids to the standard amino acid repertoire

Fig. 1.2. Classical versus new protein engineering. The discovery of site-directed mutagenesis [38] allowed for permutation of any existing gene-encoded protein sequence by codon manipulation such as TAT (Tyr) → TGG (Trp), i.e. replacement of one standard and conserved amino acid with another. The delivery of novel amino acids into the existing amino acid repertoire as prescribed by the genetic code is a novel form of protein engineering [41]. This is exemplified above by the experimental reassignment of the UGG (TGG) coding unit for Trp to aminotryprophan (AminoTrp). In this way, the interpretation of the genetic code is changed, i.e. canonical → noncanonical amino acid replacement at the level of the target protein sequence is fully achieved. Related proteins can be defined as mutants and variants. *Mutant* denotes a protein in which the wild-type sequence is changed by site-directed mutagenesis (codon manipulation at the DNA level) in the frame of the standard amino acid repertoire. *Variant* denotes a protein in which single or multiple canonical amino acids from a wild-type or mutant sequence are replaced with non-canonical ones (expanded amino acid repertoire, codon reassignment at the protein translation level).

not only novel side-chain and backbone chemistries accessible, but also open a general perspective for novel chemistry to be performed in a controlled manner exclusively inside the living cells.

Methods for the expansion of the number of amino acids that can serve as basic building blocks in ribosome-mediated protein synthesis are described in the current literature under different names, such as "expanded scope of protein synthesis" [42], "expanded amino acid repertoire" [43], "expanded genetic code" [44], "tRNA-mediated protein engineering" (TRAMPE) [45], "site-directed non-native amino acid replacement" (SNAAR) [46], etc. Their common feature is experimental re-coding, read-through or changes in meaning of coding triplets in the frame of the existing universal genetic code. The experimental read-through can be achieved either by reassignment of evolutionarily assigned coding triplets (i.e. sense codons), suppression of termination triplets (UGA, UGG and UGU) or non-triplet coding units. In the context of protein expression, this can be achieved by controlling environmental factors (i.e. selective pressure) of the intact, but genetically modified, cells, by their supplementation with redesigned translation components or by a combination of both. At the level of the universal genetic code these experiments lead to an increase in its coding capacity by expanding its amino acid repertoire. The term "engineering of the genetic code" covers all these aspects. It strictly refers to experiments aimed at changing the interpretation of the universal genetic code or changing the structure of the code by the introduction of novel coding units.

1.6 Basic Definitions and Taxonomy

A precise terminology is usually hampered in the early stages of development of any novel research field because not enough is known to permit accurate definitions. A pragmatic strategy applied in this book is to accept provisional, rough terminological characterizations which can leverage the field's early developmental stages with the taxonomic refinements emerging as the surrounding facts become clearer. For example, a great deal of knowledge gained from code engineering experiments is actually related to the chemistry of *Escherichia coli* which, to date, remains to the main cell type for protein expression experiments. There should be no doubt that genetic code engineering in eukaryotic cells will provide additional, novel and exciting facts, insights and concepts, and subsequently lead to revisions of the existing ones. By taking into account the engineering of the genetic code's integration in several disciplines, it is obvious how difficult it is to gain a comprehensive overview in this field. Molecular and systems biology, genetics, metabolic research, research on posttranslational modifications or pharmacological properties are also bringing their own rather complex terminology in this field. For example, functional genomics or proteomics cover gene actions and interactions on a genome- or proteome-wide scale, and include at least four levels of complexity: genes (the genome), messenger RNA (the transcriptome), proteins (the proteome) and metabolism (the metabolome).

In the early studies [47], the basic criterion for one substance to be regarded as an "analog" was the requirement to have a shape and size similar to a naturally occurring molecule without any dramatic differences in biophysical properties. Such amino acid analogs that are sterically almost identical to the canonical ones, e.g. Met/SeMet or Arg/canavanine, are named isosteres. Modifications that include addition/deletion of one or more side-chain methylene, e.g. Met/ethionine, or other groups resulted in amino acids termed homologs. Relatively simple terminology, already proposed in the current literature [43], that classifies α-amino acids into two general groups (canonical and noncanonical) is used throughout this book. Other terms that will occasionally be used in the text are cognate/noncognate, coded/noncoded, proteinogenic/nonproteinogenic, standard/non-standard, natural/unnatural/non-natural, special canonical amino acids (generated via cotranslational modifications) and biogenic amino acids (generated via posttranslational modifications). Indeed, such α-amino acids, e.g. selenocysteine or diiodo-tyrosine, are also "natural", "proteinogenic" or "common" and even "canonical" in the context of their physiological appearance. All these aspects are summarized in Tab. 1.1.

Nowadays, the true meaning of the nature of the universal genetic code is often confused. For example, it is often reported in the media that the human, yeast or mouse "genetic code" has been "cracked", thereby causing unnecessary confusion. In fact, this relates to the complete genome sequencing (i.e. nucleotide sequence mapping) of the human, yeast or mouse genome. The term "genetic code" specifically refers to the correspondence between nucleotide sequence and amino acid

Tab. 1.1. Taxonomy of canonical and noncanonical amino acids (these amino acids that participate in ribosome-mediated protein synthesis are classified according to their assignments/reassignments in the code table, metabolic activity and origin; posttranslational modifications are strictly separated form coding process).

Feature	***α-amino acids***[a]	
Mode of translation in the polypeptide		
sense codon-dependent incorporation	canonical	noncanonical
mostly nonsense coding unit reassignment in a context-dependent manner	special canonical[b]	special noncanonical[c]
posttranslational modifications (strictly separated from basic coding)	special biogenic	–
Other aspects		
position in the genetic code table[d]	naturally assigned	experimentally assigned
in vivo effects on cellular viability	vital	mostly toxic
origin	mostly metabolism	mostly anthropogenic

[a] Most canonical or noncanonical amino acids can indeed be built into peptides and proteins by peptide synthesis or total chemical synthesis protocols as discussed in previous sections.
[b] "Special canonical amino acid" refers to formyl-methionine, selenocysteine and pyrrolysine (see Section 3.10).
[c] Site-directed introduction of special noncanonical amino acids is extensively discussed in Chapter 5.
[d] This does not apply for special biogenic amino acids since they enter the protein structure after translation.

sequence. Canonical amino acids were defined by a three-letter code, e.g. methionine (Met), leucine (Leu), phenylalanine (Phe), tyrosine (Tyr), tryptophan (Trp), proline (Pro), etc. Noncanonical amino acids are given comparable names often denoted by a three letter code: norleucine (Nle), ethionine (Eth) or with prefixes that denote chemical functionality, e.g. 7-azatryptophan [(7-Aza)Trp]. They are all defined upon their first appearance in the text. The term "analog" defines strict isosteric exchange of canonical/noncanonical amino acids (e.g. methionine/selenomethionine) while the term "surrogate" defines nonisosteric changes (e.g. methionine/ethionine, tryptophan/thienopyrrolylalanine). Their mode of translation in proteins can be position-specific (directed by reassignments of rare codons and by suppression of termination or frameshifted coding units) or multiple-site (usually directed by reassignments of common coding triplets or codon families in the target sequence).

References

1 Koopman, G., Reutelingsperger, C., Kuijten, G., Keehnen, R., Pals, S. and van Oers, M. (1994). Annexin-V for flow cytometric detection of phosphatidylserine expression on B cells undergoing apoptosis. *Blood* **84**, 1415–1420.

2 Katre, N. V., Knauf, M. J. and Laird, W. J. (1987). Chemical modification of recombinant interleukin 2 by polyethylene glycol increases its potency in the murine Meth-A sarcoma model. *Proceedings of the National Academy of Sciences of the United States of America* **84**, 1487–1491.

3 Blundell, T. L. and Johnson, L. N. (1976). Protein Crystallography. Academic Press, New York.

4 Means, G. E. and Feeney, R. E. (1990). Chemical modifications of proteins: history and applications. *Bioconjugate Chemistry* **1**, 2–12.

5 Neet, K. E. and Koshland, D. E. (1966). The conversion of serine at the active site of subtilisin to cysteine: a chemical mutation. *Proceedings of the National Academy of Sciences of the United States of America* **56**, 1606–1616.

6 Polgar, L. and Bender, M. L. (1969). The nature of general base–general acid catalysis in serine proteases. *Proceedings of the National Academy of Sciences of the United States of America* **64**, 1335–1342.

7 Pullen, R. A., Lindsay, D. G., Wood, S. P., Tickle, I. J., Blundell, T. L., Wollmer, A., Krail, G., Brandenburg, D., Zahn, H., Gliemann, J. and Gammeltoft, S. (1976). Receptor-binding region of insulin. *Nature* **259**, 369–373.

8 Lefèvre, F., Rémy, M. H. and Masson, J. M. (1997). Alanine-stretch scanning mutagenesis: a simple and efficient method to probe protein structure and function. *Nucleic Acids Research* **25**, 447–448.

9 Hofmeister, F. (1902). Über Bau und Gruppierung der Eiweißkörper. *Ergebnisse der Physiologie* **1**, 759–802.

10 Fischer, E. (1906). Untersuchungen über Aminosäuren, Polypeptide und Proteine. *Berichte der deutschen chemischen Gesellschaft* **39**, 530–610.

11 Merrifield, R. B. (1963). Solid phase synthesis. *Journal of the American Chemical Society* **85**, 21–49.

12 Nishiuchi, Y., Inui, T., Nishio, H., Bodi, J., Kimura, T., Tsuji, F. I. and Sakakibara, S. (1998). Chemical synthesis of the precursor molecule of the Aequorea green fluorescent protein, subsequent folding, and development of fluorescence. *Proceedings of the National Academy of*

Sciences of the United States of America* **95**, 13549–13554.
13 Gutte, B. and Merrifield, R. B. (1969). Total synthesis of an enzyme with ribonuclease A activity. *Journal of the American Chemical Society* **91**, 501–507.
14 Denkewalter, R. G., Weber, D. F., Holly, F. W. and Hirschmann, R. (1969). Studies on total synthesis of an enzyme. I. Objective and strategy. *Journal of the American Chemical Society* **91**, 502–507.
15 Gaertner, H. F., Rose, K., Cotton, R., Timms, D., Camble, R. and Offord, R. E. (1992). Construction of protein analogs by site-specific condensation of unprotected fragments. *Bioconjugate Chemistry* **3**, 262–268.
16 Staudinger, H. and Meyer, J. (1919). Phosphinemethylene derivatives and phosphinimines. *Helvetica Chemica Acta* **2**, 635–646.
17 Köhn, M. and Breinbauer, R. (2004). The Staudinger ligation – a gift to chemical biology. *Angewandte Chemie* **43**, 3106–3116.
18 Dawson, P. E., Muir, T. W., Clarke Lewis, I. and Kent, S. B. H. (1994). Synthesis of proteins by native chemical ligation. *Science* **266**, 776–779.
19 Wieland, T. and Schafer, W. (1951). Synthese Von Oligopeptiden Unter Zellmoglichen Bedingungen. *Angewandte Chemie* **63**, 146–147.
20 Blake, J. (1981). Peptide segment coupling in aqueous-medium – silver ion activation of the thiolcarboxyl group. *International Journal of Peptide and Protein Research* **17**, 273–274.
21 Yamashiro, D. and Li, C. H. (1988). New segment synthesis of alpha-inhibin-92 by the acyl disulfide method. *International Journal of Peptide and Protein Research* **31**, 322–334.
22 Casi, G. and Hilvert, D. (2003). Convergent protein synthesis. *Current Opinion in Structural Biology* **13**, 589–594.
23 Muir, T. W. (2003). Semisynthesis of proteins by expressed protein ligation. *Annual Review of Biochemistry* **72**, 249–289.
24 Nilsson, B. L., Hondal, R. J., Soellner, M. B. and Raines, R. T. (2003). Protein assembly by orthogonal chemical ligation methods. *Journal of the American Chemical Society* **125**, 5268–5269.
25 Machova, Z., von Eggelkraut-Gottanka, R., Wehofsky, N., Bordusa, F. and Beck-Sickinger, A. G. (2003). Expressed enzymatic ligation for the semisynthesis of chemically modified proteins. *Angewandte Chemie-International Edition* **42**, 4916–4918.
26 Mao, H. Y., Hart, S. A., Schink, A. and Pollok, B. A. (2004). Sortase-mediated protein ligation: a new method for protein engineering. *Journal of the American Chemical Society* **126**, 2670–2671.
27 Budisa, N. (2004). Adding new tools to the arsenal of expressed protein ligation. *ChemBioChem* **5**, 1176–1179.
28 Schnolzer, M. and Kent, S. B. H. (1992). Constructing proteins by dovetailing unprotected synthetic peptides – backbone-engineered HIV protease. *Science* **256**, 221–225.
29 Kochendoerfer, G. G., Chen, S. Y., Mao, F., Cressman, S., Traviglia, S., Shao, H. Y., Hunter, C. L., Low, D. W., Cagle, E. N., Carnevali, M., Gueriguian, V., Keogh, P. J., Porter, H., Stratton, S. M., Wiedeke, M. C., Wilken, J., Tang, J., Levy, J. J., Miranda, L. P., Crnogorac, M. M., Kalbag, S., Botti, P., Schindler-Horvat, J., Savatski, L., Adamson, J. W., Kung, A., Kent, S. B. H. and Bradburne, J. A. (2003). Design and chemical synthesis of a homogeneous polymer-modified erythropoiesis protein. *Science* **299**, 884–887.
30 Muir, T. W., Sondhi, D. and Cole, P. A. (1998). Expressed protein ligation: a general method for protein engineering. *Proceedings of the National Academy of Sciences of the United States of America* **95**, 6705–6710.

31 Bordusa, F. (2002). Proteases in organic synthesis. *Chemical Reviews* **102**, 4817–4867.
32 Bergmann, M. and Fraenkel-Conrat, H. (1938). The enzymatic synthesis of peptide bonds. *Journal of Biological Chemistry* **124**, 1–6.
33 Nakatsuka, T., Sasaki, T. and Kaiser, E. T. (1987). Peptide segment coupling catalyzed by the semisynthetic enzyme thiolsubtilisin. *Journal of the American Chemical Society* **109**, 3808–3810.
34 Jackson, D. Y., Burnier, J., Quan, C., Stanley, M., Tom, J. and Wells, J. A. (1994). A designed peptide ligase for total synthesis of ribonuclease-A with unnatural catalytic residues. *Science* **266**, 243–247.
35 Chang, T. K., Jackson, D. Y., Burnier, J. P. and Wells, J. A. (1994). Subtiligase – a tool for semisynthesis of proteins. *Proceedings of the National Academy of Sciences of the United States of America* **91**, 12544–12548.
36 Mazmanian, S. K., Liu, G., Hung, T. T. and Schneewind, O. (1999). *Staphylococcus aureus* sortase, an enzyme that anchors surface proteins to the cell wall. *Science* **285**, 760–763.
37 Mullis, K. (1987). Process for amplifying nucleic acid sequences. *US Patent 4683202.*
38 Hutchison, C. A., Phillips, S., Edgell, M. H., Gillam, S., Jahnke, P. and Smith, M. (1978). Mutagenesis at a specific position in a DNA sequence. *Journal of Biological Chemistry* **253**, 6551–6560.
39 Fersht, A. R. (1985). *Enzyme Structure and Mechanism.* Freeman, San Francisco.
40 Arnold, F. H. (1997). Design by directed evolution. *Accounts of Chemical Research* **31**, 125–131.
41 Minks, C., Alefelder, S., Moroder, L., Huber, R. and Budisa, N. (2000). Towards new protein engineering: *in vivo* building and folding of protein shuttles for drug delivery and targeting by the selective pressure incorporation (SPI) method. *Tetrahedron* **56**, 9431–9442.
42 Kiick, K. L. and Tirrell, D. A. (2000). Protein engineering by *in vivo* incorporation of non-natural amino acids: control of incorporation of methionine analogues by methionyl-tRNA synthetase. *Tetrahedron* **56**, 9487–9493.
43 Budisa, N., Minks, C., Alefelder, S., Wenger, W., Dong, F. M., Moroder, L. and Huber, R. (1999). Toward the experimental codon reassignment *in vivo*: protein building with an expanded amino acid repertoire. *FASEB Journal* **13**, 41–51.
44 Noren, C. J., Anthonycahill, S. J., Griffith, M. C. and Schultz, P. G. (1989). A general method for site-specific incorporation of unnatural amino acids into proteins. *Science* **244**, 182–188.
45 Rothschild, K. J. and Gite, S. (1999). tRNA-mediated protein engineering. *Current Opinion in Biotechnology* **10**, 64–70.
46 Ibba, M. and Hennecke, H. (1994). Towards engineering proteins by site-directed incorporation *in vivo* of nonnatural amino acids. *Biotechnology* **12**, 678–682.
47 Richmond, M. H. (1962). Effect of amino acid analogues on growth and protein synthesis in microorganisms. *Bacteriological Reviews* **26**, 398–420.

2
A Brief History of an Expanded Amino Acid Repertoire

2.1
The "Golden Years" of Molecular Biology and Triplet Code Elucidation

The history of triplet code elucidation has been the subject of numerous studies, books and personal accounts. The main purpose of the brief review presented here on the crucial discoveries in the 1950s and early 1960s is to provide an appropriate background to gain insights into the research and intellectual climate in which early attempts at amino acid repertoire expansion were performed. It should also be kept in mind that the mechanism of protein synthesis was one of the most exciting research challenges in biochemistry during that time.

At the end of the 1950s the view of proteins as unique nonrandom linear arrays of only 20 amino acids began to dominate the mind of contemporary investigators [1, 2]. At the same time a direct connection between intracellular protein synthesis of a living cell and its genetic program could clearly be drawn. Zamecnik and Hoagland discovered that prior to their incorporation into proteins, amino acids are first attached to small "soluble" RNA molecules by a class of enzymes called aminoacyl-tRNA synthetases (AARS) [3]. This demonstrated that the responsibility for the fidelity of information transfer rests on AARS, which became the first translational component to be isolated and characterized [4]. "Soluble" RNA turns out to be the physical manifestation of the "adapter RNA" (since then called "transfer" tRNA) that was proposed by Crick in 1957 [5]. The adaptor role of tRNA was soon confirmed by Chapeville and coworkers (Fig. 2.1) [6], which together with the discovery of the "messenger RNA templates" (i.e. mRNA) [7] provided solid experimental evidence about the direct involvement of RNA in protein synthesis. Later, Yanofsky [8] and Brenner were the first to prove the so-called "colinearity hypothesis" which states that the linear sequence of nucleotides in DNA specifies the linear sequence of the amino acids. From the genetic studies of the T4 bacteriophage rII gene it was deduced that the code consisted of groups of three nucleotide triplets coding for a single amino acid [9]. Amino acid analysis of the capsid proteins of various tobacco mosaic virus (TMV) mutants revealed all possible mutations for a given amino acid yielded not more than 20 different amino acids [10].

The possible relations between these 20 amino acids and the triplet coding units (i.e. the rules governing the genetic code) engaged the attention of brilliant minds like the astrophysicist George Gamow. The DNA and RNA polymerizing enzymes

Engineering the Genetic Code. Nediljko Budisa

ISBN: 3-527-31243-9

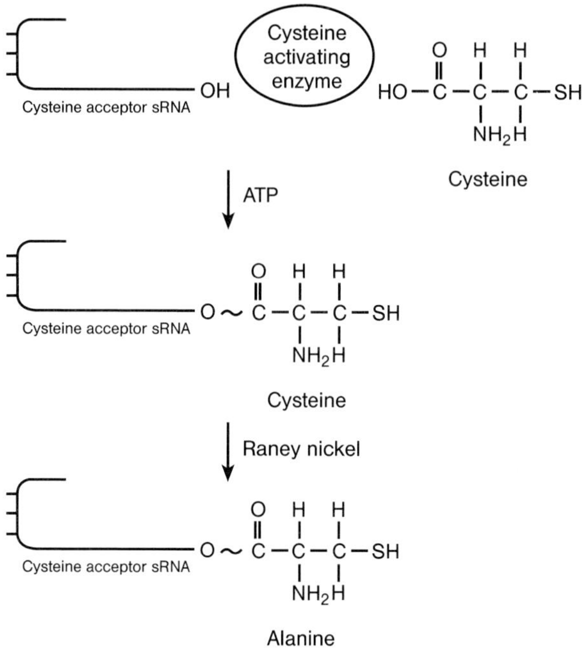

Cysteine acceptor sRNA
Cysteine activating enzyme
Cysteine
ATP
Cysteine acceptor sRNA
Cysteine
Raney nickel
Cysteine acceptor sRNA
Alanine

Fig. 2.1. The "cysteine–alanine" ("Raney nickel") experiment [6] performed in the Benzer and Lipmann laboratories established the central role of tRNA in protein synthesis. The cysteine was charged enzymatically onto its cognate $tRNA^{Cys}$. Bound Cys was then transformed into Ala by treatment of Cys-$tRNA^{Cys}$ with Raney nickel. In this reaction, the activated Cys side-chain loses its sulfur without changing its covalent bond to $tRNA^{Cys}$. Such a hybrid or misacylated Ala-$tRNA^{Cys}$ was efficiently translated into the hemoglobin sequence and the resulting tryptic peptide contains alanine side-chains at positions coded for cysteine. This experiment proved not only the adaptor hypothesis, but also yielded the discovery that the identity of the amino acid in aminoacyl-tRNA does not influence codon–anticodon interactions during mRNA translation. [Drawing reproduced from [6] with permission.]

were discovered around the same time as it became evident that knowledge of the details of the chemistry of nucleic acids was essential. This certainly cast light on how to attack the problem of the structure of the genetic code and the rules for protein synthesis. The first great breakthrough was Nirenberg and Matthei's observation in 1961 that polyuridate directs the synthesis of polyphenylalanine in the bacterial cell-free translation system [11]. Equally important was the seminal work of Ochoa [12], Khorana and coworkers that synthesized defined sequence copolymers of di-, tri- and tetranucleotides used *in vitro* to program polypeptide synthesis on ribosomes [13]. Cracking the genetic code was accomplished about 5 years after this discovery – certainly the result of strong interactions between chemistry, enzymology, theory and genetics [14]. The structure of the genetic code that emerged was in fact the result of a group effort [15].

Further development continued after the genetic code was deciphered. In the early 1970s almost all of the crucial concepts of genetics were established and are, at present, an integral part of common biochemistry. The milestone discoveries of this fascinating period relevant to the history of code engineering can be summarized as follows: (i) discovery of the standard amino acid repertoire and codons as trinucleotides, (ii) establishment of colinearity between coding polynucleotides and polypeptides, degeneracy of the genetic code, and the need for start and stop codons in translation, and (iii) discoveries of transfer (i.e. "adaptor"), messenger and ribosomal RNAs as well as AARS.

Curiously, in the late 1960s, some pioneers in this fascinating enterprise came to mistakenly believe that most of the interesting problems regarding the genetic code and mechanisms of protein synthesis had been solved, marking the end of the collaboration between theoretical studies, chemistry, enzymology and genetics. Indeed, many of them, including Francis Crick and Seymour Benzer, even decided to study nonbacterial organisms or even brain–body problems in neurology.

2.2
Early Experiments on the Incorporation of Amino Acid Analogs in Proteins

The "golden years" of molecular biology were between 1946 and 1961 when insights into the chemical nature and organization of the genetic material were gained. These years were also a period when research on the incorporation of amino acid analogs into proteins was born and subsequently rapidly developed, as shown in Fig. 2.2. Obviously, the substitution of canonical amino acids with noncanonical ones is a relatively old approach in protein chemistry. Like any other research area, it is notoriously difficult, maybe impossible, to clearly determine the starting point for the systematic attempts at amino acid repertoire expansion. Nevertheless, this research can roughly be traced back to the early studies in the 1940s regarding the inhibitory properties of amino acid structural analogs in yeasts, cell cultures, microorganisms and animals such as rats. The classical review on this topic written by Richmond in 1962 [16] provides an introduction into this topic with the following words:

> ... certain analogs are so similar to the natural compounds that they become incorporated into proteins in the place of the natural amino acids. The proteins formed in this way are sometimes altered in their specific enzyme activity, and since some of this protein will themselves be enzymes which handle the analog as a substrate, the uptake of an effective analog into a cell may have far-reaching and complex effects ...

This article by Richmond, in addition to providing a comprehensive and detailed overview of the numerous experiments performed with noncanonical analogs during the 1950s, is an undoubtedly important historic account of the early stages of this research. Generally speaking, by the end of the 1950s the study of amino acid analogs was seen as a powerful tool to determine (i) the specificity of ribosome-

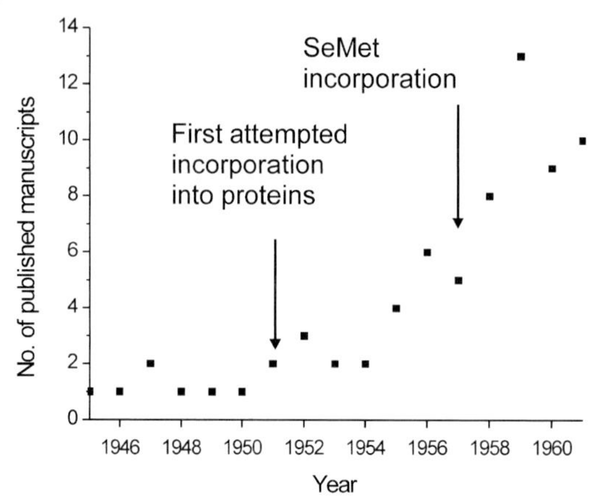

Fig. 2.2. Early experiments on noncanonical amino acid incorporation into proteins. The inhibition studies during the period 1946–1961 provided fertile ground to attempt analog incorporation into cellular proteins. The report of Levine and Traver (1951) [17] was the first to describe explicitly an attempt to substitute cellular proteins by a Met analog [^{14}C]ethionine. The number of articles counted in these statistics is based on the literature survey and reviews of Richmond (1962) [16], Hortin and Boime (1984) [18] and Kirk (1985) [19]. Only those works describing explicitly attempts to incorporate various amino acid analogs in proteins are taken into account in the statistics for 1951–1961. Although some articles might be missing from the presented statistics, the rapid growth trend in this research to the end of 1950s is obvious. Note that this 15-year period correlates well with the "golden years" of molecular biology.

mediated protein synthesis and (ii) alterations of enzyme activity upon analog incorporation. The explicit idea to change the interpretation of the genetic code, to increase its coding capacity and to even change its conserved structure had still not been conceived at that time. For example, in his recent account on experiments on the adaptor hypothesis (Fig. 2.1), Bernard Weisblum recollected [20]:

> ... Bill [Ray] considered various chemical modifications of amino acid side-chains that would yield unnatural amino acids, but Seymour [Benzer] did not consider any of these suggested modifications ... to be proof for the adaptor hypothesis. He insisted that we should convert one natural L-amino acid into another.

At that time, studies with amino acid analogs were mainly preformed in order to determine more precisely the specificity of protein synthesis [21]. The very idea of systematic amino acid repertoire expansion matured around 1990 and therefore code engineering is, strictly speaking, a very young research field. As already mentioned, these early experiments doubtless provided fertile ground for the birth of code engineering as a research field.

The experiment of Cowie and Cohen in 1957 [22] is often considered to be the point of birth of the early attempts at the expansion of the amino acid repertoire, although there are earlier reports regarding the use of amino acid analogs (Fig.

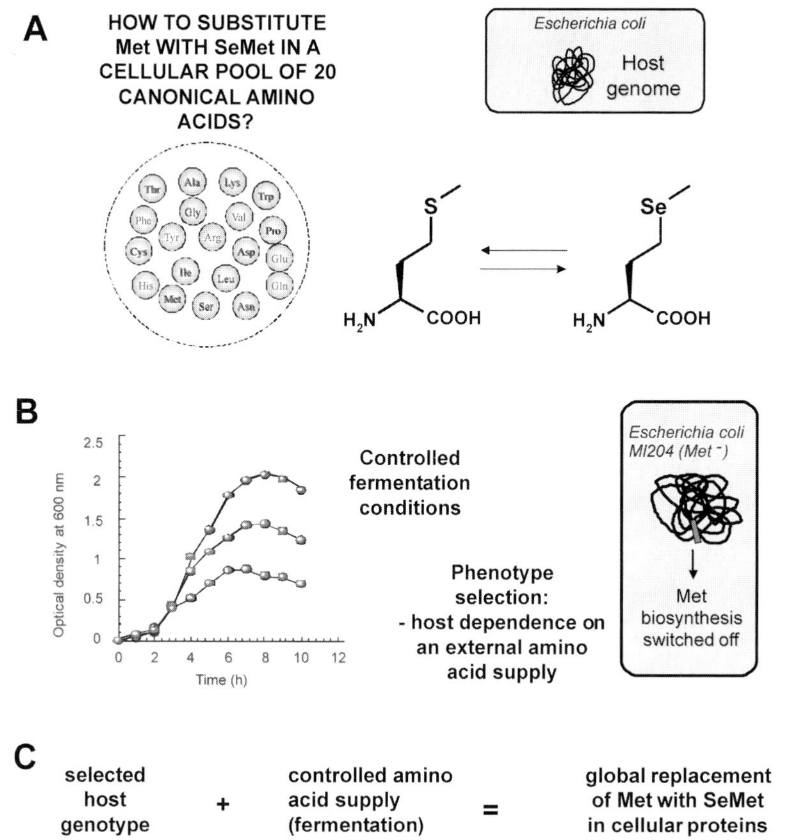

Fig. 2.3. Flow chart of the classical experiment of Cowie and Cohen [22]. (A) The basic question was how to substitute a particular amino acid from the canonical cellular pool (e.g. Met) with its noncanonical counterpart (e.g. SeMet) in cellular proteins. (B) The solution to this problem was to use selected bacterial strains deprived of their own intracellular amino acid supply (auxotrophs). In this respect, the genetic selection for such mutant bacterial strains represents a sort of metabolic engineering experiment. The availability of such bacterial strains enabled calibration of the fermentation conditions (i.e. environmental control). (C) Finally, fermentation with selected bacterial phenotypes calibrated in this way yielded global Met → SeMet substitutions in all cellular proteins (i.e. proteome-wide).

2.3). In fact, the first well-documented incorporation experiment of [^{14}C]ethionine in tissue proteins was performed by Levine and Tarver in 1951 [17]. Nevertheless, Cowie and Cohen's report [22] represents one of the most important historical landmarks in the field of the code engineering, since it has proved for the first time that "... *E. coli* can grow and synthesize active proteins under conditions where an amino acid is totally replaced by a synthetic unnatural analog". They concluded their report with the far-reaching sentence "... that the amino acid composition of proteins may be influenced by environmental changes". Their approach relies on classical genetics based on breeding, strain construction and selection of intentionally induced genetic variations in intact host expression cells that are forced to produce "mis-incorporations" by environmental control and manipulation, i.e. by externally imposed selective pressure. As a result of these pioneering experiments, it was shown that the composition of proteins can be influenced by environmental changes, i.e. by experimentally imposed selective pressure. Since then, experimental efforts have been gradually focused toward more specifically defined goals so as to shed more light on the factors that determine amino acid incorporation into proteins.

An equally important contribution in the history of the code engineering is the classical experiment of Chapeville and coworkers in 1962 [6] which demonstrated that *misacylated* tRNAs (i.e. tRNAs charged with noncognate amino acids) are able to participate in peptide bond formation at the ribosome in a fashion consistent with the adapter hypothesis (Fig. 2.1). In fact, this experiment (known also as "cysteine–alanine" or "Raney nickel") where Cys-tRNA^{Cys} was reduced to active Ala-tRNA^{Cys} was designed to answer a general question of how tRNA anticodons could be recognized by the protein-synthesizing apparatus (see Fig. 1.1). An important fact recognized later is that modified tRNA can participate in protein translation – a basic prerequisite for successful incorporation of noncanonical amino acids via chemical tRNA misacylation. Therefore, this discovery could be considered as having given birth to those experimental approaches that use mostly tRNAs with changed charging capacity, usually suppressor tRNAs.

2.3
Test Tube (Cell-free) Synthesis of Proteins and Early Incorporation Experiments

As early as the beginning of the 1950s it was shown by several investigators that ribosome-mediated protein synthesis does not require the integrity of the cell and can continue after cell disruption [23]. The cell-free translation systems developed proved to be invaluable tools in the studies of molecular mechanisms of protein biosynthesis and elucidation of the genetic code. Particularly important were the works of Zamecnik, Hoagland, Lipmann and Berg [1]. However, ribosomes in these cell extracts were programmed with endogenous DNA templates. The introduction of exogenous messengers as demonstrated by Nirenberg and Matthei has led not just to an unambiguous deciphering of the genetic code [11], but also to the development of modern cell-free systems with a dramatically improved performance. The availability of these cell-free extracts was also important for studying requirements for protein translation with amino acid analogs.

In fact, even before a clear picture of the aminoacylation reaction emerged, the activation capacities of a variety of amino acid analogs had been faithfully documented. In these initial experiments, extracts of *Escherichia coli* were used to study the activation of amino acid analogs. With these types of experiments, it was shown that norvaline, α-amino-β-chloro-butyric acid, *p*-fluorophenylalanine, Nle, SeMet and ethionine were activated by *E. coli* extracts. An important observation in these studies was that the rate of activation does not directly correlate with the translation capacity of amino acid analogs into protein sequences [24]. In other words, an amino acid analog activated by these extracts does not necessarily prove to be a suitable candidate for its incorporation into cellular proteins. For example, the extract of a Phe-requiring *E. coli* strain was capable of activating β-2-thienylalanine at the same rate as Phe [25]. On the other hand, it is a poor substrate for protein synthesis in intact *E. coli* cells. The conclusion that could be drawn from these experiments was that the activation of a particular amino acid does not automatically mean that it leads to its translational activity. Later, the extracts of microbial auxotrophic strains as working material were replaced with purified AARS and activation was directly tested in ATP–pyrophosphate (PPi) exchange reactions [26]. By the end of the 1970s impressive numbers of noncanonical analogs capable of entering an activation reaction were known [27] (Tab. 2.1).

2.4 Noncanonical Amino Acids as Tools for Studying Cell Metabolism, Physiology, Protein Processing and Turnover

Pioneering attempts to introduce variations into the amino acid composition of proteins were made by fermentation of various microorganisms in the presence of structural analogs of amino acids. At that time *E. coli* was still not used as a "model" organism for protein expression, because other microbial (*Staphylococcus*, *Bacillus*, *Lactobacillus*, *Salmonella* and *Leuconostoc*) and a variety of eukaryotic cells types (*Saccharomyces*, *Neurospora* and *Tetrahymena*, and different carcinoma cell lines like HeLa, hepatocytes, lymphocytes, etc.) were used for incorporation experiments as well [16]. However, *E. coli* soon become a prime experimental organism as it had been intensively investigated at both the biochemical and genetic level since the early 1940s. The switch to *E. coli* is understandable by taking into account the simplicity of studying its genes at the most fundamental level, its small size, nonpathogenicity, fast multiplication and uncomplicated cultivation strategies under laboratory conditions. It proved to be an excellent system to study the effects exerted by noncanonical amino acid analogs as well [21]. A remarkable property of the majority of the noncanonical amino acid analogs was their potential to serve as antimetabolites and to penetrate the amino acid pool of bacteria using the same route as their canonical counterparts [19].

Upon intracellular uptake, noncanonical amino acids may act as inhibitors of biochemical pathways or specific enzymes, precursors for analogs or surrogates of other critical biomolecules such as acetate, citrate, aminergic neurotransmitters, etc., and substrates for protein synthesis.

Fildes, in 1945 [29], had already demonstrated that 3-α-indoleacrylic acid inhibits the endogenous supply (i.e. biosynthesis) of tryptophan. Furthermore, early fermentation experiments revealed antimetabolic activity of many analogs that might be the cause of a "linear" instead of an exponential growth in the microbial cultures supplemented with analogs [24]. This linear growth was correctly attributed

Tab. 2.1. Various amino acid analogs activated by different AARS measured by the ATP–PPi exchange reaction with purified enzymes or cellular extracts between 1962 and 1983.

AARS	*Analog/surrogate*
Valyl-tRNA synthetase	penincilinamine
	allo-α-amino-β-chlorobutyric acid
	allo-isoleucine
	α-aminobutyric acid
	α-amino-β-cyclobutyric acid
	cyclobutylglycine
	cycloleucine
	cyclopropylglycine
	O-methylthreonine
	norvaline
Phenylalanyl-tRNA synthetase	β-thien-1-ylalanine
	4-thiazolealanine
	trans-2-amino-4-heptenoic acid
	2-amino-hexa-4,5-dienoic acid
	2-amino-4-methylhex-4-enoic acid
	3-furyl-3-alanine
	2-amino-5-methylhex-4-enoic acid
	2-hydroxyphenylalanine
	crotylglycine
	cyclopentenealanine
	cyclohexenealanine
	3,4-difluorophenylalanine
	ethallylglycine
	β-2-theinylalanine
	β-3-thienylalanine
	2-fluorophenylalanine
	3-fluorophenylalanine
	4-fluorophenylalanine
	methallylglycine
	4-metoxyphenylalanine
	mimosine
	β-phenylserine
	β-pyrazol-1-ylalanine
	β-pyrid-2-alanine

Tab. 2.1 *(continued)*

AARS	*Analog/surrogate*
Isoleucyl-tRNA synthetase	norvaline
	norleucine
	allo-isoleucine
	α-aminoheptanoic acid
	O-methylthreonine
	O-ethylthreonine
Arginyl-tRNA synthetase	homoarginine
	canavanine
Prolyl-tRNA synthetase	*allo*-4-hydroxyproline
	L-azetidine-2-carboxylic acid
	3,4-dehydro-proline
	N-ethylglycine
	N-methylalanine
	N-methylgycine
	trans-3-methylglycine
	N-propylglycine
	thiazolidene-4-carboxylic acid
	4-selenaproline
Lysyl-tRNA synthetase	*trans*-4,5-dehydrolysine
	5-hydrohylysine
	2,6-diamino-4-hexynoic acid
	4-oxallysine
	thiolysine
	selenalysine
Tyrosyl-tRNA synthetase	*m*-fluorotyrosine
	O-methyltyrosine
	2-hydroxytyrosine
	5-hydroxy-2-(3-alanyl)pyridine
	2,3-dihydroxyphenylalanine
	D-tyrosine
Leucyl-tRNA synthetase	azaleucine
	trifluoroleucine
	cyclopropaneglycine
	γ-hydroxynorvaline
	hypoglycine
	methallglycine
	norleucine

Tab. 2.1 *(continued)*

AARS	*Analog/surrogate*
Methionyl-tRNA synthetase	α-aminobutyric acid
	homocysteine
	ethionine
	norluecine
	norvaline
Tryptophanyl-tRNA synthetase	7-azatryptophan
	2-azatryptophan
	4-fluorotryptophan
	5-fluorotryptophan
	6-fluorotryptophan
	7-fluorotryptophan
Histidyl-tRNA synthetase	1,2,4-triazole-3-alanine
	2-thiazolealanine
	2-fluorohistidine
	4-fluorohistidine

The important feature that emerged from these experiments is the remarkable degree of nonspecificity in activation reactions with various amino acid analogs. Nowadays it is well established that the misactivated amino acids are normally substrates for editing mechanisms of cognate AARS. Data are taken from [16, 18, 27, 28]. Unless stated, all amino acids are L-chiral forms.

to amino acid analog incorporation into vital cellular enzymes which reduces or even abolishes the enzymatic activity. Almost all documented fermentation experiments with a variety of analogs (except SeMet) reported a linear growth of the culture upon addition of an analog into the synthetic media. As well as metabolic toxic effects, the reasons for such growth deviations also derive from an immediate arrest in the formation of vital proteins upon analog addition. The bactericidal analog canavanine and bacteriostatic analog *p*-fluorophenylalanine were, among others, most often used in the 1950s to study growth and physiology of yeast and other microorganisms (feedback control, metabolic interference, interruption of protein synthesis) as well as host–virus interactions, virus multiplication, bacteriostatic activity and inhibition of the growth of tumor cells [16].

In the 1960s, experimental efforts gradually became more focused towards specifically defined goals, such as the study of the effects of canonical → noncanonical amino acid replacements on the enzymatic and immunological properties of selected proteins. Studies that followed the fate of one particular protein in the cells demonstrated a diversity of effects depending on which analog was used. For example, 4-fluorotryptophan does not affect the activity of *β*-galactosidase, while 5-

fluorotryptophan and 6-fluorotryptophan induce altered kinetic parameters. Conversely, 4-fluorotryptophan incorporation into D-lactate dehydrogenase results in an enzyme mutant with higher activity, while mutants containing 5-fluorotryptophan and 6-fluorotryptophan are less active [30].

Amino acid analogs proved to be especially useful in studies on the mechanisms of protein processing and turnover (see Section 7.8.1). There are plenty of data about the processivity of substituted proteins in the context of viral infection or tumor tissues. In fact, the property of 2-fluorohistidine (see Section 5.2.1.4) to inhibit the cytopathogenic action of a variety of DNA and RNA viruses was unambiguously demonstrated. Jacobson and Baltimore discovered the mechanism of proteolytic cleavage of the viral polyprotein precursor by studying inhibition upon its substitution with amino acid analogs. Other examples include the proteolytic cleavage of the precursor forms of hormones like prosomatostatin or proinsulin, whose cleavage was inhibited by specific analogs of arginine and lysine [31] (see Section 5.2.1.9).

These works were followed by a series of extensive studies on posttranslational processing of virus-encoded proteins and viral maturation in cells. Commonly used analogs in these studies were fluorinated phenylalanines, 2-azatryptophan (tryptazan), ethionine and canavanine [21]. For example, it was demonstrated that incorporation of the arginine analog canavanine (see Section 5.2.1.9) alters the electrophoretic mobility of substituted proteins and inhibits host metabolic activity [32]. Canavanine has been intensively studied for decades because of its antitumor activity. Conversely, there are also analogs with tumorogenic activity like the Met analog ethionine (see Section 5.2.1.6), which induces liver toxicity and hepatocellular carcinomas [33]. Hydroxyproline synthesis in collagen was probed with a variety of proline analogs (see Section 5.2.1.5) in order to clarify the mechanism of posttranslational hydroxylation of proline residues. Similarly, posttranslational glycosylation of aspartic acid residues could be inhibited with their aspartic acid and threonine analogs. Numerous other examples of amino acid analog applications in biomedicine have been extensively reviewed elsewhere [18, 19, 34, 35].

2.5 Problem of Proofs and Formal Criteria for Noncanonical Amino Acid Incorporation

One of the major obstacles in this field is the lack of predictability of translational activity for the majority of noncanonical amino acids. Furthermore, it is almost impossible to conceive which effects a particular noncanonical amino acid would produce on the proteins, translational apparatus and host cells in general upon its incorporation. However, neither the methods described above nor recent developments in the field yielded any general rule that specifies the capacity to predict which changes will produce translationally active amino acid analogs – there are only empirical observations regarding the geometry, chemistry, molecular volume, etc., of the noncanonical analogs tested for translation activity.

Analyses of protein extracts of mutant bacterial strains or cell lines upon amino acid replacement with analogs should reveal their presence in cellular proteins.

Successful incorporation was reported in early experiments for amino acids such as 2-azatryptophan (tryptazan), 7-azatryptophan, fluorinated phenylalanine and tyrosine analogs, canavanine, *p*-aminophenylalanine, seleno-methionine, norleucine, *β*-2-thienylalanine, etc. However, the levels of incorporation were reported to be different, e.g. 4-fluorotryptophan was found to replace tryptophan by more than 75% in all cellular proteins, ethionine and norleucine to about 50% of all methionine residues, while *p*-aminophenylalanine only to 2–5% of all phenylalanine residues in cellular proteins [16]. Later experimental results are not consistent with the current knowledge that *p*-aminophenylalanine is quite a poor substrate for endogenous phenylalanyl-tRNA synthetase in *E. coli* [36]. Its trace presence in the analyzed cellular extracts might result from the rather limited capacities and insufficient accuracy of the contemporary analytical methods and instrumentation.

Therefore, earlier reports on noncanonical amino acid incorporation into proteins should be scrutinized closely. It should be remembered that the best available procedures at that time were microbiological assays, where quantitative analyses of particular amino acids were possible only after proper calibration of the fermentation [37]. Indeed, the idea to use metabolically engineered (mutated) bacterial strains was already employed for early amino acid analyses. These microbiological assays were based on the use of auxotrophic strains from particular organisms like *Neurospora* or *Lactobacillus* that lacked the ability to synthesize a particular amino acid. Therefore, fermentation using these microorganisms allows for a control of culture growth which is quantitatively dependent on the amount of that amino acid present in a defined (minimal) media. Proper calibration (optimization) of the fermentation conditions could provide quantitative analysis for particular amino acids. The same principles were employed by Cowie and Cohen in 1957 [22] in their experiment on global SeMet incorporation into *E. coli* proteins. Their conclusion that all Met residues in proteins were replaced with SeMet relied on a strict absence of Met from the synthetic media.

Radioactively labeled amino acids, or suitable reagents for amino acid derivatization used to facilitate their chromatographic separation like fluordinitrobenzene, ninhydrin or *N*-acyl derivatives of amino acids, were used as alternatives to the these technically demanding microbiological assays. Levine and Terver, in 1951 [17], demonstrated for the first time the incorporation of ^{14}C-labeled ethionine, an amino acid analog of Met, into *Tetrahymena* proteins. Huber and Criddle reported diagonal chromatography as a method to pinpoint the SeMet residues in partially labeled *β*-galactosidase [38]. Other approaches relied on the physicochemical properties of SeMet-containing proteins like increased/decreased hydrophobicity or altered electrophoretic mobility, e.g. canavanine-containing viral precursor proteins [32], and more recently trifluoro-containing dihydrofolate reductase [39] and green fluorescent proteins containing fluorinated tyrosines [40] having different electrophoretic mobilities as shown in Fig. 2.4. There are also other measurable parameters that can be followed upon incorporation experiments such as changes in enzymatic activity [41] or tracing fluorine atoms in protein structures using ^{19}F-nuclear magnetic resonance (NMR) spectroscopy [42].

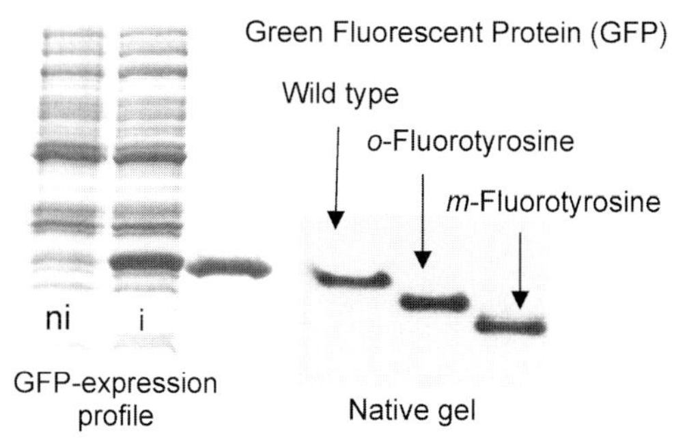

Fig. 2.4. "Gel-shift experiment" with the expression profile of native green fluorescent protein (GFP, left) in *E. coli*, and electrophoretic mobility of its *o*-fluorotyrosine- and *m*-fluorotyrosine-containing variants in 12% polyacrylamide gels (right). The difference in the electrophoretic mobility (i.e. gel shift) can be observed in both denaturing and native gels [40]. These phenomena are also well documented for canavanine-, trifluoro- and hexafluoro-leucine-containing proteins. Such experiments provide indirect evidence for the incorporation of an amino acid analog into proteins (i = induced, ni = non-induced cell lysates).

Nowadays, efficient analytical techniques to assess successful translation of amino acids into protein sequences are widely available, e.g. amino acid or mass spectrometric analyses. It is highly desirable that single substituted proteins or all cell proteins containing the desired noncanonical amino acid are isolated at least on an analytical scale. To date, there is no commonly accepted criterion for successful translation of the amino acids into proteins. In a wider perspective, defined criteria for the incorporation of noncanonical analogs and surrogates would yield a list of amino acids that are capable of entering the genetic code (see Section 6.2.2). It is for these reasons that a set of three simple, but strict, experimental criteria has been suggested recently [43]:

(i) Single proteins or all cellular proteins which are substituted with particular amino acids should be isolated at least on an analytical scale.
(ii) The incorporation must be confirmed by at least two different analytical techniques, preferably by amino acid and mass spectrometric or sequence analyses.
(iii) The protein variants should be structurally and functionally analyzed to assess whether they are properly folded, and whether there have been alternations in their activity.

One should always keep in mind that such replacement experiments often result in the loss of structural integrity and functionality of the target protein or severe impairment in the viability of the living cell. Thus, incorporation experiments make sense only if the desired effects or novel properties are generated.

2.6
Recent Renaissance – Genetic Code Engineering

The Phoenix is a fabulous golden-red bird whose body emitted pure rays of sunlight.

[Herodotus – a Greek historian]

From the beginning of the 1960s to the end of the 1980s work on systematic incorporation of noncanonical amino acids into recombinant proteins was seen as an obscure science practiced by a handful of people. Nonetheless, this period yielded important contributions, such as work on selenium metabolism on living cells, works on halogenated amino acids and their utility in biomedicine (reviewed by Kirk [19]), use of fluorinated amino acids for ^{19}F-NMR studies of proteins [42], the experiment by Wong on 4-fluorotryptophan incorporation [44], etc. These experiments are extensively covered in the section dealing with the substitution capacities for each particular canonical amino acid (see Section 5.2.1).

After the advent of recombinant DNA technology, almost two decades passed until the renaissance of noncanonical amino acid incorporation and the beginning of code engineering as a novel research field took place at the end of the 1980s and in the early part of the 1990s. Interestingly, at the same time, the protein crystallographer W. A. Hendrickson rediscovered the Cowie and Cohen approach to be an important tool for structural biology. The selenium atom of SeMet residues in proteins was recognized to be a very effective anomalous scatterer of X-rays, a main prerequisite for the multiwavelength anomalous diffraction method to solve the so-called "phase problem" in protein X-ray crystallography [45]. Nowadays, the incorporation of SeMet in proteins is an unavoidable tool for their structure elucidation [46]. The increase in the number of novel structures correlates well with the use of SeMet for protein labeling in the same manner as described by Cowie and Cohen [22], as shown in Fig. 2.5 (see also Section 7.5.1).

The possibility of facilitating structural elucidation by markers other than SeMet was first developed by Budisa and coworkers [48], in the laboratories of the crystallographer Robert Huber and bioorganic chemist Luis Moroder. This original idea was further elaborated and later extended to various fields of application [43]. In parallel, Silks and coworkers also demonstrated the possibility of telluromethionine incorporation into proteins [49]. The polymer chemist David Tirrell entered this field in the beginning of the 1990s after recognizing the potential of code engineering in materials science [50], and also made some of the most prominent contributions in the development and improvement of methodologies. Proteome-wide substitution, in addition to SeMet incorporation in *E. coli*, was demonstrated for another analog 4-fluorotryptophan in *Bacillus subtilis* by Wong and coworkers during the 1980s [51]. Wong's approach has recently been continued by Ellington and coworkers [52] who are attempting to select bacterial and viral cells with chemically distinct proteomes by using fluorinated Trp analogs. Marilere and coworkers entered the field with ambitious experimental goals to design "microbial strains with a clear-cut requirement for an additional amino acid that should be instrumental for widening the genetic code experimentally" [53].

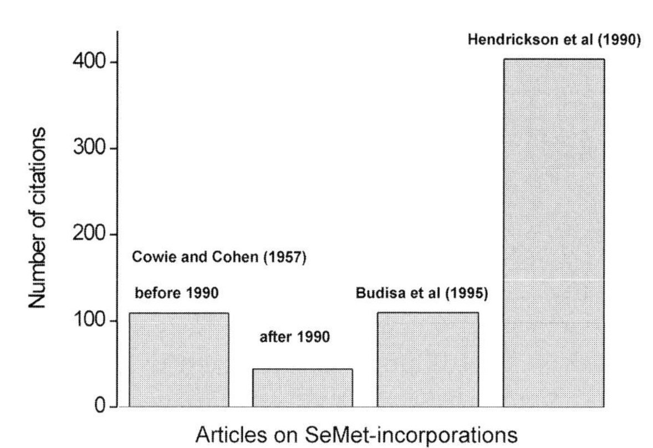

Fig. 2.5. Comparison of three most cited articles on SeMet incorporation (made in August 2004 using ISI Web of Science). The paper by Hendrickson and coworkers from 1990 [47] is one of the most frequently cited methodological papers in structural biology; that of Budisa and coworkers from 1995 [48] describes SeMet incorporation into proteins for protein X-ray crystallography. The increase in their citation frequency corresponds with an increase of the three-dimensional structures of proteins solved by the use of SeMet and reflects a renaissance of noncanonical amino acid incorporation into proteins after 1990.

Other sporadic contributions in this field are discussed with respect to individual amino acids in Section 5.2. By the use of chemically or enzymatically misaminoacylated tRNAs, missense, nonsense or frameshift mutations in a target gene sequence can be used to expand the amino acid repertoire. These methods that were pioneered by Sidney Hecht during the 1980s [54] were further developed *in vitro* and *in vivo*, in both prokaryotic and eukaryotic cells, in the laboratories of Peter Schultz, Shigeyuki Yokoyama, Masahiko Sisido, Richard Chamberlin and Dennis Dougherty during the 1990s (see Section 5.3). Finally, the idea to use complementary nucleotides in order to gain a novel Watson–Crick base pair that was pioneered by Benner and coworkers in 1992 [55] culminated recently in the findings of Kimoto coworkers that a DNA with unusual base pairing can be transcribed into an RNA molecule [56]. The race to discover new orthogonal base pairs has already begun and rapid progress, in fact a joint effort by Steven Benner, David Eaton, Eric Kool, Ichiro Hirao, Floyd Romesberg, Shigeyuki Yokoyama, Peter Schultz and other groups, was recently comprehensively reviewed by Donald Bergstorm [57].

These works sparked off a strong interaction between chemistry and genetics as it was in the time of triplet genetic code deciphering, when the mechanism of protein synthesis and the rules governing the genetic code were some of the most exciting areas in biochemistry. At present, the most exciting goal should be the ability to change these rules and subsequently reprogram protein translation. Thus, the revival or renaissance of the pioneering ideas and approaches from the middle of the last century continues. The dream is about how to re-write the genetic code experimentally and to pave the way to generate truly novel synthetic life forms.

References

1 Zamecnik, P. C. (1960). Historical and current aspects of the problem of protein synthesis. *Harvey Lectures* **54**, 256–279.

2 Fruton, J. S. (1963). Chemical aspects of protein synthesis. In *The Proteins: Composition, Structure, and Function* (Neurath, H., ed.), vol. I, pp. 189–310. Academic Press, New York.

3 Hoagland, M. B. and Zamecnik, P. C. (1957). Intermediate reactions in protein biosynthesis. *Federation Proceedings* **16**, 197–197.

4 Berg, P. and Ofengand, E. J. (1955). An enzymatic mechanism for linking amino acids to RNA. *Proceedings of the National Academy of Sciences of the USA* **44**, 78–88.

5 Crick, F. H. C. (1957). On protein synthesis. *Symposium of the Society for Experimental Biology* **12**, 138–163.

6 Chapeville, F., Ehrenstein, G. V., Benzer, S., Weisblum, B., Ray, W. J. and Lipmann, F. (1962). On role of soluble ribonucleic acid in coding for amino acids. *Proceedings of the National Academy of Sciences of the USA* **48**, 1086–1092.

7 Brenner, S., Jacob, F. and Meselson, M. (1961). An unstable intermediate carrying information from genes to ribosomes for protein synthesis. *Nature* **190**, 576–581.

8 Yanofsky, C., Carlton, B. C., Guest, J. R. and Helinski, D. R. (1964). On the colinearity of gene structure and protein structure. *Proceedings of the National Academy of Sciences of the USA* **51**, 226–272.

9 Benzer, S. (1955). Fine structure of a genetic region in bacteriophage. *Proceedings of the National Academy of Sciences of the USA* **41**, 344–354.

10 Watson, J. D., Hopkins, N. H., Roberts, J. W., Steitz, J. A. and Weiner, A. M. (1988). *Molecular Biology of the Gene*. Benjamin Cummings, Amsterdam.

11 Nirenberg, M. W. (1963). Cell-free protein synthesis directed by messenger RNA. *Methods in Enzymology* **6**, 17–23.

12 Grunberg-Manago, M. and Ochoa, S. (1955). Enzymatic synthesis and breakdown of polynucleotides; polynucleotide phosphorylase. *Journal of the American Chemical Society* **77**, 3165–3171.

13 Khorana, H. G. (1968). Synthesis in study of nucleic acids – Fourth Jubilee Lecture. *Biochemical Journal* **109**, 709–716.

14 Crick, F. H. C., Barnett, L., Brenner, S. and Watts-Tobin, J. (1961). General nature of the genetic code for proteins. *Nature*, 1227–1232.

15 Nirenberg, M. W. (2004). Deciphering the genetic code – a personal account. *Trends in Biochemical Sciences* **29**, 46–54.

16 Richmond, M. H. (1962). Effect of amino acid analogues on growth and protein synthesis in microorganisms. *Bacteriological Reviews* **26**, 398–420.

17 Levine, M. and Tarver, H. (1951). Studies on ethionine: incorporation of ethionine into rat proteins. *Journal of Biological Chemistry* **192**, 835–850.

18 Hortin, G. and Boime, I. (1983). Applications of amino acid analogs for studying co-translational and posttranslational modifications of proteins. *Methods in Enzymology* **96**, 777–784.

19 Kirk, K. L. (1991). *Biochemistry of Halogenated Organic Compounds*. Plenum, New York.

20 Weisblum, B. (1999). Back to Camelot: defining the specific role of tRNA in protein synthesis. *Trends in Biochemical Sciences* **24**, 247–249.

21 Vaughan, M. and Steinberg, D. (1959). The specificity of protein synthesis. In *Advances in Protein Chemistry* (Anfinsen, C. B., Anson, M. L., Bailey, K. and Edsall, J. T., eds), vol. XIV, pp. 116–173. Academic Press, New York.

22 Cowie, D. B. and Cohen, G. N. (1957). Biosynthesis by *Escherichia coli* of active altered proteins containing selenium instead of sulfur. *Biochimica et Biophysica Acta* **26**, 252–261.

23 Hoagland, M. B. (1960). *The*

Relationship of Nucleic Acid and Protein Synthesis as Revealed by Cell-free Systems. Academic Press, New York.

24 Cohen, G. N. and Munier, R. (1959). Effects of structural analogues of amino acids on growth and synthesis of proteins and enzymes in *Escherichia coli. Biochimica et Biophysica Acta* **31**, 347–356.

25 Janecek, J. and Rickenberg, H. V. (1964). Incorporation of beta-2-thienylalanine into beta-galactosidase of *Escherichia coli. Biochimica et Biophysica Acta* **81**, 108–112.

26 Calendar, R. and Berg, P. (1966). Catalytic properties of tyrosyl ribonucleic acid synthetases from *Escherichia coli* and *Bacillus subtilis. Biochemistry* **5**, 1690–1965.

27 Igloi, G. L., Von der Haar, F. and Cramer, F. (1977). Hydrolytic action of aminoacyl-tRNA synthetases from baker's yeast – chemical proofreading of Thr-tRNA-Val by valyl-tRNA synthetase studied with modified tRNA-Val and amino acid analogs. *Biochemistry* **16**, 1696–1702.

28 Santi, D. V. and Pena, V. A. (1973). Tyrosyl transfer ribonucleic acid from *Escherichia coli*: analysis of tyrosine and adenosine 5′ binding sites. *Journal of Medicinal Chemistry* **16**, 273–280.

29 Fildes, P. (1945). The biosynthesis of tryptophan by bact-typhosum. *British Journal of Experimental Pathology* **26**, 416–428.

30 Pratt, E. A. and Ho, C. (1974). Incorporation of fluorotryptophans into protein in *Escherichia coli*, and their effect on induction of beta-galactosidase and lactose permease. *Federation Proceedings* **33**, 1463–1463.

31 Jacobson, M. F. and Baltimorore, D. (1968). Polypeptide cleavages in formation of poliovirus proteins. *Proceedings of the National Academy of Sciences of the USA* **61**, 77–84.

32 Richmond, M. H. (1959). Incorporation of canavanine by *Staphylococcus aureus. Biochemical Journal* **73**, 261–264.

33 Yoshida, A. (1960). Studies on the mechanism of protein synthesis – incorporation of *para*-fluorophenylalanine into alpha-amylase of *Bacillus subtilis. Biochimica et Biophysica Acta* **41**, 98–103.

34 Kirk, K. L. and Nie, J. Y. (1996). Fluorinated amino acids in nerve systems. In *Biomedical Frontiers of Fluorine Chemistry* (Ojima, I., McCarthy, J. R. and Welch, J. T., eds), ACS Symposium Series 639, pp. 312–327. American Chemical Society, Washington, DC.

35 Wilson, M. J. and Hatfield, D. L. (1984). Incorporation of modified amino acids into proteins *in vivo. Biochimica et Biophysica Acta* **781**, 205–215.

36 Minks, C. (1999). In vivo *Einbau nicht-natürlicher Aminosäuren in rekombinante Proteine.* PhD Thesis, Technische Universität, München.

37 Tanford, C. and Reynolds, J. (2001). *Nature's Robots: A History of Proteins.* Oxford University Press, New York.

38 Huber, R. E. and Criddle, R. S. (1967). Isolation and properties of beta-galactosidase from *Escherichia coli* grown on sodium selenate. *Biochimica et Biophysica Acta* **141**, 587–594.

39 Tang, Y. and Tirrell, D. A. (2002). Attenuation of the editing activity of the *Escherichia coli* leucyl-tRNA synthetase allows incorporation of novel amino acids into proteins *in vivo. Biochemistry* **41**, 10635–10645.

40 Pal, P. P., Bae, J. H., Azim, M. K., Hess, P., Friedrich, R., Huber, R., Moroder, L. and Budisa, N. (2005). Structural and spectral response of *Aequorea victoria* green fluorescent proteins to chromophore fluorination. *Biochemistry* **44**, 3663–3672.

41 Ring, M., Armitage, I. M. and Huber, R. E. (1985). Meta-fluorotyrosine substitution in beta-galactosidase – evidence for the existence of a catalytically active tyrosine. *Biochemical and Biophysical Research Communications* **131**, 675–680.

42 Sykes, B. D., Weingart, H. and Schlesinger, M. J. (1974). Fluorotyrosine alkaline-phosphatase from *Escherichia coli*: preparation, properties,

and fluorine-19 nuclear magnetic-resonance spectrum. *Proceedings of the National Academy of Sciences of the USA* **71**, 469–473.

43 Budisa, N., Minks, C., Alefelder, S., Wenger, W., Dong, F. M., Moroder, L. and Huber, R. (1999). Toward the experimental codon reassignment *in vivo*: protein building with an expanded amino acid repertoire. *FASEB Journal* **13**, 41–51.

44 Bronskill, P. M. and Wong, J. T. F. (1988). Suppression of fluorescence of tryptophan residues in proteins by replacement with 4-fluorotryptophan. *Biochemical Journal* **249**, 305–308.

45 Hendrickson, W. A. and Ogata, C. M. (1997). Phase determination from multiwavelength anomalous diffraction measurements. *Macromolecular Crystallography A* **276**, 494–523.

46 Hendrickson, W. A. (1991). Determination of macromolecular structures from anomalous diffraction of synchrotron radiation. *Science* **251**, 51–58.

47 Hendrickson, W. A., Horton, J. R. and LeMaster, D. A. (1990). Selenomethionyl proteins produced for analysis by multiwavelength anomalous diffraction (MAD) – a vehicle for direct determination of 3-dimensional structure. *EMBO Journal* **9**, 1665–1672.

48 Budisa, N., Steipe, B., Demange, P., Eckerskorn, C., Kellermann, J. and Huber, R. (1995). High-level biosynthetic substitution of methionine in proteins by its analogs 2-aminohexanoic acid, selenomethionine, telluromethionine and ethionine in *Escherichia coli*. *European Journal of Biochemistry* **230**, 788–796.

49 Boles, J. O., Lewinski, K., Kunkle, M., Odom, J. D., Dunlap, R. B., Lebioda, L. and Hatada, M. (1994). Bio-incorporation of telluromethionine into buried residues of dihydrofolate reductase. *Nature Structural Biology* **1**, 283–284.

50 Yoshikawa, E., Fournier, M. J., Mason, T. L. and Tirrell, D. A. (1994). Genetically engineered fluoropolymers – synthesis of repetitive polypeptides containing *p*-fluorophenylalalanine residues. *Macromolecules* **27**, 5471–5475.

51 Wong, J. T. F. (1983). Membership mutation of the genetic code – loss of fitness by tryptophan. *Proceedings of the National Academy of Sciences of the USA (Biological Sciences)* **80**, 6303–6306.

52 Bacher, J. M. and Ellington, A. D. (2001). Selection and characterization of *Escherichia coli* variants capable of growth on an otherwise toxic tryptophan analogue. *Journal of Bacteriology* **183**, 5414–5425.

53 Lemeignan, B., Sonigo, P. and Marliere, P. (1993). Phenotypic suppression by incorporation of an alien amino acid. *Journal of Molecular Biology* **231**, 161–166.

54 Heckler, T. G., Chang, L. H., Zama, Y., Naka, T., Chorghade, M. S. and Hecht, S. M. (1984). T4 RNA-ligase mediated preparation of novel chemically misacylated phenylalanine transfer-RNA. *Biochemistry* **23**, 1468–1473.

55 Bain, J. D., Switzer, C., Chamberlin, A. R. and Benner, S. A. (1992). Ribosome-mediated incorporation of a nonstandard amino acid into a peptide through expansion of the genetic-code. *Nature* **356**, 537–539.

56 Kimoto, M., Endo, M., Mitsui, T., Okuni, T., Hirao, I. and Yokoyama, S. (2004). Site-specific incorporation of a photo-crosslinking component into RNA by T7 transcription mediated by unnatural base pairs. *Chemistry and Biology* **11**, 47–55.

57 Bergstrom, D. E. (2004). Orthogonal base pairs continue to evolve. *Chemistry and Biology* **11**, 18–20.

3
Basic Features of the Cellular Translation Apparatus

3.1
Natural Laws, Genetic Information and the "Central Dogma" of Molecular Biology

The main feature of living systems as autonomous and self-replicating entities is the high level of organization and integration of their basic physicochemical processes. These interactions are emergent features of life as a process and distinguish it from anything else in the physical world. This book follows the traditional biology (and philosophy of biology) viewpoint that such processes are purely physical and casual [1]. Thus, life processes are shaped by (i) deterministic laws of chemistry and physics ("natural laws"), and (ii) the genetic program which determines all biological activities and phenomena. This program operates such that nucleic acids encode information and proteins execute it. At first glance there is almost nothing in common between the chemistry of nucleotides and that of amino acids. It is the genetic code itself that brings together the two realms – that of protein chemistry with that of nucleic acids chemistry – through a set of optimized and finely tuned interactions [2].

For a long time it was believed that proteins were the hereditary material. The reasons are probably historical, since by the end of the last century protein chemists had already identified most of the amino acids, which were known to be the main building blocks of all proteins. At the same time, Fisher and Hofmeister conceptually solved the problem of amino acid linkage in proteins with the "peptide hypothesis" (i.e. the concept of the peptide bond) [3]. The development of a variety of analytical techniques during the following almost half a century was required until this hypothesis was fully confirmed with the successful sequencing of insulin by Sanger [4]. However, that DNA might be key genetic molecule emerged unexpectedly from the studies on nonvirulent → virulent transformation of pneumonia-causing bacteria by Griffith [5]. After DNA was unambiguously identified as the hereditary molecule ("transforming principle") by Avery, Hershey and Chase [6, 7], it captured the attention of biologists, chemists and physicists alike. The mechanisms of synthesis of polypeptide chains from amino acid sequences were fully elucidated only upon experimental demonstration that the genes are nucleic acids and that a nucleic acid template is required to direct the synthesis of each protein [8]. The basic problem was how the synthesis of a precisely ordered sequence of amino acids in polypeptide chains occurs. It was fully understood after

Engineering the Genetic Code. Nediljko Budisa

ISBN: 3-527-31243-9

experimental demonstration that the RNA template is required to direct the synthesis of each protein. At this point, the inevitable connection between intracellular protein synthesis of living beings and their genetic program became clear. Tatum and Beadle's "one gene → one enzyme" concept [9] certainly paved the way for Jacob and Monod's discoveries [10] that led to the first description of gene expression regulation. Although the "one gene → one enzyme" concept does not fit into the recent picture of the multidimensionality of a proteome, is still a common practical guide in everyday laboratory routine.

The syntax of the genetic information is reflected in a precise sequence, either of bases in nucleic acids or amino acids in proteins. However, the semantic (meaning) of the genetic information stored in the DNA molecule is realized via two processes: transcription and translation. During transcription the sequence of nucleotide bases along one of DNA strand is copied onto a complementary strand of various RNA forms [messenger (mRNA), transfer (tRNA), ribosomal (rRNA), small nuclear (snRNA), transfer messenger (tmRNA), etc.]. In some viruses reverse transcription from RNA back to DNA (by reverse transcriptase) is also possible as well. Translation can be defined as the conversion of information stored in the nucleic acid language into the protein language. In particular, mRNA is used as a template for protein synthesis on ribosomes, i.e. the genetic program of a cell is "read" from mRNA into a complementary amino acid sequence which, once assembled, spontaneously assumes a particular three-dimensional form and begins to perform its specific biological function.

The connection between these processes was recognized very early and unified in the concept termed the "central dogma of molecular biology" [11]. The information flow between nucleic acids and protein is unidirectional, i.e. a polynucleotide is never determined by a protein template. This early working hypothesis predicts that DNA functions as the template for RNA molecules, which subsequently move to the cytoplasm where they determine the arrangement of amino acids within polypeptides. This transfer is governed by the genetic code as a set of rules which precisely relates the sequence of base triplets in nucleic acid (polynucleotide) with a sequence of amino acid residues in protein (polypeptide). In order to get a full picture of the genetic information flow, the "central dogma" should be extended as follows (Fig. 3.1). After transcription and before translation, the RNA transcripts are processed (e.g. through editing, splicing, cap addition, modifications, polyAAA

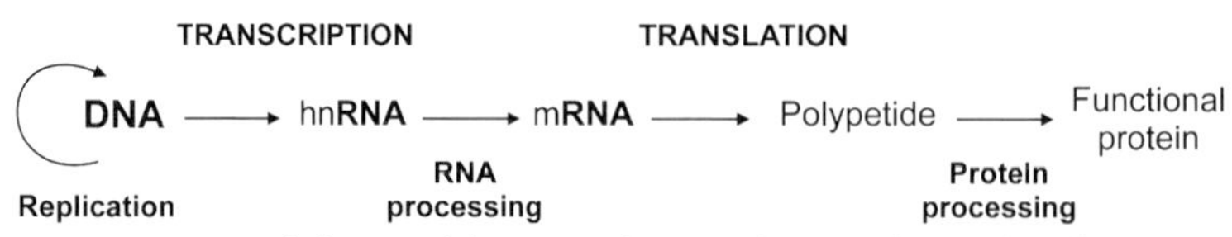

Fig. 3.1. An extended "central dogma" – the informational flow from genotype to phenotype in living beings. The arrow encircling DNA indicates that this molecule is the template for its own replication. The arrow between DNA and RNA indicates that all RNAs are made on DNA templates ("transcription"), whereas all proteins are determined by RNA templates ("translation"). hnRNA = homonuclear RNA, mRNA = messenger RNA.

addition) to produce mature mRNA. Polypeptides as translational products must be processed in many cases (see Section 3.11), producing the mature and active proteins capable of acting as structural components of the cell or serving as functional biochemical machines.

3.2
Cellular Investments in Ribosome-mediated Protein Synthesis

Protein synthesis, a key step in realizing templated genetic information, is performed by ribosomes in all living organisms. The genes involved in replication, transcription, aminoacylation and translation are almost universally conserved among the different kingdoms of life [12]. Protein synthesis requires two steps: transcription and translation. The high interdependence, synchronization, integration and efficiency of these processes are hallmarks of living cells. This is obvious when, for example, the timescale of these processes is considered: the rate of DNA replication is 50–100 times faster than the rate of transcription, which is thought in bacteria to roughly match the rate of translation. Three nucleotides are transcribed during the time one canonical amino acid is incorporated into the elongating polypeptide chain. Immediately after mRNA is copied from a DNA strand it is captured by the ribosome which scans for initiation codons, and is further followed by mRNA uncoiling and translation. In this way, each mRNA is translated 10–20 times [12]. Not surprisingly, the protein translation process consumes approximately 5% of the human caloric intake or about 30–50% of the energy generated by rapidly growing *Escherichia coli* cells (in the form of energy-rich amino acid–tRNA bonds and GTP hydrolysis in the elongation process) [13, 14]. It was also argued that the minimum amount of the genome assigned to the protein synthesis apparatus might be as high as 35–45%, which is not surprising in light of the high proportion of a cell's resources and energy budget that is allotted to translation [15]. Finally, the extent of cellular investment in protein synthesis can be measured by the amount of RNA in living cells. All RNA types together contribute about 21% of the total biomass in an average cell population of *E. coli* during balanced growth at 37 °C (Table 3.1).

The protein translation process, by which the correct sequence of amino acids is assembled to form a protein, encompasses two distinct molecular recognition events: codon–anticodon interaction between tRNAs and mRNA on the ribosome A site (translation of the genetic information in *sensu strictu*), and a specific selection of the amino acid substrates by enzyme actions of the aminoacyl-tRNA synthetases (AARS) (interpretation of the genetic code). It is this pairing of amino acids and tRNAs which defines the genetic code. While the translation of the genetic message relies on nucleic acid–nucleic acid interactions (i.e. base pairing), the interpretation of the genetic message relies on the enzyme catalytic actions (i.e. enzyme–substrate interactions) of the AARS (Fig. 3.2) and, exceptionally, on other enzymes as well. In the other words, the accuracy of the basic translational process depends largely on the precision of aminoacylation (i.e. covalent linkage of cognate

Tab. 3.1. Basic molecular building blocks of *E. coli*.

Components	*Percent total dry weight*	*Amount (g, 10^{-15}) per cell*	*Molecular weight*	*Molecules per cell*	*No. of different kinds of molecules*
Protein	55.0	156	4.0×10^4	2350000	1850
RNA	20.5	58			
23S rRNA		31.0	1.0×10^6	18700	1
16S rRNA		15.5	5.0×10^5	18700	1
5S rRNA		1.2	3.9×10^4	18700	1
tRNA		8.2	2.5×10^4	198000	60
mRNA		2.3	1.0×10^6	1380	600
DNA	3.1	8.8	2.5×10^9	2.1	1
Lipid	9.1	25.9	705	22000000	
Lipopolysaccharide	3.4	9.7	4070	1430000	1
Peptidoglycan	2.5	7.1	$(904)_n$	1	1
Glycogen	2.5	7.1	1.0×10^6	4300	1
Polyamines	0.4	1.1			
putrescine		0.83	88	5600000	1
spermidine		0.27	145	1100000	
Metabolites, ions, cofactors	3.5	9.9			800+

Chemical composition of an average *E. coli* strain B/r was determined in aerobic glucose minimal medium with a mass doubling time of 40 min during balanced growth at 37 °C. (Reproduced from [15] with permission.)

amino acids with cognate tRNAs) and subsequent noncovalent interactions between charged tRNA with ribosome-bound mRNA (translation in *sensu strictu*) [16].

3.3
Molecular Architecture of AARS

The tRNA aminoacylation is catalyzed by corresponding AARS which are responsible for specific esterification of tRNAs with their cognate amino acids (Fig. 3.3). They constitute a family of 20 cellular enzymes, and operate at the interface between nucleic acids and proteins. They are also essential in maintaining the fidelity of the protein biosynthetic process. Therefore, AARS are at the heart of the translation process. It is believed that they belong to the oldest class of enzymes, i.e. they appeared very early in evolution [17]. The first enzymes from these groups

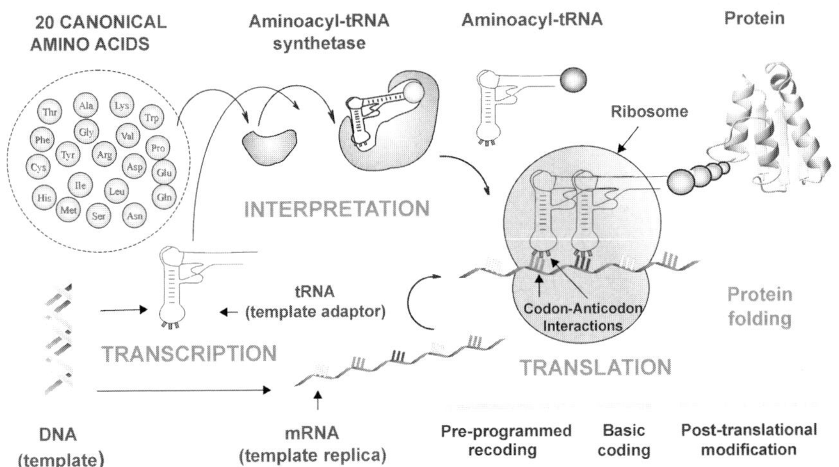

Fig. 3.2. Protein translation and genetic code interpretation in the context of genetic message transmission. The basic coding is achieved via conservative mRNA–tRNA codon–anticodon specific base pairing; the meaning of each codon is interpreted mainly through the stringent substrate specificity of AARS in the aminoacylation reaction. This reaction is an interpretation level in the context of the flow of genetic information transmission. Preprogrammed re-coding (see Section 3.10) and posttranslation modifications (see Section 3.11) are the means to diversify protein functionality beyond its basic coding. DNA replication, RNA transcription and protein translation are mutually interdependent and synchronized processes followed by protein folding and posttranslation modifications.

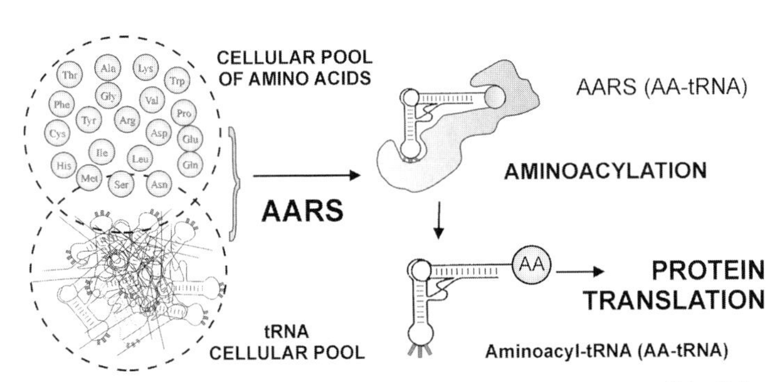

Fig. 3.3. The central role of AARS in protein translation. In well-studied model organisms such as *E. coli*, yeast or various mammals, 20 AARS catalyze via the aminoacylation reaction the amino acid activation and accurate biosynthesis of aminoacyl-tRNA, the immediate precursors for encoded proteins. Note that each AARS should select its "own" (i.e. cognate) amino acid (AA) which is subsequently covalently linked to the cognate tRNA isoacceptor "fished" from the cellular pool. Immediately after their dissociation from AARS, the aminoacyl-tRNAs are shuttled or channeled to the ribosome where the anticodon is matched to the mRNA codon and the tRNA is deacylated, with the amino acid being added as the next residue of a nascent protein chain [AARS(AA-tRNA)–enzyme–aminoacyl-adenylate complex].

are discovered 1950s, while the first partially purified AARS was TrpRS from pancreas [18]. Prokaryotic AARS appear as free cytosolic enzymes seldom bound with other proteins to form complexes. The eukaryotic AARS enzymes are usually larger than their prokaryotic counterparts, due to the presence of the C- and N-terminal extensions that are unnecessary for aminoacylation, but their full function is still unclear [19]. It is now also well documented that eukaryotic AARS participate as part of multienzyme complexes in a broad repertoire of cellular activities not just restricted to protein biosynthesis, e.g. tRNA processing, RNA splicing, RNA trafficking, apoptosis, and transcriptional and translational regulation [20]. On the other hand, in some eukaryotes (e.g. yeast) particular AARS are shared between the mitochondria and cytoplasm [21]. Finally, some AARS such as ProRS or GlyRS, according to sequence analysis, occur in different organisms with one of two quite distinct structural architectures: prokaryote like and eukaryote/archaeon like [22]. Both types might even occur in the same organism, most probably as a result of horizontal gene transfer [23].

Although AARS catalyze the same basic reaction and share a common substrate (ATP) and Mg^{2+} as cofactor, they have long been known to differ in their size, amino acid sequence and subunit structure. For example, their molecular weight is in the range between 51 (CysRS) and 384 kDa (AlaRS), their primary polypeptide structures can be composed from "only" 334 (TrpRS) to 1112 amino acids (PheRS), while their atomic crystal structures are mainly monomeric (α: Arg, Cys, Glu, Gln, Ile, Leu, Val) or dimeric (α_2: Met, Tyr, Trp, Asn, Asp, His, Lys, Pro, Ser, Thr). PheRS and GlyRS are functional heterotetramers ($\alpha_2\beta_2$), while AlaRS is believed to be a functional homotetramer (α_4) (Fig. 3.4). Prokaryotic MetRS are dimers, whereas yeast cytoplasmic and mitochondrial enzymes behave as monomers. In addition the Archaea and eukaryotes possess "eukaryote-like" synthetases (the prototype being the yeast or human cytoplasmic enzymes), whereas eubacteria and mitochondria possess more or less distinct "prokaryote-like" synthetases (the prototype being the *E. coli* enzyme) [24].

In spite of the fact that the two classes of AARS catalyze the same reaction, earlier sequence analyses have suggested the existence of two distinct classes of AARS (class I and II) of 10 members each on the basis of exclusive sets of sequence motifs [25]. Subsequent elucidation of their three-dimensional structures in free and complex forms with various substrates and substrate analogs revealed a well-defined structural basis of the two more or less independent classes.

These studies provided mechanistic insights into the specificity of substrate recognition and catalysis. Both AARS classes also differ as to where the terminal adenosine of the tRNA the amino acid is placed: class I enzymes link cognate amino acids to the 2′-hydroxyl group, whereas the class II enzymes prefer the 3′-hydroxyl group [26]. It is generally assumed that an AARS of a given specificity will always belong to the same class – this assignment is conserved very early in evolution [17]. In other words, regardless of its origin, a particular AARS will always belong to the same class. The only known exceptions to this rule are lysyl-tRNA synthetases which are composed of two unrelated families, class I enzymes in certain Archaea and bacteria, and class II enzymes in all other organisms [27] (class I:

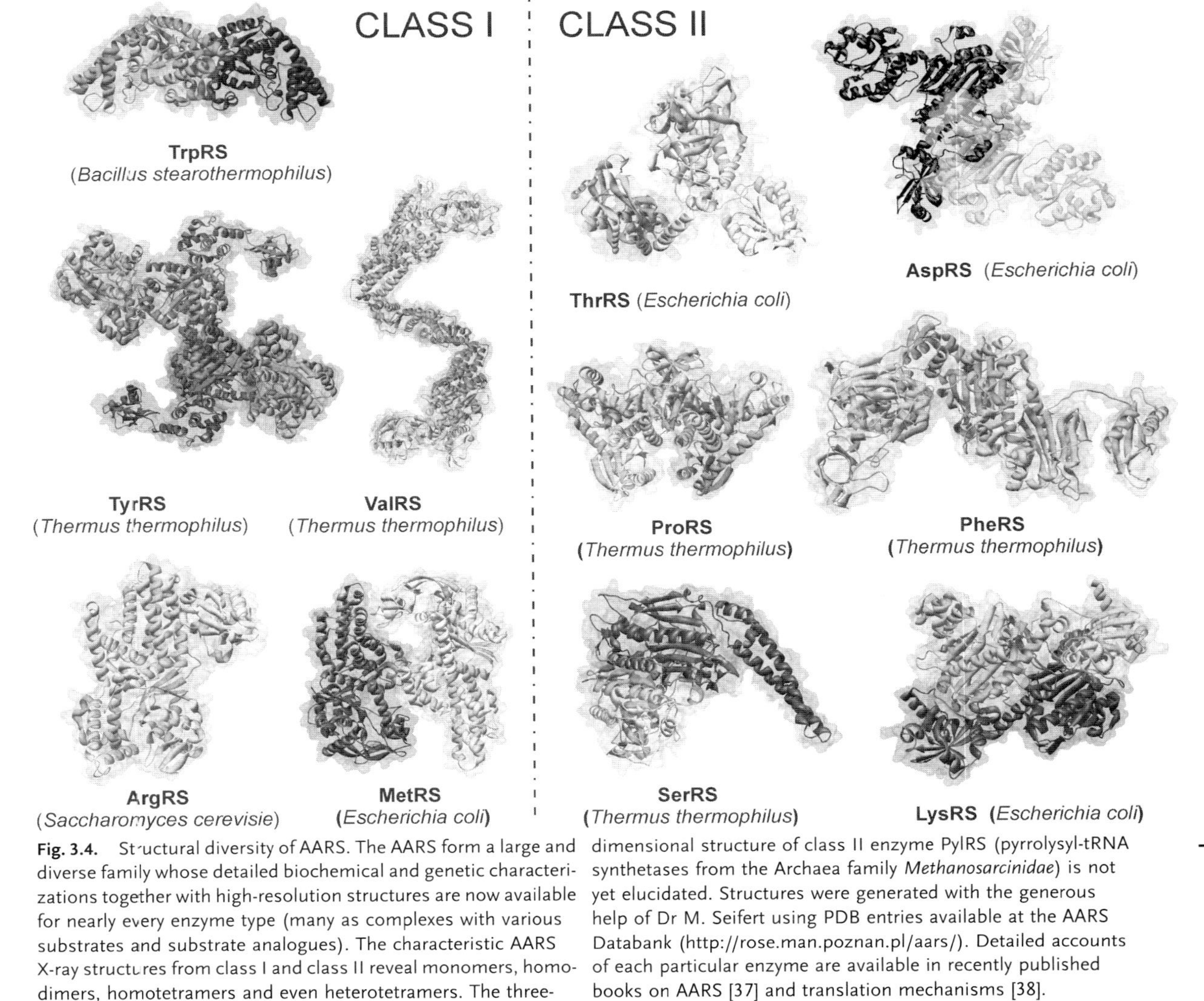

Fig. 3.4. Structural diversity of AARS. The AARS form a large and diverse family whose detailed biochemical and genetic characterizations together with high-resolution structures are now available for nearly every enzyme type (many as complexes with various substrates and substrate analogues). The characteristic AARS X-ray structures from class I and class II reveal monomers, homodimers, homotetramers and even heterotetramers. The three-dimensional structure of class II enzyme PylRS (pyrrolysyl-tRNA synthetases from the Archaea family *Methanosarcinidae*) is not yet elucidated. Structures were generated with the generous help of Dr M. Seifert using PDB entries available at the AARS Databank (http://rose.man.poznan.pl/aars/). Detailed accounts of each particular enzyme are available in recently published books on AARS [37] and translation mechanisms [38].

MetRS, ValRS, LeuRS, IleRS, CysRS, ArgRS, GluRS, GlnRS, LysRS(1), TyrRS and TrpRS; class II: SerRS, ThrRS, AlaRS, GlyRS, ProRS, HisRS, AspRS, AsnRS, LysRS(2) and PheRS).

The overwhelmingly monomeric enzymes from class I contain highly conserved HIGH and KMSKS sequence motifs, which are critical elements in the structure of the active site for aminoacyl-adenylate synthesis. X-ray studies on the first three class I AARS (MetRS [28], TyrRS [29] and GlnRS [30]) revealed a classic nucleotide binding fold called the Rossmann fold (five-handed parallel β-sheet), already found in NADH-binding dehydrogenases and ATP-binding kinases. The X-ray structure of TyrRS from *Bacillus stearothermophillus* [31] and subsequent mutagenesis studies provided an insight into the mechanisms of amino acid recognition and activation in class I enzymes [32]. For example, the mutagenesis of the KMSKS motif in TyrRS, MetRS and TrpRS has shown that these residues are part of a mobile loop that stabilizes the transition state of the activation reaction [33]. A similar role was assigned for the HIGH motif in MetRS and TrpRS [34]. The amino acid and ATP binding and activation are accompanied by distinct conformational changes via movement of small domains that contain these motifs. In addition, the topology of the binding site is further influenced by the presence of insertion segments (CP1 and CP2) that contribute to the intrinsic diversification in the frame of class I (see Fig. 3.5 below).

The first X-ray structure of one AARS from class II (SerRS from *E. coli*) [35] revealed a rather unique topology completely different from those of class I enzymes. The dominant structural feature of overwhelmingly dimeric class II enzymes is a seven-stranded β-sheet fold with three sequence motifs marked with numbers 1 (18 amino acids), 2 (23–31 amino acids) and 3 (29–34 amino acids). Motif 1 contains an invariant Pro residue involved in dimmer formation, while motives 2 and 3 are involved in aminoacyl-adenylate formation and binding of the 3′-end of tRNA. The specificity of amino acid recognition in these enzymes is further enhanced by an induced fit mechanism (i.e. substrate-induced ordering of the active site) [36]. In some class II AARS (e.g. AspRS and SerRS) this process is mainly local, while in others such as HisRS, LysRS and ProRS more global changes including a relative movement of other domains are observed [33].

The general molecular architecture of these enzymes is based on the existence of (i) a catalytic domain (group I: Rossmann fold; group II: unique fold, built around a six-stranded antiparallel β-sheet), (ii) tRNA-binding domains which can be represented by full motifs or by insertions scattered though the whole structure and (iii) oligomerization domains (Fig. 3.5) [37–40]. Several AARS contain "insertion peptides" relative to the canonical sequence for that class. In the class I AARS the active site contains a Rosmann dinucleotide binding fold, whereas this fold is absent from class II enzymes, in which the active site has a seven-stranded antiparallel β-sheet feature. For that reasons, class I enzymes bind ATP in an extended conformation, while class II bind ATP in a bent conformation [21]. There is even greater diversity inside the class II than in the class I. Many idiosyncratic domains are attached to and/or inserted in the "class-defining" catalytic core, and are responsible for binding and recognition of cognate RNAs. Such vast structural diver-

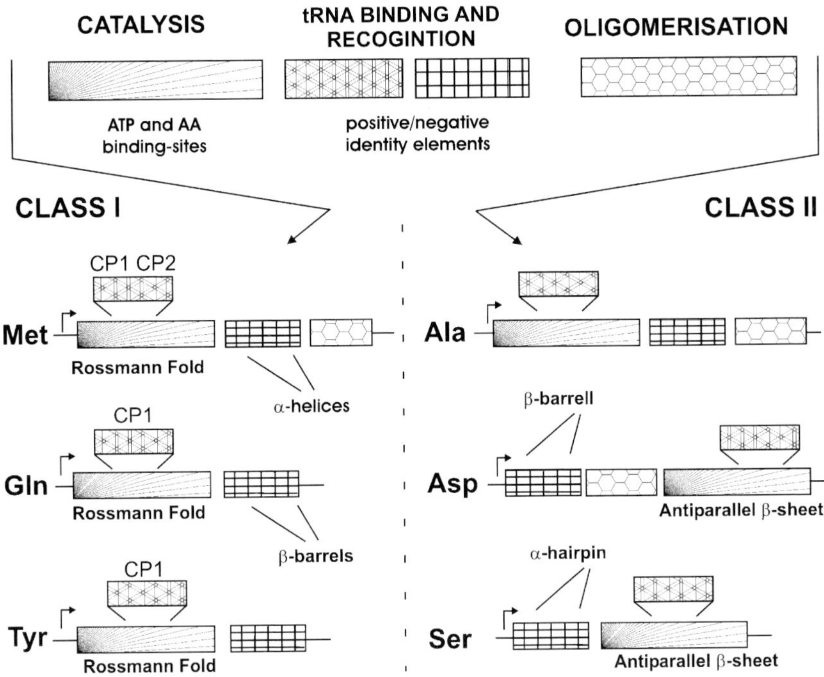

Fig. 3.5. Principal organization into functional domains for catalyses; tRNA binding and oligomerization is a constant feature of AARS. In spite of their conserved mechanisms of catalysis, the AARS are divided into two unrelated classes based on mutually exclusive sequence motifs that reflect distinct active site morphologies. The recently solved crystal structure of the 453-amino-acid catalytic fragment of *Aquifex aeolicus* AlaRS has shown that this enzyme contains a catalytic domain characteristic of class II synthetases, a helical domain with a hairpin motif critical for acceptor-stem recognition and a C-terminal domain of a mixed α/β fold [40]. Other principal differences are reflected in the binding of tRNA. CP1 and CP2 = insertion elements. Detailed accounts on particular systems are available in numerous reviews and recently published books [37, 38].

sity in topologies within the active site and two different structural frameworks reflects the facts that different AARS evolved independently to catalyze the aminoacylation reaction [23]. There are other division criteria as well, e.g. AARS can be distinguished according to their ability to bind Zn^{2+} ions [39].

Recent mapping of entire genomes revealed a surprise in that some organisms do not have genes for all 20 AARS, but they use all 20 amino acids to construct their proteins. It is now well documented that many Archaea possess only 16 out of the 20 canonical AARS [21]. The absence of a particular AARS (like AsnRS and GlnRS) in many taxa of the Archaea and bacteria is often compensated for by indirect pathways, i.e. by the recruitment of the enzymes of intermediary metabolism [23]. For instance, some bacteria do not have an enzyme for charging glutamine onto its $tRNA^{Gln}$. Instead, a single enzyme adds glutamic acid to all $tRNA^{Glu}$ and

$tRNA^{Gln}$ from the cellular tRNA pool; a second enzyme then converts the Glu-$tRNA^{Gln}$ into Gln-$tRNA^{Gln}$, which is than ready for translation at the ribosome [41]. Another impressive example includes the absence of CysRS, which is compensated for by the dual function of ProRS (which obviously recognizes both substrates), i.e. ProRS is an example of a bifunctional AARS [42]. The reasons for the variations in the number of AARS might also be due to the doubling of some AARS genes, postaminoacylation enzymatic modification of aminoacylated tRNA (AsnRS and GlnRS) as well an alternative decoding mechanism (e.g. Sec and Pyl). In contrast, a full set of AARS genes has been found in all eukaryotes [43]. These surprising findings will most probably be further extended by future genome and proteome mapping among different life kingdoms.

3.4
Structure and Function of the tRNA Molecule

It is now well established that tRNA plays the role of an adaptor between amino acids and mRNA molecules because it links, in a colinear manner, the genetic information coded in the nucleotide sequence with the amino acid sequence of the corresponding protein. In 1955, Zamecnik and coworkers found that amino acids were first attached to a low-molecular-weight RNA in the soluble fraction of a cell-free tube and were subsequently transferred to proteins in ribosomes [44]. They named these intermediary molecules "soluble" RNA (sRNA) [45]. Soon afterwards, Watson recognized that sRNA meets the requirements of Crick's adaptor hypothesis, i.e. each of these sRNA pairs with a specific amino acid and ferries it to the ribosome for protein synthesis. After that, the name "transfer" RNA (tRNA) rapidly displaced the original descriptive term sRNA [46].

Once the amino acid has been added to the tRNA, it will be used for protein synthesis according to the specificity dictated by the anticodon sequence in the tRNA (i.e. tRNA decodes the template). This was demonstrated by the classical Raney nickel experiment [47] (see Fig. 2.1). The successful sequencing of yeast $tRNA^{Ala}$ reported by Holley was the first time that a nucleic acid had had its sequence determined [48]. tRNAs are small molecules (MW $\sim$ 25 000–30 000 Da) composed of a single polynucleotide chain with 72–93 ribonucleotides and constitute 10–15% of the total RNA in *E. coli* (Tab. 3.1).

In most organisms there are more codons than tRNAs and more distinct tRNAs than AARS. The genetic code redundancy or degeneracy (i.e. the existence of more than one codon for most amino acids) is the main reason for the presence of more tRNAs for charging of only one amino acid. Different tRNAs chargeable with the same amino acid are called isoaccepting tRNAs. The number of isoaccepting tRNA species varies with the organism – while cells with minimal genomes have only one tRNA for each amino acid, eubacteria have about 40–55 tRNA species in cells. They all have a CCA nucleotide sequence at the 3′-end with terminal adenine 3′-OH or 2′-OH groups as amino acid-binding sites, whereas the 5′-end is phosphorylated (usually pG). On the ribosome, the aminoacyl residues are positioned only at

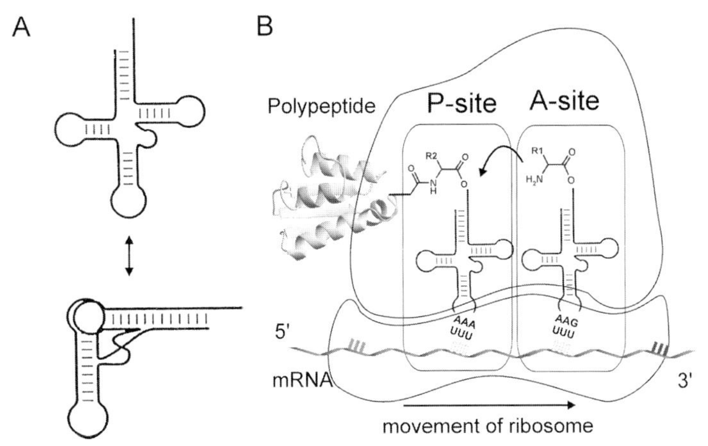

Fig. 3.6. (A) General representations for tRNA secondary (cloverleaf) and tertiary (L-shaped) structures. The basic tRNA architecture consists of two main parts: the acceptor and anticodon arms enclosing mutually an angle of about 90°. The acceptor stem is capable of being loaded with one amino acid, while the anticodon loop provides complementarity for coding triplets at mRNA. Such a design predisposes the tRNA molecule to participate in all recognition events crucial for correct translation of particular mRNA into the target protein sequence. (B) Schematic representation of tRNA placement in the ribosome in the context of its A and P sites during protein synthesis [50].

3′-OH groups since transesterification between the 2′- and 3′-position at the ribose is possible. The 2′-OH group plays an essential role in tRNA translocation from the A to the P site in the ribosome (Fig. 3.6) [49].

The structures and shapes of all known tRNAs are similar since they have to be recognized by different components of the polypeptide formation machinery. However, they should also have rather strong individual differences that predispose them toward specific interactions with cognate AARS in order to be charged with specific amino acids (see Section 3.6). The tRNA sequences contain, besides canonical A, U, G and C, more that 90 different "unusual" nucleotides such as inozine, pseudouridine, dihydrouridine, ribothymidine, methylated derivatives of guanosine and inosine, etc. Almost 25% of tRNA nucleotides are modified enzymatically during maturation [51]. These modifications include ribose/base methylations, base isomerization, base reduction and thiolation as well as base deamination. The physiological functions of such modifications are related to the diverse physiological roles of tRNAs in living cells. For example, methylations prevent pairing with some nucleotides, but enable other interactions, and contribute additionally to hydrophobicity which certainly influences tRNA interactions with AARS and ribosomal proteins. There are also numerous tertiary interactions that are not based in complementary principles (e.g. A–A, G–G and C–C) as well as hydrogen bonds between bases and riboses, bases and phosphates or riboses and phosphates [49]. In general, such interactions are attributed to tertiary structure modifications, specificity in interactions with other translational components and tRNA decoding on the ribosome. For example, deamination of the adenine in the anticodon loop results in inosine, which is capable of pairing with any of the three bases A, C and U (wobbling rules). In fact, Crick's wobble hypothesis explained much of the degeneracy of the genetic code by taking into account such base modifications – a single tRNA anticodon can recognize multiple codons by nonstandard base pairing (see Section 4.7.1). Interestingly, for *in vitro* experiments tRNAs are usually generated by run-off transcription from a DNA template using T7 RNA polymerase. Transcripts have generally been found to be efficient substrates for cognate AARS despite the fact that they lack base modifications [51].

It was found that a planar cloverleaf structure portrays the best representation of the tRNA secondary structure (Fig. 3.6). In this two-dimensional model approximately half the nucleotides in the tRNA are paired in four helices with three loops. The acceptor helix (or acceptor stem) with a conserved N^{73}CAA-3′, contains a specific nucleotide at position 73 functioning as a "discriminatory" base (Fig. 3.7) that is important for aminoacylation specificity [52]. There is a correlation between the nucleotide type at position 73 and the chemical nature of the cognate amino acid (tRNA with A73 is an acceptor with hydrophobic amino acids, while those with G73 are acceptors with hydrophilic amino acids). The CCA end (which docks at the A and P sites of the ribosome) is encoded by a tRNA gene sequence in eubacteria, whereas in higher organisms the CCA end is added posttranscriptionally. The D-stem–loop (so-called due to the presence of unusual dihydrouridine bases) is a stem–loop structure whose helical region consists of 3–4 bp and a loop length of 8–11 nucleotides. The anticodon stem–loop with a universal length of 7 nucleotides contains an anticodon in the middle of its loop. Similarly, the "T-loop" or "TΨC-loop" (T stands for ribozyl-thymidine and Ψ for pseudouridine) also has a conserved length of 7 nucleotides. Between them there is a variable loop of length between 4 and 24 nucleotides. Based on the structures of their variable loops, there are two tRNAs types. A vast majority of tRNAs belongs to type I with 4–5 nucleotides in the variable loops, whereas type II tRNAs have more that 5 nucleotides that form a stem–loop structure (e.g. eubacterial $tRNA^{Leu}$, $tRNA^{Ser}$ and $tRNA^{Tyr}$) [51, 52]. The U33 in the anticodon loop is universally conserved since it is instrumental for what is known as a "U-turn" – a sharp 180° turn which provides a proper geometry for tRNA interactions to the P and A sites as well as translocation in the ribosome cycles of mRNA decoding [49].

The first three-dimensional structures of $tRNA^{Phe}$ from yeast by Klug [53], Rich [54] and their associates revealed a rigid and stable structure that assumes an L-shape with the anticodon loop and the CCA-3′-OH at opposite ends of the L (Fig. 3.7). There are two helical mutually perpendicular segments, with numerous noncanonical base pairings and interactions as described above. The crucial role of Mg^{2+} ions and polyamines is reflected in their interaction with the tRNA sugar-phosphate backbone in order to stabilize the rather unusual folding of the tRNA molecule. Although an adaptor function is assigned as a central role to tRNA, it also dictates the functional and conformational states of the ribosome during the elongation cycle of protein synthesis. Finally, in addition to their central role in protein synthesis, tRNAs are involved in a series of other cellular processes ranging from viral infection to metabolism [51].

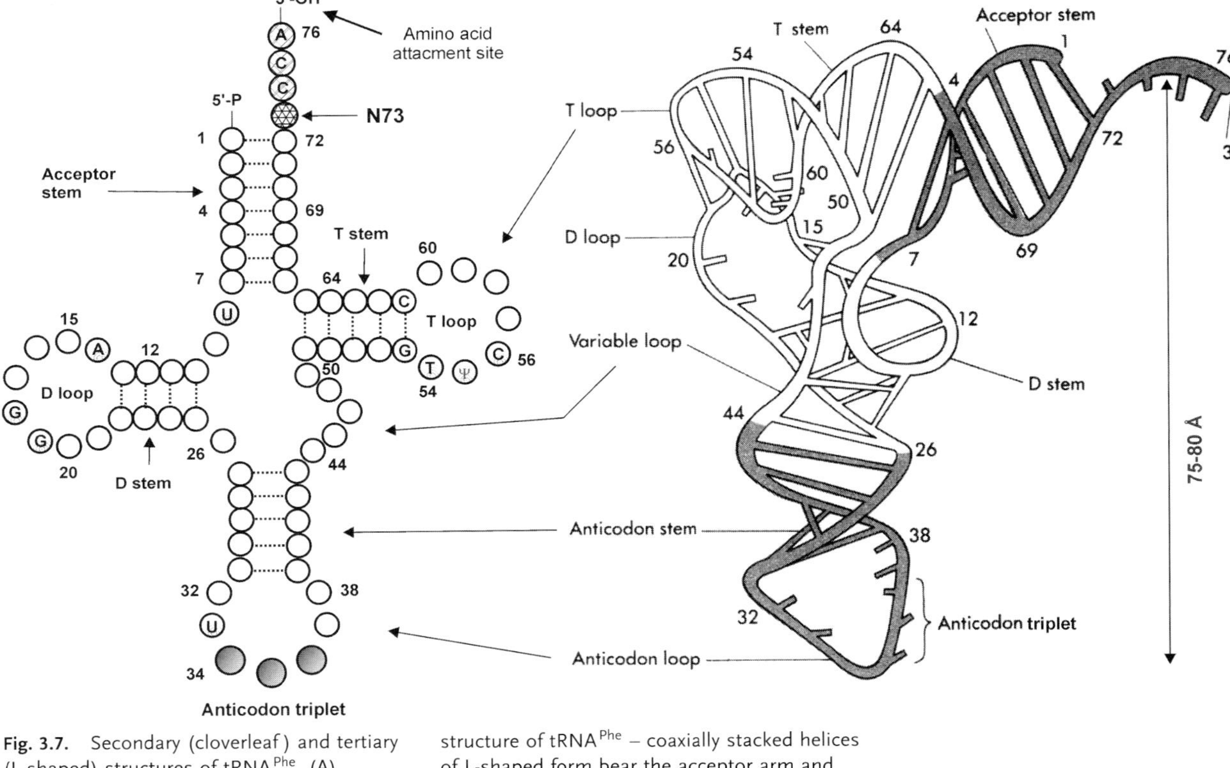

Fig. 3.7. Secondary (cloverleaf) and tertiary (L-shaped) structures of $tRNA^{Phe}$. (A) Cloverleaf model of secondary structure of $tRNA^{Phe}$ from yeast with marked nucleotides that are present in nearly all tRNA species. For example, U33 is essential for translocation in the ribosome, while N73 is the so-call "discriminatory base". (B) Three-dimensional structure of $tRNA^{Phe}$ – coaxially stacked helices of L-shaped form bear the acceptor arm and the anticodon loop are separated from each other by a distance of 75–80 Å. This value perfectly correlates with the distance from the coding center on the 30S ribosomal subunit to the peptidyl transferase center on the 50S ribosomal subunit.

3.5 Aminoacylation Reaction

In neutral aqueous solution with free amino acids together with cognate tRNA, Mg^{2+} ions and ATP, the AARS catalyze an aminoacylation reaction which results in the generation of aminoacyl-tRNA (AA-tRNA). The energy required to form the aminoacyl bond comes from a high-energy pyrophosphate linkage in ATP. In the cellular milieu, these enzymes have to recognize the correct amino acid and tRNA from a large cellular pool of similar molecules (Fig. 3.3). Therefore, this reaction is very precisely performed via preferential binding of the enzyme to amino acids and ATP with tRNA charging as the rate-limiting step.

Aminoacylation of the tRNA proceeds through a two-step reaction mechanism: both steps occur in the active site of the enzyme. The recognition of the correct amino acid occurs in a manner analogous to that by which all enzymes recognize their substrates. Each amino acid will fit into an active-site pocket in the AARS where it will bind through a network of hydrogen bonds, and electrostatic and hydrophobic interactions. Only amino acids with a sufficient number of favorable interactions will bind. The amino acid is activated by attacking a molecule of ATP at the α-phosphate, giving rise to a mixed anhydride intermediate, aminoacyl-adenylate, and an inorganic pyrophosphate leaving group. The ATP as substrate is complexed with Mg^{2+} ions which act as an electrophilic catalyst by binding to γ- and β-phosphates. An amino acid-activated carboxylic group bound to AMP (as a good leaving group) is an easy target for nucleophilic attack by the ribose 2′- or 3′-OH group in the terminal adenosine of tRNA (Fig. 3.8). The first step takes place in the absence of tRNA; however, in some cases (GlnRS, GluRS and ArgRS) amino acid activation is tRNA dependent [55]. There is no dissociation of the aminoacyl-adenylate intermediate from the active site during the reaction and in some cases enzyme–aminoacyl-adenylate complex is stable enough to be isolated chromatographically. The activation reaction is usually examined by pyrophosphate exchange where radioactively labeled pyrophosphate (usually [^{32}P]PPi) is distributed among total PPi in the reaction mixture and γ- and β-phosphates in ATP upon equilibration [18, 56]. In this way, the determination of dissociation and velocity constants is possible, and reaction profiles accompanied with free energy changes can be derived.

In the second step, the activated amino acid moiety is transferred to the 2′- or 3′-OH of the terminal ribose in the cognate tRNA, yielding the specific aminoacyl-tRNA and AMP (Fig. 3.8). The mechanisms by which the AARS recognizes the correct tRNA involve processes that proceed through an initial broad specificity interaction followed by more precise recognition and conformational changes of both AARS and cognate tRNA. Selection of the correct tRNA is assumed to occur as a result of preferential reaction kinetics for cognate protein–RNA complexes [55]. The tRNA first binds the AARS (either as a free enzyme or as an enzyme–aminoacyl-adenylate complex) with recognition depending on sequence-specific protein–RNA interactions. Then, this transient protein–RNA complex catalyzes attachment of the amino acid to the 2′- or 3′-OH-terminal adenosine of the tRNA.

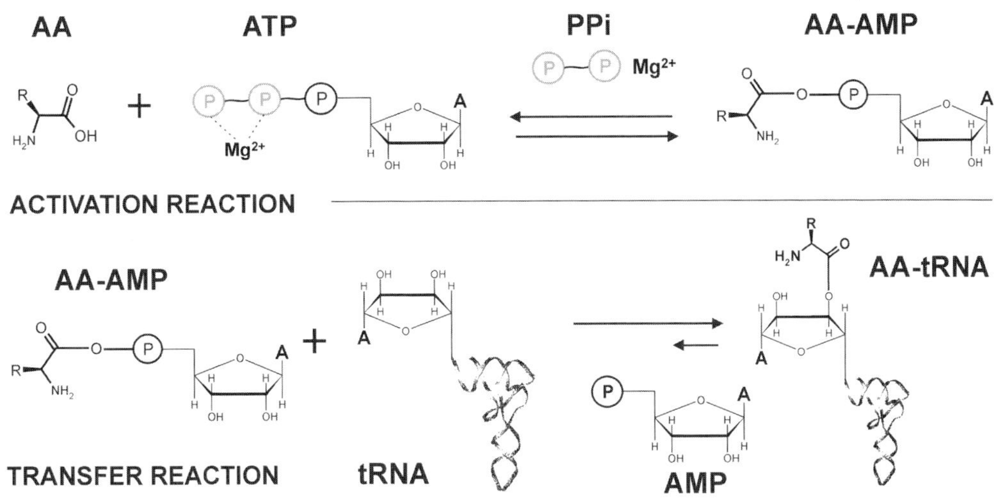

Fig. 3.8. Chemical mechanism of the aminoacylation reaction. The majority of aminoacyl-tRNAs are generated in a two-step reaction, whereby the amino acid (AA) is first recognized, activated and subsequently uncharged tRNA aminoacylated with the appropriate amino acid. The aminoacyl-tRNA then interacts with a translation elongation factor which allows its delivery to the ribosomal A site. P = phosphate, PPi = pyrophosphate, AMP = adenosine monophosphate, AA-AMP = aminoacyl-adenylate. A detailed description of the aminoacylation reaction is available in [33].

The mode of tRNA binding, especially in interactions with acceptor parts, differs between two AARS classes [51].

The two classes of AARS catalyze the same global reaction of attachment of cognate amino acid to the tRNA, but differ as to where the terminal adenosine of the tRNA of the amino acid is placed: class I enzymes act on the 2′-hydroxyl group, whereas the class II enzymes prefer the 3′-hydroxyl group. However, the aminoacyl group attached to the 2′-hydroxyl group of the terminal ribose cannot be used during protein synthesis: it is subsequently transferred by a transesterification reaction from the 2′-hydroxyl to the 3′-hydroxyl of the tRNA terminal adenosine [49].

3.6 AARS:tRNA Interactions – Identity Sets

tRNA is a central player in the process of protein translation. In this process, the meaning of each particular codon can easily be altered by mutation of the tRNA. This is because the ribosome matches codons with anticodons, but never checks whether tRNAs are charged with cognate, noncognate, canonical or noncanonical amino acids. The responsibility for linking specific amino acids with specific codons (via anticodon-bearing tRNAs) lies with AARS which can employ sophisticated proofreading mechanisms, but do not always recognize the anticodon specifically [57]. In this context, the universal genetic code is malleable [58, 59].

The DNA-binding proteins can recognize and form hydrogen bonds with suitable donor or acceptor groups in the major and minor grooves of a double helix. However, the tRNA has no such uniformity in structure – its conformation is a combination of single- and double-strand regions forming stems and loops. For these reasons the AARS:tRNA interactions are very complex and appear to be different for different AARS:tRNA pairs. To a limited extent, it is possible to distinguish between class I and class II AARS based on the way in which they recognize tRNA: most of the class I enzymes (with the exception of LeuRS) require the anticodon for proper recognition, whereas some class II enzymes (SerRS and AlaRS) recognize only features in the acceptor stem [60].

In the investigations on tRNA recognition by AARS it was first postulated that there should be a general set of rules for these interactions. They are defined as the "tRNA identity" – a sum of properties that enable recognition through cognate AARS and prevent binding of other enzymes [61, 62]. Since tRNA molecules are involved in multiple interactions with other diverse components of the translational apparatus, they must have very similar secondary and tertiary structures. On the other hand, all variations in each individual tRNA that enable interactions with specific AARS have to be kept in this frame [60].

It was first believed that the tRNA anticodon is the main recognition element for AARS [63]. Many excellent examples of changes of the anticodon leading to tRNA identity changes *in vitro* are provided in the contemporary literature [62]. For example, the transplantation of valine anticodon nucleotides into fMet-tRNA^{Met} changed it such that tRNA^{Met} was a substrate for ValRS. Similarly, the "identity" of this fMet-tRNA^{Met} was just as successfully changed to the "identity" of Tyr, Trp, Cys, Phe, Met, Ile, Gly and Gln as well [61]. In all these cases, noncognate amino acids were incorporated at the N-terminal position of a reporter protein. This *in vitro* system provided an invaluable model to study contributions of anticodon nucleotides to overall the identity of the tRNA. Later, the crystal structures of *E. coli* tRNA^{Gln}:GlnRS [64] and *Saccharomyces cerevisiae* tRNA^{Asp}:AspRS [65] fully confirmed the role of anticodon nucleotides in tRNA recognition. For most of the tRNA species (17 of 20), anticodon nucleotides mainly dictate the identity of the amino acid to be charged onto an acceptor stem.

However, full tRNA recognition does not solely rely to anticodons. There are other important recognition elements, especially bases at position 73 ("discriminatory base") [52] (Fig. 3.7) and, in some systems, at the variable arm of tRNAs. At least three well-documented cases exist where the anticodon is not important in aminoacylation: Ser, Leu and Ala acceptor tRNAs [55]. Therefore, it should be possible to change *in vitro* the tRNA identity without exchange of anticodon loops. An example is the identity switches in *E. coli* of the amber suppressor tRNA^{Leu} though the transplantation of only 2 nucleotides from tRNA^{Ser} into the variable stem [66]. Suppressor tRNA^{Leu} changed in this way served as a serine isoacceptor tRNA. Even in systems where an anticodon is the dominating identity element, the acceptor stem still plays an important role: in the first 4 bp the acceptor stems in $tRNA^{Ala}$, $tRNA^{Gln}$, $tRNA^{Gly}$, $tRNA^{Met}$, $tRNA^{Pro}$, $tRNA^{Ser}$ and $tRNA^{Thr}$ from *E. coli* are markers of their own identity [55]. In addition, the "discriminatory" base 73 proved to be important in $tRNA^{Cys}$, $tRNA^{His}$ and $tRNA^{Leu}$ from *E. coli* and yeast [51].

For these reasons, the concept of an "operational RNA code" for the relationship between the structures and sequences of tRNA acceptor stems and amino acids was proposed [67]. These interactions are a sort of "precursor" to the present genetic code. It was assumed that primitive AARS:tRNA recognition was primarily based on interactions with the acceptor stem or variable loop nucleotides, which is today remains rudimentary only in some AARS:tRNA systems (Ser, Leu and Ala). During the evolution of the genetic code, these amino acid recognition functions were most probably gradually shifted to codon–anticodon interactions. In fact, the "operational RNA code" might have limited the number of variations in tRNA recognition by AARS and these limitations are believed to shape the interpretation of the universal genetic code [17].

In living cells, possible competition reactions between particular AARS and non-cognate tRNAs are prevented through the presence of positive and negative recognition elements. Together, they build up an "identity set" for each tRNA isoacceptor [52]. A considerable body of experimental data has accumulated on the identity elements essential for many tRNA aminoacylation reactions over the past few decades [51, 55]. In this light, attempts to change the "identity" in most cases would require an exchange of the whole set of elements (either positive or negative) rather than exchange of single nucleotides [68].

The fidelity of aminoacylation is also controlled by positive (identity) and negative regulatory elements in tRNAs and AARS which permit both recognition and productive binding of the cognate pairs as well as discrimination against nonproductive binding of noncognate pairs [55]. The accuracy of this process is considerably high, i.e. the error rate in tRNA selection is of order of 10^{-6} or even higher, and constitutes one of the fundamental phenomena in protein–nucleic acid molecular recognition [69]. Identity residues, properly located in several regions of the tRNAs, trigger specific recognition and charging by the cognate AARS. Numerous currently known identity elements in tRNA charged by class I and class II AARS have already been systematically investigated, and even their tables have been established [51].

Studies on conservation of aminoacylation among different species from all kingdoms revealed paradoxical results – although the genetic code is nearly universal, there are many examples of species-specific aminoacylation reactions [70]. Cross-aminoacylation experiments reported by Soma and Himeno have shown that SerRS and LeuRS from *E. coli* are not able to aminoacylate cognate yeast tRNAs, whereas yeast enzymes recognize *E. coli* cognate tRNAs [71]. In addition, prokaryotic and eukaryotic TyrRS, IleRS and GlyRS are not able to recognize the corresponding tRNA species from other organisms [72]. In several cases, cytoplasmic AARS cannot aminoacylate the corresponding mitochondrial or chloroplastic tRNA *in vitro*, which at least is in part due to the natural variation in the genetic code of these organelles [17] (see Section 4.7). This is also a consequence of the large functional diversity around the catalytic sites of the AARS.

3.7 Translational Proofreading

There are remarkable similarities between the members of the encoded amino acids. They can have the same isosteric shape (Val–Thr and Cys–Ser), the same molecular volume (Ile–Val and Pro–Cys) or very similar chemistry (Ser–Thr). Since these amino acids have very similar side-chains, a proofreading mechanism must exist in many cases to make sure that the correct amino acid is chosen. To select the correct amino acid, AARS must be able to discriminate or distinguish between all intracellular canonical and noncanonical amino acids (Fig. 3.3).

In the simplest case, the activation of the larger amino acids might be kept down to an insignificant level by steric hindrance of their binding [73]. Such discrimination is primarily achieved by preferential binding of a cognate amino acid to the correct active site, and by exclusion of larger and chemically different substrates. If the enzyme side-chain can be distinguished adequately from any other similar side-chains, then proofreading is not necessary. For example, among the 20 canonical amino acids TyrRS is almost absolutely specific for tyrosine. Discrimination against Phe is achieved exclusively through the differences in initial binding energies [74]. Therefore, not all AARS have, or need, proofreading mechanisms.

However, when differences in binding energies of a particular amino acid to AARS are inadequate, editing is used as a major determinant of enzyme selectivity. Aminoacyl-adenylate formation is the least accurate step in the tRNA aminoacylation pathway and in many cases AARS misactivate similar amino acids (e.g. Leu/Ile, Ser/Thr, etc.) at such high frequencies that formation of AARS-bound noncognate aminoacyl-adenylates is inevitable. Such noncognate adenylates are destroyed via two alternative pathways: (i) pretransfer, by hydrolysis of the noncognate aminoacyl-adenylates, or (ii) posttransfer, by the hydrolysis of the mischarged tRNA. Overall, by including an editing step in aminoacylation, considerably increased AARS selectivity (by a factor over 100) for amino acids can be obtained [75].

AARS for branched-chain amino acids like ThrRS, ValRS, LeuRS and IleRS have evolved additional domains or insertions in their structures, where each amino acid (cognate, noncognate, noncanonical, biogenic) is checked though a "double sieve" [21]. This means that a single (catalytic) active site is supplemented with an additional editing (hydrolytic) site that dissociates noncognate amino acids (Fig. 3.9). There is solid structural evidence for the presence of a second binding site, as found in the crystal structure of IleRS [76]. Recently, structure elucidation of class II ThrRS enzymes from *E. coli* and *S. aureus* revealed the recruitment of Zn^{2+} ions in the process of cognate amino acid recognition [77]. Zinc ions interact with hydroxyl and amino groups of the cognate amino acid Thr; due to presence

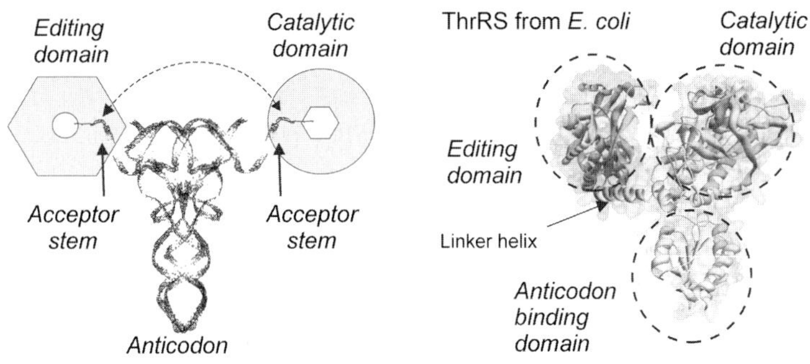

Fig. 3.9. The general concept of double-sieve (two-step) substrate selection in isome AARS classes (left), and the three-dimensional structure of class II ThrRS with marked catalytic (aminoacylation), editing and anticodon-binding domains (right). The editing mechanism in both classes is based on the dynamics of acceptor end binding: the aminoacylated 3′-end of cognate tRNA should be faster in movement towards the editing site than aminoacyl-tRNA dissociation from the enzyme. For more detailed account, see [76, 77]. The ThrRS structure was generated by MolMol using PDB entries available at the AARS Databank (http://rose.man.poznan.pl/aars/).

of Zn^{2+} in the active center, the isosteric, but noncognate, amino acid valine is rejected. However, this reaction is not sufficient to discriminate the sterically different, but chemically similar, noncognate amino acid serine. Serine is indeed charged onto $tRNA^{Thr}$; this misacylated Ser-$tRNA^{Thr}$ is then translocated from the catalytic to the editing site through a shuttle of the acceptor (CCA) part and noncognate serine is hydrolyzed on the basis of steric size. In summary, the chemistry of discrimination is achieved by inclusion of Zn^{2+} in the catalytic site; discrimination based on stereochemistry is gained by addition of the editing domain to an enzyme. The generally accepted model of functional editing is that either the tRNA accepting end or the noncognate adenylate shuttles from the catalytic to the editing site (Fig. 3.9). Such unimolecular substrate translocations are general mechanisms involved in editing steps at other levels of genetic message transfer as well, i.e. in DNA replication as well as in translation on the ribosome [78]. The most recent finding of tRNA-dependent editing by AlaRS indicates that editing of misactivated glycine or serine in this enzyme requires a tRNA cofactor [40]. This class-specific editing domain was found to be essential for cell growth in the presence of elevated concentrations of glycine or serine.

In general, there are three error-correcting levels in protein translation: (i) amino acids selection (pretransfer editing), (ii) tRNA charging (posttransfer editing) and (iii) ribosome synthesis. These processes together ensure that the error frequency in translation does not exceed on average around 0.03% [79]. The need for discrimination goes beyond the canonical 20 amino acids, since the cellular milieu is also full of metabolic intermediates and other biogenic amino acids. For example, homocysteine, a metabolic precursor of Met, is the substrate for activation via MetRS, but is efficiently edited *in vitro* and *in vivo* in a way that enzyme-bound homocysteinyl adenylate is cyclized to yield homocysteine lactone [80]. Others Met analogs, norleucine and selenomethionine, are recognized in aminoacylation reactions and subsequently loaded onto $tRNA^{Met}$ [81]. Obviously, *E. coli* and other organisms were not under the selective pressures to evolve such a mechanism to prevent their activation, tRNA loading and translation into protein sequences. In other words, editing has not evolved against specific features of chemically or sterically similar synthetic amino acids which living cells have never encountered in their evolutionary paths [82]. Such a lack of absolute substrate specificity of AARS has been well documented in many systems for decades (for more details, see the excellent reviews of Jakubowski [75], Fersht [73] and Cramer [83], and references cited therein) and was very early recognized as a route for the expansion of the amino acid repertoire *in vivo*.

3.8
Ribosomal Decoding – A Brief Overview

Subsequent to their aminoacylation, aminoacyl-tRNAs, although they might be subject to further modifications, are usually ready for translation. Aminoacyl-tRNAs are channeled to the ribosomal A and P sites where elongation takes place. The events between the initial binding of aminoacyl-tRNA and amino acid translation into the polypeptide sequence can be summarized as the ribosomal elongation cycle. This cycle is divided into three basic reactions: (i) occupation of the A site by aminoacyl-tRNA, (ii) peptide bond formation where the nascent polypeptide chain is transferred from the P site tRNA onto the A site tRNA (this leaves deacylated tRNA at the P site and a peptidyl-tRNA at the A site) and (iii) translocation involving movement of mRNA:tRNA on the ribosome by one codon length, thus freeing the A site for the next incoming aminoacyl-tRNA (Fig. 3.6). An aminoacyl-tRNA is delivered to the ribosomal A site as the ternary complex AA-tRNA:EF-Tu:GTP where the displayed codon is specific for the cognate tRNA. Cognate codon–anticodon interactions result in the dissociation of EF-Tu from the ribosome accompanied by GTP hydrolysis and accommodation of the acceptor end of tRNA into the A site of the 50S subunit. Since many other tRNA competitors exist (in *E. coli* there are 41 anticodon species alone; in eukaryotic cells even more) the ribosome must discriminate between relatively large ternary complexes (selection pathway) on the basis of an anticodon (which represents a relatively small discrimination area) [50].

During the initial decoding reaction the A site is in a low-affinity state which restricts reaction of the ternary complex to mainly codon–anticodon reactions, thus enabling discrimination between cognate and noncognate interactions. Such decoding is able to sense the stereochemical correctness of the base pairing in the anticodon–codon duplex. Simple mistakes in the decoding of a codon are thought to be prevented by ribosomal proofreading, also known as "kinetic" proofreading (for a more detailed discussion, see recent extensive books and reviews [50, 84–86]). It was observed that the second letter of the codon is most important (see Section 4.5; Fig. 4.3) due to its lowest detectable frequency in misreading, whereas the

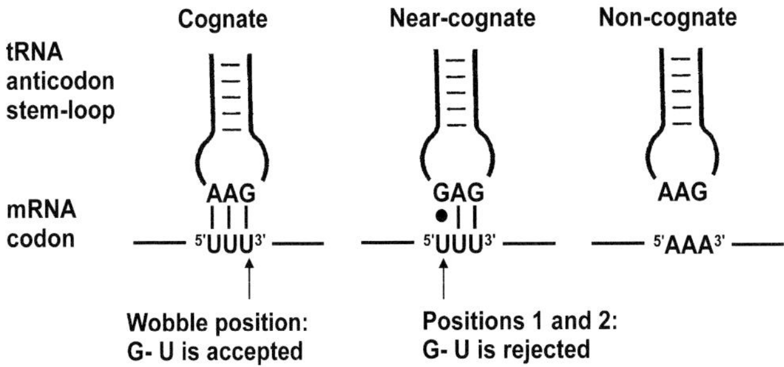

Fig. 3.10. Codon–anticodon base pairing in the ribosomal 30S A site ("decoding center"). In addition to the cognate aminoacyl-tRNA there are numerous noncognate as well as a few "near-cognate" aminoacyl-tRNA species that must be efficiently discriminated. The anticodon must correctly interact with first two nucleotides of the codon following the Chargraff pairing rules, whereas Crick's "wobbling" is allowed at the third position. Near-cognate aminoacyl-tRNA is rejected initially or during proofreading after accommodation in the A site of 50S subunit. (Modified from [89].)

third codon position ("wobble" position) allows some tRNAs to recognize several different codons (Fig. 3.10). The pairing at the wobble positions is also influenced by the nature of tRNA nucleoside modifications.

Loss of the correct reading frame (frameshift) usually means loss of genetic information due to the appearance of a premature stop codon in the A site and will result in the release of the peptide fragment from the ribosome. In other cases, such as the release of prematurely short peptidyl-tRNA from the ribosome, processing errors take place. For example, proteins whose synthesis on a ribosome has stalled (e.g. due to the frequent appearance of rare codons) are marked for degradation by the addition of peptide tags to their C-termini in a reaction-mediated *ssrA* (for small stable RNA) RNA (also known as tmRNA or 10Sa RNA), which functions as both a tRNA and mRNA [87]. Aminoacylation of tmRNA by AlaRS is an emergency mechanism to recycle stalled ribosomes in *E. coli*; it results in tagging of unfinished peptide and release of the ribosome from the unproductive complex. The control mechanisms of ribosomal translation, initiation and termination are discussed in detail elsewhere [88, 89], and are still the subject of intensive research.

3.9 Codon Bias and the Fidelity of Protein Synthesis

Protein sequences are reproduced many times over in the cell ontogenetic and phylogenetic history by a highly accurate mechanism that guarantees the invariance of their basic structure. The fidelity at all levels of protein synthesis is guaranteed by the already discussed proofreading (i.e. error-correcting) mechanisms [69]. During replication of DNA, which is a crucial process for the preservation of species, error rates are in the range 10^{-6} to 10^{-10}, while translation errors are usually about 2×10^{-3} to 2×10^{-4} in normal *E. coli* cells [75]. The most common errors during protein translation are frameshifts, premature truncations, low expression rates and misincorporations [90]. The rate of these errors is, among others, dependent on the nature and the context of the particular codons in a target sequence. For example, the error rate increases by about 100 times when the supply of charged tRNAs is reduced [91]. Indeed, it is possible to demonstrate *in vitro* that the fidelity of the translation is directly dependent on relative concentrations of isoacceptor tRNAs, i.e. each living cell has its own set of biases for the use of the 61 codons for the amino acid and tRNA population which is in direct correlation with the codon composition of total mRNA [92]. The frequency of appearance of particular synonymous codons (i.e. codon bias or codon usage) can be significantly different between genes, cell compartments, cells and organisms. Codon usage as a direct result of the degeneration of the genetic code is correlated to the abundance of a particular tRNA species in the cytosol.

It is well known that evolutionary advanced, i.e. specialized, eukaryotic cells, like erythrocytes, translate hemoglobin mRNA sequences with high fidelity. Indeed during ontogenic development of highly specialized erythrocytes, its amino acid pools and supply routes are regulated and properly balanced. However, the imbalances in amino acid pools lead to errors in tRNA charging. These considerations should be particularly taken into account in the case of heterologous gene expression. For example, *E. coli* that produces high levels of recombinant proteins which constitute up to 30% of total cell protein mass [93] can have problems with the intracellular balance and amino acid supply. It is not surprising since such *E. coli* is not an evolutionary specialized cell like the erythrocyte. Thus, efficient routine heterologous gene expression in *E. coli* might produce target proteins with considerable higher error rates [94].

A relatively high expression level of a cloned heterologous gene might impose a request for charged tRNAs on a host cells far above the capacity of their normal tRNA population ("stringent response") [15]. For example, the Arg codon AGA contributes on average 40% of the total $tRNA^{Arg}$ pool in yeast, while in *E. coli* it is the rarest codon (contributing less than 4% to total $tRNA^{Arg}$ population) [92]. There are well-documented cases of the expression of yeast genes in *E. coli* as host cells, where the higher rate of Arg $\rightarrow$ Lys side-chain substitutions in target proteins was observed [95]. This is possibly due to pronounced differences in codon usage between yeast and *E. coli*, and could even be enhanced through proper growth media manipulation. Thus, the presence of rare codons in heterologous mRNA resulted in the rapid clearance of specific charged isoacceptor tRNA, thus limiting higher rates of expression. Rare codons in heterologous mRNA might be responsible not only for errors in aminoacyl-tRNA selection on ribosomes and decreased rates of protein expression, but also might lead to mRNA stalling at ribosomes, and induce tagging and subsequent degradation of the partial translation product [87].

Finally, codon usage also dictates the strength of expression of a particular gene in a particular organism and serves as a signal for the control of gene expression

[96]. That means that the relative abundance of intracellular tRNA species correlates directly with gene expression. The most abundant tRNA species are mainly a function of expression of those genes that are important for cell growth and metabolism [97].

3.10 Preprogrammed Context-dependent Recoding: fMet, Sec, Pyl, etc.

The proteome of the living cell is multidimensional, i.e. a single gene can specify a number of different protein products. Indeed, organisms of all taxa are provided with a whole repertoire of pathways for molecular variations between a gene and its corresponding active protein (or RNA) product, like suppression, promotion, splicing, cotranslational and posttranslational modifications. In this way, the stem–loops, pseudoknots or specific 5′ or 3′ sequence elements in the mRNA structure can facilitate in-frame changes of decoding (frameshift, bypass) or even induce re-coding (read-through, suppression) or reassignment of a particular codon. For example, viruses such as retroviruses have evolved strategies to bend host machinery to favor their replication by various mechanisms that result in "hijacking" normal translation. These include programmed ribosome slipping (backward or forward frameshifting), ribosome hopping (bypassing a long sequence of a message) and suppression (read-through, reassignments) in specific sequence contexts. These phenomena are not restricted to viruses; in a minority of mRNA of most organisms diverse sequence-specific instructions are preprogrammed: (i) for altering the linear mechanism of readout and (ii) for sense or nonsense codon reassignments [98].

The AUG codon is the principal signal for initiation of protein synthesis in both eukaryotes and prokaryotes. However, this initiation event is context dependent. For example, in addition to AUG, there is a requirement for the Shine–Dalgarno sequence for proper initiation in prokaryotes; in eukaryotes it is the feature of the 5′-end of mRNA that is important for initiation of protein synthesis. The context-dependent alternative reading or re-coding of the same sense AUG codon should be principally possible only if two adaptors exist. Indeed, two distinct methionyl-tRNA^{Met} exist as substrates for the same cellular MetRS: one exclusively for initiation codons and the other for internal codons. In addition, prokaryotic Met-tRNA^{Met} is formylated [99] (Met-tRNA^{Met} $\rightarrow$ fMet-tRNA^{Met}) by intracellular enzymes in a similar manner to asparagines and glutamines (i.e. Asp-tRNA^{Asn} $\rightarrow$ Asn-tRNA^{Asn}; Glu-tRNA^{Gln} $\rightarrow$ Gln-tRNA^{Gln}) [41].

The extension of this strategy for in-frame stop codon suppression is also possible only in a given context. For this, special signals are required; these should allow the translation of machinery to distinguish between internal (those that should be reassigned) and termination coding triplets. In a limited number of genes, the in-frame UGA and UAG codons, normally termination triplets, specify the special noncanonical amino acid selenocysteine (Sec) [100] and pyrrolysine (Pyl) [101, 102]. For example, several enzymes (e.g. glutathione peroxidase, deiodinase and

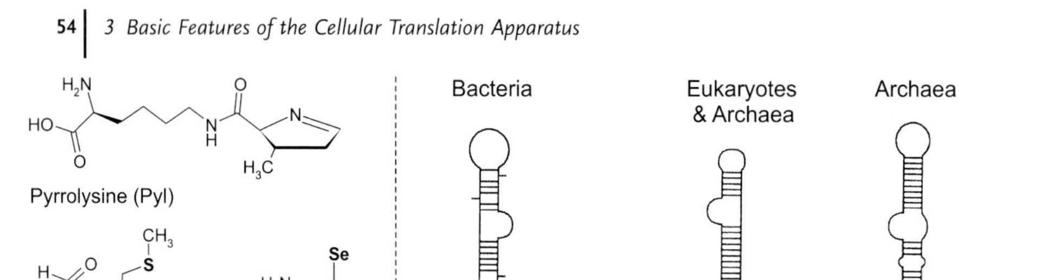

Fig. 3.11. Preprogrammed context-dependent re-coding of AUG, UGA and UAG codons. Structures of special canonical amino acids are presented on the left side. On the right side are RNA structures that serve as context signals for in-frame UGA (Sec) and UAG (Pyl) codon reassignment (modified from [107] with permission). The initial and AUG internal codons are distinguished by specific features of the N-terminal mRNA sequence (e.g. Shine–Dalgarno sequence element). The internal and terminal UGA (Sec) and UAG (Pyl) coding triplets are differentiated by specific (both upstream and downstream) local arrangements of the mRNA structure around the stop codon preprogrammed for suppression. In bacteria, SelB binds to Sec-tRNASec and SECIS, resulting in suppression of UGA which is in this context cognate to Sec (in Archaea and eukaryotes this process is slightly different) [103]. In other words, the immediate context of the UGA stop signal facilitates, for example, Pyl insertion at greater that 75% efficiency [105]. This is much more than for engineered suppression of orthogonal AARS:tRNA pairs, which are usually less than 20% efficient in incorporation (see Section 5.5.4). The existence of the Pyl insertion element (PYLIS), which acts in a manner similar to that of SECIS, is also postulated [107].

formate dehydrogenase) contain UGA codons almost in the middle of their coding regions and correspondingly Sec in their polypeptide sequences. In the deiodinase mRNA, a 250-bp region in the 3′-untranslated region (about 1 kbp removed from the UGA) is also required for selenocysteine incorporation. This local mRNA structure is known as the "Sec insertion element" (SECIS) that allows Sec-tRNA^{Sec} to find and translate the target UGA codon (Fig. 3.11). Sec has it own tRNA^{Sec} which is first charged with Ser by SerRS. Subsequently, the Ser-tRNA^{Sec} is enzymatically modified to form Sec-tRNA^{Sec}. Upon binding on a special EF-Tu-like elongation factor (SelB) Sec-tRNA^{Sec} competes directly with release factor 2 (RF2) for binding to the P site of the ribosome. Briefly, the cotranslational incorporation requires concerted interactions between specialized mRNA structures (SECIS), elongation factors (SelB) and accessory proteins (selenophosphate synthase, Sec synthase) [103].

The evolutionary rationale behind Sec context-dependent translational integration into proteins might be explained by chemical argumentation – there are significant physicochemical differences between Cys and its analog Sec [e.g. $pK_a(Sec) = 5.2$; $pK_a(Cys) = 8.3$]. At physiological pH 7.0 thiol function is not deprotonated, while selenols exist almost exclusively as selenolates and these are better nucleophiles than the thiolates [82]. Sec occurs as the active center of a few enzymes such as formate dehydrogenase; the Sec → Cys replacement in this enzyme lowers its turnover number by two orders of magnitude [103]. The complex chemistry of selenium requires Sec to be protected from the bulk solvent, i.e. water exclusion. Indeed, in the crystal structure of the Sec enzyme glutathione peroxidase the single active site Sec is located in a flat depression caged with aromatic and hydrophobic amino acids that sequesters the residue from the bulk solvent [104]. In the other words, an efficient cotranslational system that allows Sec protection during its translational integration into proteins evolved.

While Sec is metabolically generated by a pretranslational modification of Ser-$tRNA^{Sec}$, i.e. via indirect tRNA-dependent amino acid biosynthesis, Pyl is directly attached to $tRNA^{Pyl}{}_{CUA}$ by pyrrolysyl-tRNA synthetase (PylRS) as a response to a single in-frame UAG codon in *Methanosarcina barkeri* monomethylamine methyltransferase. The PylRS is truly the 21st AARS, belonging to an archaeal class II family which charge specifically Pyl to pyrrolysyl-tRNA ($tRNA^{Pyl}{}_{CUA}$); lysine itself and its cognate $tRNA^{Lys}$ are not substrates of this enzyme. Krzycki and coworkers have recently demonstrated that PylRS activates Pyl with ATP and charges it to $tRNA^{Pyl}{}_{CUA}$ *in vitro* and in *E. coli* cells [105]. The biochemical necessity of Pyl in the versatile methane-producing metabolism in *Methanosarcinidae* is easy to understand – the crystal structure of monomethylamine methyltransferase indicates a vital role of the Pyl side-chain in methylamine activation [101]. The questions of Pyl abundance and integration into metabolism and protein structure of other species remain to be clarified in future experiments.

Although versatile and multifaceted mechanisms for Sec, Pyl or fMet insertions in protein sequences exist, context dependency is the ubiquitous feature of these rather exceptional phenomena. They include recruitment of intermediary metabolism enzymes (fMet and Sec) and even an increase in the number of AARS:tRNA pairs (Pyl). They can be taxonomically assigned as species-specific local changes in the codon meaning or preprogrammed modifications of canonical decoding rules or simply context-dependent re-coding (Fig. 3.11). In this context, translated fMet, Sec and Pyl can be assigned as special canonical amino acids (see Section 1.6; Tab. 1.1), although contemporary literature almost unanimously treats them as the 21st (Sec) [100] and 22nd (Pyl) [106, 107] canonical amino acids. Future mapping of genomes and proteomes might yield yet other unpredictable mainly species-specific surprises, which (predictably) will be baptized as the 23rd, 24th, 25th, etc., amino acids.

The processes of the local changes in codon meaning are not strictly separated from the coding process; thus, they are probably as old as the basic coding itself. The extent of these phenomena is still the subject of speculation, i.e. whether such preprogrammed re-coding events are widespread mechanisms to produce minor products in addition to standard proteins from mRNA [98]. On the other hand, it is also unclear which intrinsic limitations in such variations are imposed. For example, which factors limit the variability of specific recognition mechanisms in AARS:tRNA interactions? Maybe these phenomena could be better understood in the light of growing experimental evidence which shows that aminoacylation and editing reactions play an important role in the control of diverse cellular processes [87]. However, it seems that during the evolution of advanced cellular forms, these progressed a step further and bypassed those ancestral machineries in order to replace them with more efficient enzymatic mechanisms which are strictly separated from basic coding (posttranslational modifications) as better strategies to compensate for restrictions in the expansion of the coding repertoire. Indeed, the chemical diversity of the proteins achieved by local changes in codon meaning could not rival those gained via posttranslational modifications.

3.11 Beyond Basic Coding – Posttranslational Modifications

Translation from the language of nucleotides to the language of amino acids takes place when the appropriate codon–anticodon pairs are linked together on the ribosome, while all reactions beyond this point are in fact posttranslational. It is well known that only a few proteins have a final covalent structure which is a simple accurate translation of mRNA. Indeed, there are strong reasons why the release of a completed polypeptide chain from a ribosome is usually not the last chemical step in the formation of a protein. For example, the regulation of protein activity is usually accomplished at the level of its gene expression, which determines the amount to be synthesized by the cell. However, other levels of regulation also exist; they are achieved by interactions with other proteins and small molecules or by enzymatically controlled posttranslational modifications.

Various covalent modifications (e.g. disulfide bridge formation, carboxypeptidase processing of basic amino acids, γ-carboxylation of Glu residues, phosphorylation, glycosylation, C-terminal amidation, etc.) often occur, either during or after assembly of the polypeptide chains. In fact, protein maturation is a result of processing that is inevitably connected with transport, noncovalent folding and covalent modification of peptide bonds and amino acid side-chains. With the delivery of new inorganic or organic groups for modifications, living cells provide protein structures with an additional versatility able to profoundly influence/modulate/change their properties and functions. Such modifications are normally enzyme-catalyzed, reversible (e.g. phosphorylation, nucleotidylation, methylation, acetylation or ADP rybosylation) or irreversible processes (e.g. limited proteolyses of zymogens like chymotrypsinogen, proelastase, procarboxypeptidase and prophospholipase) [108]. The spontaneous formation of mixed disulfides or cross-linking reactions in proteins such as collagen and elastin are examples of nonenzymatic posttranslational reactions. Protein maturation also includes cleavage of signal sequences in protein trafficking or polypeptide targeting for degradation. Other mature forms of proteins are generated by precursor cleavage (e.g. polyprotein processing, proteolytic cleavage, intein splicing), by specific modifications (e.g. L → D amino acid conversions) or by addition of various chemical groups and molecules (e.g. phosphoryl, methyl, hydroxyl, carboxyl, myristyl, prenyl, glycophospholipid, sugars, ADP-ribose, dipthamide and ubiquitin). The most versatile posttrans-

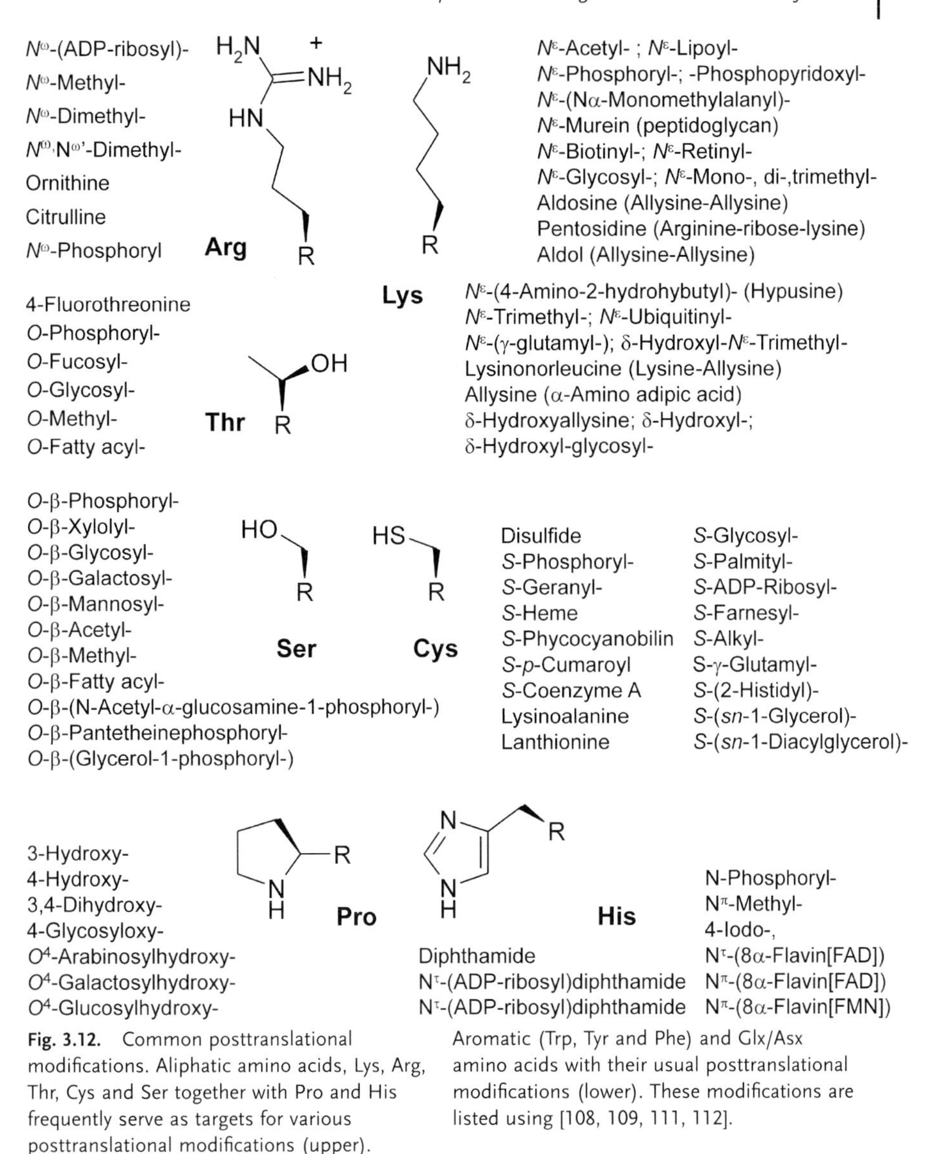

Fig. 3.12. Common posttranslational modifications. Aliphatic amino acids, Lys, Arg, Thr, Cys and Ser together with Pro and His frequently serve as targets for various posttranslational modifications (upper). Aromatic (Trp, Tyr and Phe) and Glx/Asx amino acids with their usual posttranslational modifications (lower). These modifications are listed using [108, 109, 111, 112].

lational chemistry is achieved at the functional groups of Tyr, Ser, Lys and Cys (Fig. 3.12) [109].

Although the "Anfinsen dogma" [110] predicts signals for the proper folding of a polypeptide to be contained within its amino acid sequence (i.e. principally no other proteins are needed for proper folding), some proteins do indeed require other proteins (i.e. chaperones) that assist in their folding in order to prevent aggregations or fibril formation, or cofactors such as flavins, pyridoxal and metal

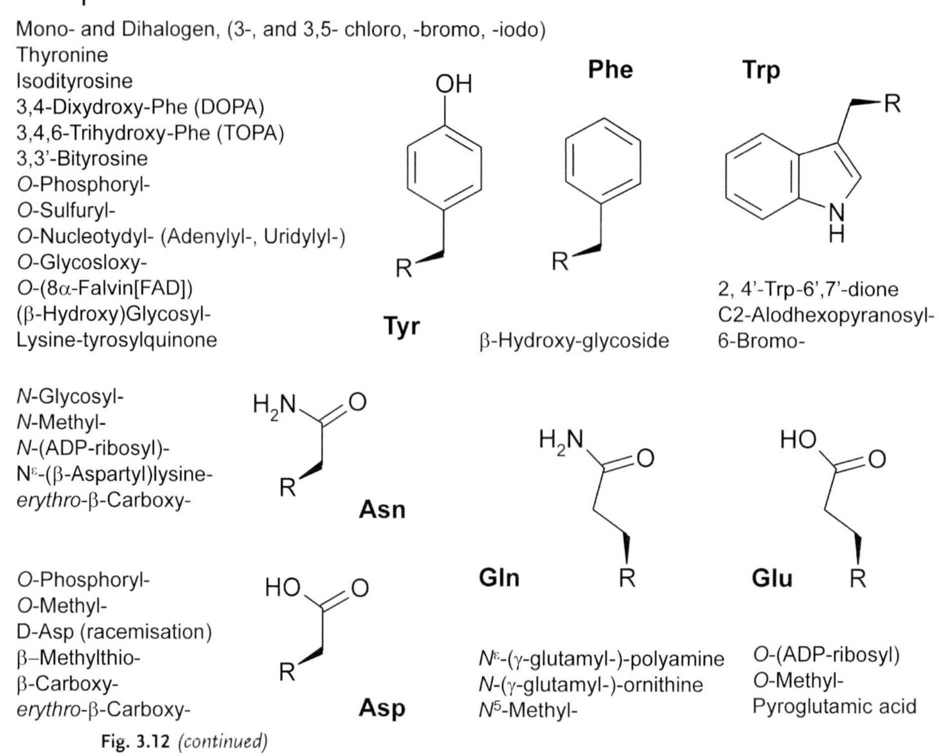

Fig. 3.12 *(continued)*

ions. However, disulfide bond formation, glycosylation, hydroxylation, proteolytic processing and *cis/trans* prolylpeptidyl isomerization can also influence the attainment of the native-folded state. In addition, sulfation, phosphorylation–dephosphorylation and a variety of other reactions might alter the conformation of a whole protein. Not surprisingly, reversible protein phosphorylation, principally on serine, threonine or tyrosine residues, plays critical roles in the regulation of many cellular processes, including the cell cycle, growth, apoptosis and signal transduction pathways [111].

A greatly expanded functional diversity achieved through posttranslational modifications of proteins can be found, especially in the evolutionarily advanced metazoans. The chemistry of these modifications is strictly separated from basic coding processes and includes selective attachment of different classes of chemical groups on defined residues in specific proteins. Protein glycosylation with carbohydrates like pentoses, hexosamines, *N*-acetylhexosamines, deoxyhexoses, hexoses and sialic acid, that are usually attached at residues like serine, threonine or asparagines, is one of the major classes of posttranslational modifications. Carbohydrates in the form of asparagine-linked (*N*-linked) or serine/threonine (*O*-linked) oligosaccharides are crucial structural components of many cell-surface and secreted proteins.

Glycoproteins on cell surfaces are important for communication between cells, for maintaining cell structure and for self-recognition by the immune system. Therefore, alteration of cell-surface glycoproteins can generate significant effects on cellular physiology and activity [112].

Primary amines from the protein N-terminus or lysine residues are targets for the following posttranslational chemical reactions: methylation, acetylation, farnesylation, biotinylation, stearoylation, formylation, myristoylation, palmitoylation, geranylgeranylation and lipoic acid additions [109]. Sulfhydryl group of cysteine which participates in a rather large number of posttranslational reactions (Fig. 3.12) is also a target for glutationylation and oxidation. The appearance of Met-sulfoxide in proteins is the result of an undesirable, spontaneous oxidation. This process (that might lead to inactivation of the protein) is reversed *in vivo* by Met(O) reductase, an enzyme with a repair function [113] (see Section 5.2.1.6, Fig. 5.13). Asparagines and glutamines are often deamidated, especially after longer protein storage in various buffers [114].

Amino acids modified in this way are assigned as "biogenic" (see Section 1.6; Tab. 1.1). This class of amino acids includes metabolic products like urea cycle intermediates (ornithine, citrulline and arginosuccinate), homocysteine, β-alanine or amino acids with a special physiologic function like γ-butyric acid, creatine or thyroxine, or they can arise as secondary metabolites often found in large quantities in plants and microorganisms [115]. Enzyme-mediated production of novel biogenic amino acids, accompanied by large protein functional diversification, is mainly a "privilege" of eukaryotes, especially metazoans. This is expected, since these advanced and mostly specialized cell forms with networks of membrane structures and many compartments require additional chemistries to perform their sophisticated functions. Although these operations are energetically expensive, eukaryotes can afford them by switching-off many of their basic synthetic activities. Prokaryotes indeed have a simple synthetic machinery that uses raw chemical building blocks and are thus independent of any other life form [116]. Many eukaryotes can do away with the task of completing some of the basic biological syntheses and, by retaining the basic cytoplasmic chemistry of prokaryotes, can use them as a source of essential chemicals like fats, essential amino acids, co-enzymes and minerals [117].

References

1 Mayr, E. (2001). *What Evolution Is.* Perseus Books, New York.

2 Budisa, N. (2004). Prolegomena to future efforts on genetic code engineering by expanding its amino acid repertoire. *Angewandte Chemie International Edition* **43**, 3387–3428.

3 Fischer, E. (1906). Untersuchungen über Aminosäuren, Polypeptide und Proteine. *Berichte der deutschen chemischen Gesellschaft* **39**, 530–610.

4 Sanger, F. and Thompson, E. O. P. (1952). The amino-acid sequence in the glycyl chain of insulin. *Biochemical Journal* **52**, R3–R3.

5 Griffith, F. (1928). The significance of pneumococcal types. *Journal of Hygiene* **27**, 113–159.

6 Avery, O. T., MacLeod, C. M. and McCarty, M. (1944). Studies on the chemical nature of the substance inducing transformation of pneumococcal types. *Journal of Experimental Medicine* **79**, 137–157.

7 Hershey, A. D. and Chase, M. (1952). Independent functions of viral protein and nucleic acid in growth of bacteriophage, *Journal of General Physiology* **36**, 39–56.

8 Brenner, S., Jacob, F. and Meselson, M. (1961). An unstable intermediate carrying information from genes to ribosomes for protein synthesis. *Nature* **190**, 576–581.

9 Beadle, G. W. (1957). Genetic basis of biological specificity. *Journal of Allergy* **28**, 392–400.

10 Jacob, F. and Monod, J. (1961). Genetic regulatory mechanisms in the synthesis of proteins. *Journal of Molecular Biology* **3**, 318–356.

11 Crick, F. H. C. (1957). On protein synthesis. *Symposium of the Society for Experimental Biology* **12**, 138–163.

12 Danchin, A., Guerdoux-Jamet, P., Moszer, I. and Nitschke, P. (2000). Mapping the bacterial cell architecture into the chromosome. *Philosophical Transactions of the Royal Society of London Series B Biological Sciences* **355**, 179–190.

13 Dupont, C. (2003). Protein requirements during the first year of life. *American Journal of Clinical Nutrition* **77**, 1544S–1549S.

14 Meisenberg, G. and Simmonis, W. H. (1998). Principles of Medical Biochemistry. Mosby, St Louis, MO.

15 Neidhardt, F. C. and Umbarger, E. N. (1996). Chemical composition of Escherichia coli. In *Escherichia coli and Salmonella: Cellular and Molecular Biology* (Neidhardt, F. C. et al. eds), pp. 13–16. 2nd Ed. ASM Press, Washington, DC.

16 Ibba, M. and Söll, D. (1999). Quality control mechanisms during translation. *Science* **286**, 1893–1897.

17 de Pouplana, L. R. and Schimmel, P. (2004). Aminoacylations of tRNAs: record-keepers for the genetic code. In *Protein Synthesis and Ribosome Structure* (Nierhaus, K. and Wilson, D. N., eds), pp. 169–184. Wiley-VCH, Weinheim.

18 Vaughan, M. and Steinberg, D. (1959). The specificity of protein synthesis. *Advances in Protein Chemistry* **XIV**, 116–173.

19 Doolittle, R. F. and Handy, J. (1998). Evolutionary anomalies among the aminoacyl-tRNA synthetases. *Current Opinion in Genetics and Development* **8**, 630–636.

20 Szymanski, M., Deniziak, M. and Barciszewski, J. (2000). The new aspects of aminoacyl-tRNA synthetases. *Acta Biochimica Polonica* **47**, 821–834.

21 Ibba, M. and Söll, D. (2000). Aminoacyl-tRNA synthesis. *Annual Review of Biochemistry* **69**, 617–650.

22 Yaremchuk, A., Cusack, S. and Tukalo, M. (2000). Crystal structure of a eukaryote/archaeon-like prolyl-tRNA synthetase and its complex with tRNAPro(CGG). *EMBO Journal* **19**, 4745–4758.

23 Ibba, M. and Söll, D. (2003). Aminoacyl-tRNA synthetase and evolution. In *Translation Mechanisms* (Lapointe, J. and Brakier-Gingras, L., eds), pp. 25–37. Landes Biosciences, Georgetown, TX.

24 Mirande, M. (2003). Multi-AARS complexes. In *Aminoacyl-tRNA Synthetases* (Ibba, M., Francklyn, C. and Cusack, S., eds), pp. 298–309. Landes Bioscience, Georgetown, TX.

25 Eriani, G., Delarue, M., Poch, O., Gangloff, J. and Moras, D. (1990). Partition of tRNA synthetases into two classes based on mutually exclusive sets of sequence motifs. *Nature* **347**, 203–306.

26 Henkin, T.M. (2003). Regulation of Aminoacyl-tRNA Synthetase gene expression in bacteria. In *Aminoacyl-tRNA Synthetases* (Ibba, M., Francklyn, C. and Cusack, S., eds), pp. 309–313. Landes Bioscience, Georgetown, TX.

27 Ibba, M., Morgan, S., Curnow, A. W., Pridmore, D. R., Vothknecht, U. C., Gardner, W., Lin, W., Woese, C. R. and Söll, D. (1997). A euryarchaeal

Lysyl-tRNA synthetase: resemblance to class I synthetases. *Science* **278**, 1119–1122.

28 Blanquet, S., Crepin, T., Mechulam, Y. and Schmitt, E. (2003). Methionyl-tRNA synthetases. In *Aminoacyl-tRNA Synthetases* (Ibba, M., Francklyn, C. and Cusack, S., eds), pp. 47–58. Landes Biosciences, Georgetown, TX.

29 Bedouelle, H. (2003). Tyrosyl-tRNA synthetases. In *Aminoacyl-tRNA Synthetases* (Ibba, M., Francklyn, C. and Cusack, S., eds), pp. 111–125. Landes Biosciences, Georgetown, TX.

30 Dubois, D. Y., Lapointe, J. and Shun-ichi, S. (2003). Glutamyl-tRNA synthetases. In *Aminoacyl-tRNA Synthetases* (Ibba, M., Francklyn, C. and Cusack, S., eds), pp. 89–94. Landes Biosciences, Georgetown, TX.

31 Brick, P. and Blow, D. M. (1987). Crystal-structure of a deletion mutant of a tyrosyl-transfer RNA-synthetase complexed with tyrosine. *Journal of Molecular Biology* **194**, 287–297.

32 Gay, G. D., Duckworth, H. W. and Fersht, R. A. (1993). Modification of the amino-acid specificity of tyrosyl-tRNA synthetase by protein engineering. *FEBS Letters* **318**, 167–171.

33 First, E. A. (2003). Catalysis of the tRNA aminoacylation reaction. In *Aminoacyl-tRNA Synthetases* (Ibba, M., Francklyn, C. and Cusack, S., eds), pp. 328–347. Landes Bioscience, Georgetown, TX.

34 Carter, C. W. (2003). Trytophanyl-tRNA synthetases. In *Aminoacyl-tRNA Synthetases* (Ibba, M., Francklyn, C. and Cusack, S., eds), pp. 99–108. Landes Bioscience, Georgetown, TX.

35 Cusack, S., Colominas, C., Härtlein, M., Nassar, N. and Leberman, R. (1990). A second class of synthetase structure revealed by X-ray-analysis of *Escherichia coli* seryl-transfer RNA synthetase at 2.5 Å. *Nature* **347**, 249–255.

36 Weygand-Durasevic, I. and Cusack, S. (2003). Seryl-tRNA synthetases. In *Aminoacyl-tRNA Synthetases* (Ibba, M., Francklyn, C. and Cusack, S., eds), pp. 177–190. Landes Bioscience, Georgetown, TX.

37 Ibba, M., Francklyn, C. and Cusack, S. (eds) (2003). *Aminoacyl-tRNA Synthetases.* Landes Bioscience, Georgetown, TX.

38 Lapointe, J. and Brakier-Gingras, L. (2003). *Translation Mechanisms*, Landes Bioscience, Georgetown, TX.

39 Nureki, O., Kohno, T., Sakamoto, K., Miyazawa, T. and Yokoyama, S. (1993). Chemical modification and mutagenesis studies on zinc-binding of aminoacyl-transfer RNA-synthetases. *Journal of Biological Chemistry* **268**, 15368–15373.

40 Swairjo, M. A. and Schimmel, P. R. (2005). Breaking sieve for steric exclusion of a noncognate amino acid from active site of a tRNA synthetase. *Proceeding of the National Academy of Sciences of the USA* **102**, 988–993.

41 Feng, L., TumbulaHansen, D., Min, B., Namgoong, S., Salazar, J., Orellana, O. and Söll, D. (2003). Transfer RNA-dependent amidotransferases: key enzymes for Gln-tRNA and Asn-tRNA synthesis in nature. In *Aminoacyl-tRNA Synthetases* (Ibba, M., Francklyn, C. and Cusack, S., eds), pp. 314–320. Landes Bioscience, Georgetown, TX.

42 Stathopoulos, C., Li, T., Longman, R., Vothknecht, U. C., Becker, H. D., Ibba, M. and Söll, D. (2000). One polypeptide with two aminoacyl-tRNA synthetase activities. *Science* **287**, 479–482.

43 Ibba, M., Becker, H. D., Stathopoulos, C., Tumbula, D. L. and Söll, D. (2000). The Adaptor hypothesis revisited. *Trends in Biochemical Sciences* **25**, 311–316.

44 Zamecnik, P. C. (1960). Historical and current aspects of the problem of protein synthesis. *Harvey Lectures* **54**, 256–279.

45 Hoagland, M. B. and Zamecnik, P. C. (1957). Intermediate reactions in protein biosynthesis. *Federation Proceedings* **16**, 197–197.

46 Watson, J. D., Hopkins, N. H., Roberts, J. W., Steitz, J. A. and Weiner, A. M. (1988). *Molecular Biology of the Gene.* Benjamin Cummings, Amsterdam.

47 Chapeville, F., Ehrenstein, G. V., Benzer, S., Weisblum, B., Ray, W. J. and Lipmann, F. (1962). On the role of soluble ribonucleic acid in coding for amino acids. *Proceedings of the National Academy of Sciences of the USA* **48**, 1086–1092.
48 Holley, R. W., Apgar, J., Everett, G. A., Madison, J. T., Merrill, S. H., Penswick, J. R. and Zamir, A. (1965). Structure of an alanine transfer RNA. *Federation Proceedings* **24**, 216.
49 Marquez, V. and Nierhaus, K. (2003). tRNA and synthesis. In *Protein Synthesis and Ribosome Structure* (Nierhaus, K. and Wilson, D. N., eds), pp. 145–183. Wiley-VCH, Weinheim.
50 Nierhaus, K. and Wilson, D. N. (eds) (2004). *Protein Synthesis and Ribosome Structure.* Wiley-VCH, Weinheim.
51 Giege, R. and Frugier, M. (2003). Transfer RNA structure and identity. In *Translation Mechanisms* (Lapointe, J. and Brakier-Gingras, L., eds), pp. 1–24. Landes Biosciences, Georgetown, TX.
52 Giege, R., Sissler, M. and Florentz, C. (1998). Universal rules and idiosyncratic features in tRNA identity. *Nucleic Acids Research* **26**, 5017–5035.
53 Robertus, J. D., Ladner, J. E., Finch, J. T., Rhodes, D., Brown, R. S., Clark, B. F. and Klug, A. (1974). Structure of yeast phenylalanine tRNA at 3 Å resolution. *Nature* **250**, 546–551.
54 Kim, S. H., Suddath, F. L., Quigley, G. J., McPherson, A., Sussman, J. L., Wang, A. H. J., Seeman, N. C. and Rich, A. (1974). 3-Dimensional tertiary structure of yeast phenylalanine transfer-RNA. *Science* **185**, 435–440.
55 Beuning, P. J. and Musier-Forsyth, K. (1999). Transfer RNA recognition by aminoacyl-tRNA synthetases. *Biopolymers* **52**, 1–28.
56 Calendar, R. and Berg, P. (1966). Catalytic properties of tyrosyl ribonucleic acid synthetases from *Escherichia coli* and *Bacillus subtilis. Biochemistry* **5**, 1690–1695.
57 Schimmel, P. R. and Söll, D. (1979). Aminoacyl transfer RNA-synthetases – general features and recognition of transfer-RNAs. *Annual Review of Biochemistry* **48**, 601–648.
58 Schultz, D. W. and Yarus, M. (1994). Transfer-RNA mutation and the malleability of the genetic code. *Journal of Molecular Biology* **235**, 1377–1380.
59 Schultz, D. W. and Yarus, M. (1996). On malleability in the genetic code. *Journal of Molecular Evolution* **42**, 597–601.
60 Hendrickson, T. L. and Schimmel, P. (2003). Transfer RNA-dependent amino acid discrimination by aminoacyl-tRNA synthetase. In *Translation Mechanisms* (Lapointe, J. and Brakier-Gingras, L., eds), pp. 35–69. Landes Bioscience, Georgetown, TX.
61 Schulman, L. H. A. J. (1988). Recent excitement in understanding transfer RNA identity. *Science* **240**, 1591–1592.
62 McClain, W. H. (1993). Transfer-RNA identity. *FASEB Journal* **7**, 72–78.
63 Shimizu, M., Asahara, H., Tamura, K., Hasegawa, T. and Himeno, H. (1992). The role of anticodon bases and the discriminator nucleotide in the recognition of some *Escherichia coli* transfer-RNAs by their aminoacyl-transfer RNA-synthetases. *Journal of Molecular Evolution* **35**, 436–443.
64 Rould, M. A., Perona, J. J., Söll, D. and Steitz, T. A. (1989). Structure of *E. coli* glutaminyl-tRNA synthetase complexed with tRNA(Gln) and ATP at 2.8 Å resolution. *Science* **246**, 1135–1142.
65 Cavarelli, J., Rees, B., Ruff, M., Thierry, J. C. and Moras, D. (1993). Yeast transfer RNA (Asp) recognition by its cognate class-II aminoacyl-transfer RNA synthetase. *Nature* **362**, 181–184.
66 Saks, M. E., Sampson, J. R. and Abelson, J. N. (1994). The transfer-RNA identity problem – a search for rules. *Science* **263**, 191–197.
67 de Pouplana, L. R. and Schimmel, P. (2001). Operational RNA code for amino acids in relation to genetic code

in evolution. *Journal of Biological Chemistry* **276**, 6881–6884.

68 Schimmel, P. and Söll, D. (1997). When protein engineering confronts the tRNA world. *Proceedings of the National Academy of Sciences of the USA* **94**, 10007–10009.

69 Jakubowski, H. (2003). Accuracy of aminoacyl-RNA synthetases: proofreading of amino acids. In *Aminoacyl-tRNA Synthetases* (Ibba, M., Francklyn, C. and Cusack, S., eds), pp. 384–396. Landes Bioscience, Georgetown, TX.

70 Wakasugi, K., Quinn, C. L., Tao, N. J. and Schimmel, P. (1998). Genetic code in evolution: switching species-specific aminoacylation with a peptide transplant. *EMBO Journal* **17**, 297–305.

71 Soma, A. and Himeno, H. (1998). Cross-species aminoacylation of tRNA with a long variable arm between *Escherichia coli* and *Saccharomyces cerevisiae. Nucleic Acids Research* **26**, 4374–4381.

72 Nair, S., de Pouplana, L. R., Houman, F., Avruch, A., Shen, X. Y. and Schimmel, P. (1997). Species-specific tRNA recognition in relation to tRNA synthetase contact residues. *Journal of Molecular Biology* **269**, 1–9.

73 Fersht, A. R. (1981). Enzymic editing mechanisms and the genetic code. *Proceedings of the Royal Society of London Series B Biological Sciences* **212**, 351–379.

74 Fersht, R. A., Shindler, J. S. and Tsui, W. C. (1980). Probing the limits of protein-amino acid side chain recognition with the AARS: discrimination against Phe by tyrosyl-tRNA synthetases. *Biochemistry* **19**, 5520–5524.

75 Jakubowski, H. and Goldman, E. (1992). Editing of errors in selection of amino acids for protein synthesis. *Microbiological Reviews* **56**, 412–429.

76 Fukai, S., Nureki, O., Sekine, S., Shimada, A., Tao, J. S., Vassylyev, D. G. and Yokoyama, S. (2002). Structural basis for double-sieve discrimination of L-valine from L-isoleucine and L-threonine by the complex of tRNA(Val) and valyl-tRNA synthetase. *Cell* **103**, 353–362.

77 Dock-Bregeon, A. C., Rees, B., Torres-Larios, A., Bey, G., Caillet, J. and D. M. (2004). Achieving error-free translation: the mechanism of proofreading of threonyl-tRNA synthetase at atomic resolution. *Molecular Cell* **16**, 375–386.

78 Fersht, A. R. (1980). Enzymic editing mechanisms in protein synthesis and DNA-replication. *Trends in Biochemical Sciences* **5**, 262–265.

79 Francklyn, C. (2003). tRNA synthetase paralogs: evolutionary links in the transition from tRNA-dependent amino acid biosynthesis to *de novo* biosynthesis. *Proceedings of the National Academy of Sciences of the USA* **100**, 9650–9652.

80 Jakubowski, H. (2000). Translational incorporation of *S*-nitrosohomocysteine into protein. *Journal of Biological Chemistry* **275**, 21813–21816.

81 Fersht, A. R. and Dingwall, C. (1979). Editing mechanism for the methionyl-tRNA synthetase in the selection of amino acids in protein synthesis. *Biochemistry* **18**, 1250–1256.

82 Budisa, N., Minks, C., Alefelder, S., Wenger, W., Dong, F. M., Moroder, L. and Huber, R. (1999). Toward the experimental codon reassignment *in vivo*: protein building with an expanded amino acid repertoire. *FASEB Journal* **13**, 41–51.

83 Igloi, G. L., von der Haar, F. and Cramer, F. (1977). Hydrolytic action of aminoacyl-tRNA synthetases from baker's yeast – chemical proofreading of Thr-tRNA-Val by valyl-tRNA synthetase studied with modified tRNA-Val and amino acid analogs. *Biochemistry* **16**, 1696–1702.

84 Wilson, D. N. and Nierhaus, K. H. (2003). The ribosome through the looking glass. *Angewandte Chemie International Edition* **42**, 3464–3486.

85 Ramakrishnan, V. (2002). Ribosome structure and the mechanism of translation. *Cell* **108**, 557–572.

86 Moore, P. B. and Steitz, T. A. (2003). The structural basis of large ribosomal

subunit function. *Annual Review of Biochemistry* **72**, 813–850.

87 Keiler, K. C., Waller, P. R. H. and Sauer, R. T. (1996). Role of a peptide tagging system in degradation of proteins synthesized from damaged messenger RNA. *Science* **271**, 990–993.

88 Rodnina, M. V. and Wintermeyer, W. (2001). Ribosome fidelity: tRNA discrimination, proofreading and induced fit. *Trends in Biochemical Sciences* **26**, 124–130.

89 Ogle, J. M., Carter, A. P. and Ramakrishnan, V. (2003). Insights into the decoding mechanism from recent ribosome structures. *Trends in Biochemical Sciences* 128.

90 Emilsson, V. and Kurland, C. G. (1990). Growth rate dependence of transfer-RNA abundance in *Escherichia coli*. *EMBO Journal* **9**, 4359–4366.

91 Sorensen, M. A., Kurland, C. G. and Pedersen, S. (1989). Codon usage determines translation rate in *Escherichia coli*. *Journal of Molecular Biology* **207**, 365–377.

92 Osawa, S., Jukes, T. H., Watanabe, K. and Muto, A. (1992). Recent evidence for evolution of the genetic code. *Microbiological Reviews* **56**, 229–264.

93 Studier, F. W., Rosenberg, A. H., Dunn, J. J. and Dubendorff, J. W. (1990). Use of T7 RNA-polymerase to direct expression of cloned genes. *Methods in Enzymology* **185**, 60–89.

94 Kane, J. F. (1995). Effects of rare codon clusters on high-level expression of heterologous proteins in *Escherichia coli*. *Current Opinion in Biotechnology* **6**, 494–500.

95 Calderone, T. L., Stevens, R. D. and Oas, T. G. (1996). High-level misincorporation of lysine for arginine at AGA codons in a fusion protein expressed in *Escherichia coli*. *Journal of Molecular Biology* **262**, 407–412.

96 Kurland, C. G. (1991). Codon bias and gene expression. *FEBS Letters* **285**, 165–169.

97 Kurland, C. G. (1992). Translational accuracy and the fitness of bacteria. *Annual Review of Genetics* **26**, 29–50.

98 Poole, E. S., Major, L. L., Cridge, A. G. and Tate, W. P. (2003). The mechanism of recoding in pro- and eukaryotes. In *Protein Synthesis and Ribosome Structure* (Nierhaus, K. and Wilson, D. N., eds), pp. 397–426. Wiley-VCH, Weinheim.

99 Dury, A. (1968). Formyl-methionine tRNA and valine tRNA available. *Science* **162**, 621.

100 Bock, A., Forchhammer, K., Heider, J., Leinfelder, W., Sawers, G., Veprek, B. and Zinoni, F. (1991). Selenocysteine – the 21st amino-acid. *Molecular Microbiology* **5**, 515–520.

101 Srinivasan, G., James, C. M. and Krzycki, J. A. (2002). Pyrrolysine encoded by UAG in Archaea: charging of a UAG-decoding specialized tRNA. *Science* **296**, 1459–1462.

102 Hao, B., Gong, W. M., Ferguson, T. K., James, C. M., Krzycki, J. A. and Chan, M. K. (2002). A new UAG-encoded residue in the structure of a methanogen methyltransferase. *Science* **296**, 1462–1466.

103 Bock, A., Thanbichler, M., Rother, M. and Resch, A. (2003). Selenocysteine. In *Aminoacyl-tRNA Synthetases* (Ibba, M., Francklyn, C. and Cusack, S., eds), pp. 320–328. Landes Bioscience, Georgetown, TX.

104 Ladenstein, R., Epp, O., Bartels, K., Jones, A., Huber, R. and Wendel, A. (1979). Structure analysis and molecular model of the selenoenzyme glutathione peroxidase at 2.8 Å resolution. *Journal of Molecular Biology* **134**, 199–218.

105 Blight, S. K., Larue, R. C., Mahapatra, A., Longstaff, D. G., Chang, E., Zhao, G., Kang, P. T., Green-Church, K. B., Chan, M. K. and Krzycki, J. A. (2004). Direct charging of tRNA(CUA) with pyrrolysine *in vitro* and *in vivo*. *Nature* **431**, 333–335.

106 Atkins, J. F. and Gesteland, R. (2002). The 22nd amino acid. *Science* **296**, 1409–1410.

107 Schimmel, P. and Beebe, K. (2004). Genetic code sizes pyrrolysine. *Nature* **431**, 257–258.

108 Graves, D. J., Martin, B. L. and Wang, J. H. (1994). *Co- and Post-Translational Modification of Proteins – Chemical Principles and Biological Effects*. Oxford University Press, Oxford.

109 Kannicht, C. (2002). *Posttranslational Modification of Proteins Tools for Functional Proteomics*. Humana Press, Totowa, NJ.

110 Baldwin, R. L. (1996). Protein folding from 1961 to 1982. *Nature Structural and Molecular Biology* **6**, 814–817.

111 Clemens, M. J. (1997). *Protein Phosphorylation in Cell Growth Regulation*. Gordon & Breach, London.

112 Brockhausen, I. and Kuhns, W. (1997). *Glycoproteins and Human Disease*. Landes Bioscience, Georgetown, TX.

113 Vogt, W. (1995). Oxidation of methionyl residues in proteins – tools, targets, and reversal. *Free Radical Biology and Medicine* **18**, 93–105.

114 Wright, H. T. (1991). Nonenzymatic deamidation of asparaginyl and glutaminyl residues in proteins. *Critical Reviews in Biochemistry and Molecular Biology* **26**, 1–52.

115 Rosenthal, G. (1982). *Plant Non-protein Amino and Imino Acids. Biological, Biochemical and Toxicological Properties*. Academic Press, New York.

116 Budisa, N., Moroder, L. and Huber, R. (1999). Structure and evolution of the genetic code viewed from the perspective of the experimentally expanded amino acid repertoire *in vivo*. *Cellular and Molecular Life Sciences* **55**, 1626–1635.

117 Williams, R. J. P. and Fraústo da Silva, J. J. R. (1996). *The Natural Selection of the Chemical Elements: The Environment and Life's Chemistry*. Oxford University Press, New York.

4
Amino Acids and Codons – Code Organization and Protein Structure

Numerum combinationum in terminis etiam numero finitis esse infinitum; si rite omnia expendantur. In ipso immenso combinatorium numero in immensum plures esse combinationes inordinatas, quam ordinatas. [Careful examination reveals the infinite number of combinations between the finite numbers of terms. In this large number of combinations there are far more without order than with it.]

R. J. Boscovich (1763). *Theoria Philosophie Naturalis Redacta ad Unicam Legem Virium in Natura Existenttium*, pp. 254–256. Ex Typographia Remondiana, Venice.

4.1
Basic Features and Adaptive Nature of the Universal Genetic Code

The genetic code arrangement as presented in Fig. 4.1 is also known as the "universal code" and has at least three basic features: (i) it has a strict codon synonymy, (ii) it specifies a definite number of chemical species that serve as protein building blocks and (iii) it is redundant in the values of physicochemical properties (e.g. the extent of encoded hydrophobicity and hydrophilicity) [3]. These features can be elaborated as follows. First, since most of the amino acids are represented by more than one codon, the genetic code is said to be degenerate or even "redundant" in its structure. Coding triplets that encode the same amino acid are termed synonymous codons. Second, the chemical diversity encoded in this way includes a limited number of side-chain functional groups: carboxylic acids (Glu and Asp), amides (Gln and Asn), thiol (Cys) and thiol ester (Met), alcohols (Ser and Thr), basic amines (Lys), guanidine (Arg), aliphatic (Ala, Val, Leu and Ile) and aromatic side-chains (Phe, Tyr, His and Trp), and even the cyclic imino acid Pro (see Section 5.2.1.5). While Met (AUG) and Trp (UGG) are assigned only with single coding triplets, the rest of the 18 amino acids have at least two codons. On average, an amino acid is encoded by three different codons. As a consequence, any gene can be expressed using different codons, i.e. each gene has a specific codon usage. Third, the increased redundancy in coding of strictly polar and strictly apolar amino acids reflects highly synonymous quotas for Arg, Ser (XGX group of codons), Leu and other hydrophobic amino acids (XUX group of codons). Such an

Engineering the Genetic Code. Nediljko Budisa

ISBN: 3-527-31243-9

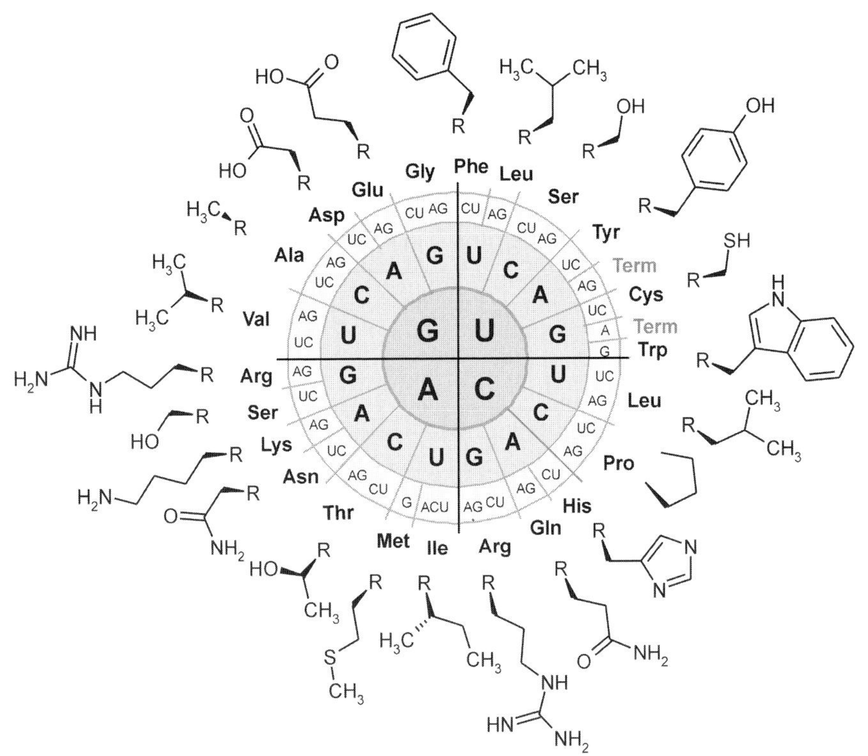

Fig. 4.1. Codons (mRNA format) and related amino acids as basic building blocks for protein synthesis. Although there are 10^{84} possible codes containing at least one codon for each of the 20 amino acids and termination signal, the structure of the genetic code is such that it assigns 20 amino acids to 61 codons and remaining three are stop codons: UAG (amber), UGA (opal) and UAA (ochre). (R denotes an amino carboxyl group). More details about the history of genetic code deciphering and its general features are available in standard textbooks [1, 2].

arrangement is directly related to protein folding and building as well as translational error minimization (Section 4.5). Despite numerous reports about variations in codon assignments, mainly in mitochondria and ciliate protozoa (Section 4.7), the genetic code can be regarded as conserved and universal.

This book will follow the line of reasoning that explains patterns within the genetic code as a result of optimization processes during evolution. Such a code is adaptive, i.e. it results from adaptation that optimizes its function such that the number of errors during translation is minimized and the effects of mutations reduced. For code engineering as a research field this is indeed plausible since it would be highly disadvantageous for living cells to accumulate a lot of errors in proteins. Such reasoning can be traced back to the pioneering work of Woese [4], who was among first to suggest that patterns within the code reflect the physicochemical properties of amino acids (Section 4.5). However, some recent theoretical considerations indicate that the genetic code itself is not optimal. For example, Di Giulio and Medugno [5] calculated that the code has achieved 68% minimization of polarity distance (which is a normalized chemical scale derived from water-to-ethanol distribution interactions). This is not surprising since the code itself is not the result of any intelligent design, but rather the result of evolutionary interplay between chance and necessity. In other words, the genetic code as a product of evolutionary optimization cannot considered as optimal (see also Section 6.4.2). One might therefore expect that a major challenge for code engineering as a new research field might be the design of target-engineered genetic codes with a much higher degree of optimization than natural ones.

4.2 Metabolism and Intracellular Uptake of Canonical Amino Acids

Amino acid metabolism in all living beings is characterized by a balance between the rates of protein synthesis and degradation. As a result of this balance, amino acids are found in living organisms in both their free (modified or unmodified) forms, covalently integrated into peptides and proteins or conjugated into other structures. Since life is based on protein chemistry, through its evolution on Earth, a large variety of organisms have developed idiosyncratic ways of handling amino acids. Therefore, many findings regarding (i) the mechanism of the formation of amino acids, (ii) their chemical nature, and (iii) their intracellular transport, turnover and discharge properties in *Escherichia coli* apply equally well to mammalian and other cell types [6].

In the context of cellular metabolism, canonical amino acids are "multitasking" i.e. they act as substrates perfectly integrated into the fine network of cellular biosynthetic, metabolic and energetic functions. Amino acids may undergo different metabolic pathways, serving as catabolic substrates for energy production by deamination that results in keto-acid residues capable of entering the glycolytic pathway at the level of pyruvate or acetyl CoA or else enter the Krebs cycle [7]. In addition, they are basic building blocks for protein synthesis, and important components in the biosynthesis of numerous biological molecules like nucleic acids, glutathione, nicotinamide sphingosins, porphyrin derivatives, various metabolic cofactors, neurotransmitters and pigments. For example, the majority of Tyr unused in proteins synthesis is catabolized for energy production or enters a pathway for conversion to catecholamines (neurotransmitters: dopamine, norepinephrine and epinephrine), while Trp serves as the precursor for the synthesis of serotonin (5-hydroxytryptamine) [8].

Most prokaryotes have functional biochemical pathways for the synthesis of all 20 amino acids. In contrast, mammalian cells generate only nonessential amino acids (Ala, Arg, Asp, Asn, Cys, Glu, Gln, Gly, Pro, Ser and Tyr), while the essential ones are provided via an exogenous supply [9]. However, even nonessential amino acids are imported from the surrounding medium when available. Even *E. coli* cells that have the ability to synthesize all amino acids from inorganic salts and glucose maintain the capacity to actively accumulate external amino acids [10]. There are three main pathways by which amino acids enter cells: simple or free diffusion, facilitated diffusion and active transport [11]. In many cases, simple diffusion is probably too slow for essential life processes. Like passive diffusion, facilitated transport does not require energy, but is brought about by protein molecules that bind to a particular amino acid or its derivative (e.g. dopamine) and facilitate its movement down a concentration gradient into the cell. Systems for active transport or cotransport deliver amino acids into the cells against steep concentration gradients. Examples include the re-uptake of Glu from the synapse back into the presynaptic neuron by sodium-driven pumps [12] or coupled amino acid/H^+ cotransporters for the cellular import of neutral and charged amino acids that can exhibit low or higher affinity [13]. Metazoan Na^+-dependent and -independent carrier systems transporting amino acids comprise $B0^+$, LAT1, LAT2 and TAT secondary transporters and ATP-binding cassette (ABC) pumps [14]. Over the past decades, numerous proton-coupled amino acid and peptide transporters have been discovered from different plant and animal species. Compared with single-cell organisms, higher plants have evolved a larger proportion of genes involved in energy-dependent transport. This apparent redundancy could be explained by tissue- and development-specific requirements in these organisms. For example, larvae or embryos cannot tolerate dietary deprivation of any one of the basic (Arg, Lys and His) and neutral L-amino acids (Leu, Met, Ile, Thr and Val) [15]. Aromatic amino acids (Phe and Trp) are exceptionally critical in mammalian development since they are essential precursors for the synthesis of neurotransmitters [12]. At present, the molecular identities of essential amino acid transporters as well as mechanisms of accumulation and redistribution of essential amino acids in pools are poorly understood in metazoans. However, one general feature of these transport systems is that they exhibit no strict substrate specificity, i.e. in addition to their "own" group of amino acids, they can also carry amino acids from other groups, although with lower affinity [8]. For example, a Na^+-dependent system ASC preferentially transports Ala, Ser and Cys, but also transfers other aliphatic amino acids [8, 16].

4.3 Physicochemical Properties of Canonical Amino Acids

Canonical amino acids can be divided according to their physicochemical properties (e.g. positively and negatively charged and uncharged, acidic, basic, etc.), and metabolic and physiological roles. Regarding their physicochemical properties, it is indeed quite difficult to put all amino acids of the same type into an invariant group. For example, the same amino acid can behave differently in the context of different solvents and protein structure microenvironments. Nevertheless, amino acids have been arranged into several groups by using various criteria such as size and volume of their side-chains, isoelectric point, solubility in water, free energy of transfer between different phases (vapor/aqueous) and solvent systems (polar/apolar), optical properties such as refractive index, optical rotation or chromatographic migration, effects on surface tension, and their solvent accessibility in known protein structures. Some of these features are presented in Table 4.1. Such

Tab. 4.1. Some properties of canonical amino acids (such tables can be also regarded as an index where the columns refer to a particular amino acid, and the rows define the biological, chemical and physical properties).

	Gly	*Ser*	*Cys*	*Pro*	*Ala*	*Val*	*Leu*	*Phe*	*Ile*	*Met*	*Thr*	*Trp*	*Tyr*	*His*	*Asp*	*Asn*	*Glu*	*Gln*	*Lys*	*Arg*
Abundance in *E. coli*[a]	582	205	87	210	488	402	428	176	276	146	241	54	131	90	229	229	250	250	326	281
Occurrence in proteins[b]	7.10	7.25	2.44	5.67	6.99	6.35	9.56	3.84	4.50	2.23	5.68	1.38	3.13	2.35	5.07	3.92	6.82	4.47	5.71	5.28
van der Waals volume (Å^3)	48	73	86	90	67	105	124	135	124	124	93	163	141	118	91	96	109	114	135	148
Isoelectric point (pI)[c]	6.06	5.68	5.05	6.30	6.11	6.00	6.01	5.49	6.05	5.74	5.60	5.89	5.64	7.60	2.85	5.41	3.15	5.65	9.60	10.76
pK_a (side-chain (R))	–	–	8.37	–	–	–	–	–	–	–	–	–	8.37	6.04	3.90	–	4.07	–	10.54	12.48
Hydropathy index[d]	−0.4	−0.8	2.5	−1.6	1.8	4.2	3.8	2.8	4.5	1.9	−0.7	−0.9	−1.3	−3.2	−3.5	−3.5	−3.5	−3.5	−3.9	−4.5
Solubility in water[e]	24.99	5.023	FS	162.3	16.7	8.85	2.426	2.965	4.117	3.381	FS	1.136	0.0453	4.19	0.778	3.53	0.864	2.5	66.6	15.0
$\Delta G_{(chx \rightarrow water)}$ (kJ mol^{-1})[f]	0.94	−3.40	1.28	–	1.81	4.04	4.92	2.98	4.92	2.35	−2.57	2.33	−0.14	−4.66	−8.72	−6.64	−6.81	−5.54	−5.55	−14.92

[a] Determined as described in [30] and expressed as μmol g^{-1} of dried cells.
[b] Data taken from [31], based on an analysis of 1490 human genes.
[c] The isoelectric point (pI) is defined as the pH at which a molecule carries no net electrical charge.
[d] These values are derived from the hydropathy scale of Kyte and Doolittle [22], which assigns a hydropathy index to each amino acid, based on its relative hydrophobicity (positive value) or hydrophilicity (negative value). It is an index of solubility characteristics in water which combines hydrophobic and hydrophilic tendencies, and can be used to predict protein structure (i.e. propensities to occur in different secondary structures).
[e] Water-solubility is expressed as g 100 g^{-1} at 25 °C (data from [32]) (FS = fully water soluble).
[f] Nonpolar (chx; cyclohexane) → polar (water) distributions of amino acid side-chains expressed in free energies (kcal mol^{-1}) from Radzicka and Wolfenden [20].

indexing might be useful in predicting favorable substitution patterns in proteins or in an attempt to understand the solvation effects in molecular recognition. Moreover, the property which has appeared to be the major driving force in protein folding is hydrophobicity [17, 18]. Not surprisingly, the various methods employed over the past decades have yielded numerous hydrophobicity scales in order to quantify this property at the level of a single amino acid [19].

One class of hydrophobicity scale which is based on the measurement of the distribution coefficients between water and apolar solvents such as 1-octanol or cyclohexane revealed that this parameter is correlated with solvent accessibility in globular proteins [20]. In an ideal case a solvent that would be a good representative of (or at least mimic) the interior of a protein is desirable. Conversely, no single reference phase capable of serving as a general model for the interior of proteins has yet been found. Another class of hydrophobic scale is constructed by examining proteins with known three-dimensional structures and defined hydrophobic character (i.e. the tendency for a residue to be found inside the protein, rather than on its surface) [21]. Finally, correlations from the various types of data presented by the scales can also be modeled mathematically [22–24].

By analyzing 43 different hydrophilic/hydrophobic scales, Trinquier and Sanejounad [19] found a fairly good consensus in assigning strictly hydrophilic characters to Asp, Asn, Glu and Gln. They are seldom buried in the interior of a folded protein, and are normally found on the surface of the protein where they interact with water and other important biological molecules. Curiously, Arg and Lys (see Section 5.2.1.9) are in some scales regarded to be the most hydrophilic, whereas in others only as "moderately" hydrophilic. This is not surprising since both Arg and Lys can be considered hydrophobic by virtue of the long hydrocarbon side-chain, whereas the end of the side-chain is positively charged.

Lys and Arg are almost exclusively distributed on protein surfaces and have part of the side-chains buried, and only the charged portion is solvent exposed. Similarly, Trp occupies the most hydrophobic rank in some scales, whereas it is considered "moderately" hydrophilic in others (Fig. 4.2). Indeed, Trp and Met are relatively rare amino acids, not strictly polar or apolar, but combine both properties – as reflected in their "promiscuous" topological distribution in proteins [28, 29]. Residues such as Ala, Thr and Ser are located at the mid-scale, which is not surprising since they might be both buried as well as on the surface of the protein (such amino acids are sometimes called "ambivalent" or "amphipathic" in the literature) (see Figure 7.2).

They are capable either of forming hydrogen bonds with other polar residues in the protein interior or with water on the surface. The strict hydrophobic character of Phe, Leu, Ile, Val and Met is plausible since the interior of most proteins is composed almost exclusively of them, stabilized by numerous van der Waals interactions. The most extreme case is Cys, which in some scales can be found at the most hydrophobic positions (probably due to its overwhelming tendency to create disulfide bonds in protein interiors), while in other scales it occupies the most hydrophilic positions.

The difficulties in such positioning are understandable by taking into account

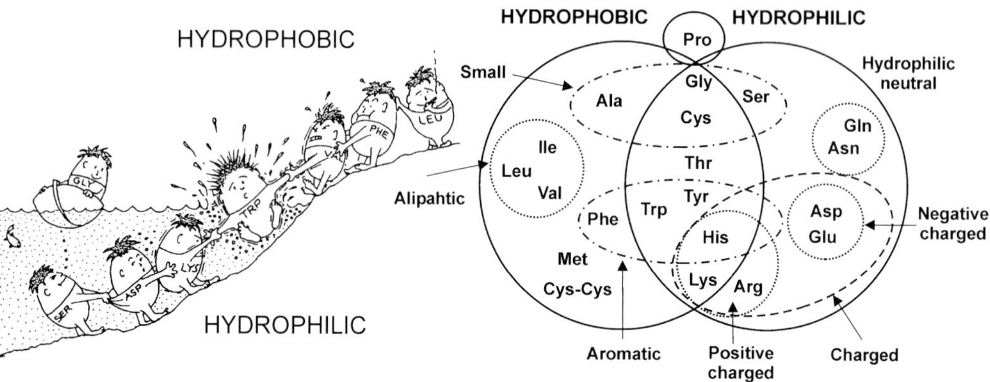

Fig. 4.2. Left: cartoon illustrating the scientific dispute (Fauchere and Pliska [25] versus Wolfenden and Radzicka [20]) over the question "How hydrophilic is tryptophan?" [26]. There is indeed a wide diversity of positions for many amino acids in different hydrophobic scales, especially for Arg, Trp, Cys and the imino acid Pro. (Right) Venn diagram presenting some basic physicochemical features (also shown in Table 4.1) and relations between 20 canonical amino acids. This is especially useful in showing the multiple properties of an amino acid, e.g. positive–negative charge, neutral–hydrophilic, aromatic–hydrophobic, aliphatic–hydrophobic, etc. The amino acids that possess the dominant hydrophobic and hydrophilic properties are defined by their set boundaries. Subsets contain amino acids with the properties aliphatic, aromatic, charged, positive, negative and small. There are certainly areas which should define sets of properties possessed by none of the cannonical amino acids – these areas can be "occupied" by noncanonical amino acids delivered to the code by its engineering (Section 6.2.2). (The carton is reproduced from [26] with permission from Elsevier, copyright 1986; the Venn diagram was drawn according to [27]).

the fundamentally different methods used for constructing such scales (Fig. 4.2). These difficulties are further enhanced thought the complex nature of hydrophobicity itself, which can be defined in several ways. In fact, Charton and Charton [33] wrote:

> ... no single hydrophobicity parameter ... can represent the complete range of amino acid behavior. There is no special phenomenon denoted by hydrophobicity in amino acids. It is the natural and predictable result of differences in the intermolecular forces between water and the amino acid and those between the amino acid and some other medium.

For example, the aromatic amino acids Phe, Tyr and Trp harbor a combination of two properties, i.e. hydrophobicity and polarity (ability to bind ions through cation–π interactions), that are often considered to be mutually exclusive [34]. This combination of properties makes them suitable candidates for both the interior of a protein or for interactions with cell membranes. For that reason, aromatic side-chains are often part of the rigid protein interior, where they participate in networks of three or more interacting side-chains that are supposed to serve as nucleation sites in protein folding and as a main stabilizing force of tertiary structures.

On the other hand, these amino acids are crucial in mediating processes such as receptor–ligand interactions, enzyme–substrate binding and antigen–antibody recognition [29].

4.4 Reasons for the Occurrence of Only 20 Amino Acids in the Genetic Code

Evolution is based on a changing chemical environment. These changes drive the cells to optimize their chemistry for survival purposes. Only those cells whose chemistry fits the newly emerged environmental conditions will survive such a natural selection process and transmit their genes to the following generations. From that point of view, entry in the code repertoire would be reserved only for those novel amino acids that bring superior cellular functionality (i.e. increase survival fitness). However, it is reasonable to suppose that in most cases the integration of novel amino acids into the proteome or smaller numbers of proteins takes place without any substantial advantage for host cells and might even be harmful. Such cells are usually subjected to negative selection, i.e. elimination from the life tree.

According to the early proposal of Woese [4], primitive cells or proto-cells, nowadays usually described in the framework of the "last universal common ancestor" (LUCA) [35] or as "progenote" local populations [36], started with a completely random, highly ambiguous set of codon assignments with very inaccurate translation. Such early systems were capable of adopting a novel amino acid much easier due to the relatively minor deleterious effects on the proto-cell. This is indeed plausible in the context of smaller genomes and proteomes, simple metabolism, rather modest biosynthetic capacity, and their low level of integration. Once such an entry of a novel amino acid brings about a great advantage in a given chemical environment, it would be "cemented" in the code by selection processes. Such code enrichment with novel amino acid building blocks would lead to more advanced and sophisticated life forms to the point when any further repertoire expansion would not be possible without causing large harmful effects and even lethality. At that point "freezing" or conservation of the code took place [37], as already elaborated in Crick's "frozen accident" concept [38]. According to this hypothesis, the current code repertoire reached an evolutionary dead-end, since cells were no longer capable of assimilating new amino acids in their genetic code. When life had evolved to a certain level of complexity, the proteome of the cell was further functionally diversified by other means: context-dependent preprogrammed re-coding (see Section 3.10) and posttranslational modifications (see Section 3.11). Therefore, the answer to the question as to why only 20 amino acids represent the standard repertoire of the universal genetic code is an evolutionary one. This is the point when the standard repertoire of the 20 amino acids was established either in the context of the LUCA [39], "progenote" local populations [36] or something else. This establishment was a historical process, a sort of "bottleneck" event to which all coding properties could be traced back. Another line of reasoning was followed by Miller and Weber [40]. They were considering the hypothetical scenario of independent life origin on another planet in which proteins were chosen as catalysts. Their main question was: which basic building blocks would then be used for their synthesis? Applying physicochemical, genetic and historical argumentation for such a scenario, Miller and Weber concluded that for an independent origin of life on another planet using protein catalysis, about two-thirds of the basic building blocks would have to be of the same physicochemical nature as those on Earth. They speculated that certain classes of amino acids were excluded from the code through selection against adverse effects that they might cause on protein structure, stability and function. The explanation for the exclusion of other amino acid types that do not violate these rules is most probably historical, such as their unavailability or insufficient evolutionary time to produce them metabolically [3]. In fact, these rules and determinants are at least in part responsible for the fixation of 20 canonical amino acids in the standard repertoire of the universal genetic code (see Section 6.3.2).

4.5
What Properties of Amino Acids are Best Preserved by the Genetic Code?

In its very essence, known life is based on self-organized polymeric matter, i.e. proteins, nucleic acids and carbohydrates. Such polymeric structures in present-day organisms are the result of the selection process during evolution. The basic feature of these functional biopolymers is their nonlinear (all-or-nothing) response to external stimuli. In other words, small changes happen in response to a varying parameter until a critical point is reached – when a large change occurs over a narrow range of the varying parameter. After such a transition is completed, there is no further significant response of the system [41]. These nonlinear responses of biopolymers are caused by highly cooperative interactions between separate monomeric subunits. In the case of proteins, this is indeed plausible by taking into account their prominent feature – foldability [42]. This means that functional protein sequences fold into conformations that possess a pronounced energy minimum separated by a large energy gap from the bulk of structurally unrelated misfolded conformations [43]. Sequences that satisfy these requirements are able to fold fast and to exhibit a cooperative folding transition [44]. They are stable against mutations as well as against variations in solvent conditions and temperature. Indeed, characteristic folds and stable conformations in proteins are largely preserved across the various phylogenetic trees despite their great sequence variations [18].

A simple system that operates on a few basic principles should be devised to build up such structures. The simplest rule to be obeyed for protein building is the general "apolar in–polar out" principle, allowing them to fold into tight particles that have an internal hydrophobic core shielded from the surrounding solvent [45]. This binary partitioning of polar and nonpolar amino acids is indeed an intrinsic part of the genetic code (Fig. 4.3). In its structure, these relations are visibly regular: all codons with a central U are cognate to amino acids with chemically relatively uniform, hydrophobic side-chains ("convergent types"). On the other hand, coding triplets with a central A are cognate to amino acids with chemically variable, polar side-chains ("divergent types") [46] (see Section 7.2 and Fig. 7.2). Not only is such a chemical order apparent by careful inspection, it can also be quantitatively measured in a simple chromatographic experiment. Woese and coworkers [4] measured the chromatographic mobility of amino acids on paper in various solvents and interpreted the observed mobility as the number of water molecules required for solvation. On this basis, the "polar requirement" index was developed. This index reflects remarkably well an order in the coding table, clearly indicating that the genetic code assigns amino acids to related triplets on the basis of polarity/hydrophobicity. Later studies by Wolfenden and coworkers [47] confirmed that the relative distribution of amino acids between the surface and interior of a native globular proteins is indeed shaped with a sharp bias imposed by the genetic code see also Section 6.2.2 (Fig. 6.2).

As a consequence of its degeneracy, there is an increased redundancy in the code structure, which allows for the existence of larger synonymous quotas for some amino acids. Interestingly, they are either strictly polar or strictly apolar. For example, 15 codons are allocated to apolar amino acids Ile, Val and Leu, which shows the extent of encoded hydrophobicity. On the other hand, the number of codons allocated to strictly polar amino acids is even higher (Fig. 4.3), which means that hydrophilicity is equally well represented in the genetic code structure. Obviously, such allocation of codons in the genetic code has consequently rather forced the distinction of hydrophobic versus hydrophilic residues in the protein structures. Protein interiors are dominantly populated with hydrophobic residues (XUX group of codons), which are rarely found at the surfaces of proteins. Conversely, hydrophilic residues (mainly XAX and AGX groups of coding triplets) are seldom in the interior since their spontaneous appearance would be seriously disruptive to the overall stability of a protein. Such distributions are associated with large differences between free energies of solvation between polar/apolar residues [47].

This general globular protein organization (polar out–apolar in) indicates first additions to the code during its evolution that were either strictly apolar or strictly polar. This rigid mechanism for maintaining protein structure can also be recognized in the frame of an "operational RNA code" [51] that is believed to be a relic from primitive proto-cells [52]. This means that the discriminatory base N73 in the acceptor stem of tRNA reflects well such binary relationships: A73 is the acceptor for hydrophobic amino acids, while those with G73 are acceptors with hydrophilic amino acids (see Section 2.4). Therefore, primitive cells with relatively simple organization started code expansion first through additions that were strictly apolar (in the core or more probably as the integral transmembrane part) and strictly polar (at the surface) [37]. Later additions were more "promiscuous", of which Trp and Met are excellent examples – both are almost equally distributed in the protein core, at surfaces and minicores, and have increased chemical variability [53]. They are also characterized by smaller synonymous quotas and, correspondingly, a greater probability that they may be substituted, but without significant disturbance of the protein structure.

In the evolution of proteins, amino acid substitutions are found to occur more frequently between similar amino acids than between dissimilar ones [38]. Such functional state by certain selective pressures that resulted in structurally and functionally meaningful proteins. The interplay of such selective pressures in evolution is the main driving force for convergent processes (e.g. preservation of structural types, such as protein folds or body plans in nature) as well as divergence that increase variety in functionality or performance. This structure can be seen as an equilibrium between two opposing extremes, i.e. invariance and mutability. Such a dual nature of the genetic code consequently has the possibilities that genes can simultaneously be fragile (mutable) and robust (invariant) to mutations.

This concept was introduced for the first time by Maeshiro and Kimura [55], and has recently been elaborated in great detail in an excellent study by Judson and Hydon [56]. While mutability promotes diversity in the functionality of proteins, invariance is the barrier against which such mutations would be restricted, especially those that would compromise the active structure of proteins. The single coding triplets of the rare amino acids Met (AUG) and Trp (UGG) are fragile, and any point mutation or translation error leads to nonsynonymous (amino acid altering) substitution in gene sequences like AUG (Met) and AUA (Ile). Mutability is not only the main source of protein sequence diversification, but also of evolution of living beings in general. It is a basic prerequisite for the development of more complex life forms and provides enough flexibility to allow adaptation to a changing environment. Conversely, invariance is responsible for preventing amino acid substitutions in proteins (synonymous or silent codon substitutions), making them robust to mutations [e.g. AUU (Ile) and AUA (Ile)]. In other words, invariance is evolutionarily neutral, because many synonymous codon changes and conservative amino acid changes occur without deleterious effects. This property is the main source of the characteristic protein folding and structure conservation in all living beings. It has been argued that nonsynonymous versus synonymous substitution ratios might be used as a measure of the selective pressure at the level of the protein [57].

Conservatism in protein structure is related to structural stability and folding kinetics rather that function. Clearly, function does not require a particular sequence, but rather particular amino acids at defined sites, while there are other possibilities of variation at different sites. A comparison of protein homologs (i.e. proteins that are functionally and evolutionary related) with protein analogs (those that are structurally similar proteins, but evolutionary unrelated) shows that such analogous proteins in most cases share folding, but not function [58]. On a more general level, such balanced genetic code structure allows for convergence in the structural types, since protein folds found in nature represent a finite set of build-in forms. Other levels of complexity of the living beings obey these principles as well, e.g. concept of the finite number of body plans in nature as proposed more than two centuries ago by Göethe and Cuvier [59]. In fact, all the properties of living beings rest on the fundamental mechanism of molecular invariance. Extant cells are characterized by their invariant basic chemical organization. Therefore, evolution is not a property of living beings; it stems from imperfections in the conservation of the physiological/molecular mechanisms. Such an "imperfection" is indeed a unique privilege of living beings [60].

4.7
Natural Variations in Assignment of Codons of the Universal Genetic Code

Deciphering of the triplet genetic code by the end of the 1960s revealed a nearly identical assignment of amino acids to coding triplets in "standard" organisms (e.g. *E. coli*, *Bacillus subtilis*, *Saccharomyces cerevisiae*, *Arabidopsis*, *Drosophila*, etc.), indicating its universality as well as the possibility of a common origin for all life forms. A decade later, the sequencing of some genes from the mitochondrial genome and their comparison with nuclear counterparts revealed that there are deviations in codon assignments between nuclear and mitochondrial genes [61]. Changes in codon meaning can be classified into two general categories. The first category includes reassignment of standard termination codons (UGA, AAA and UAG) to Trp or Gln in some prokaryotes, Archaea, mitochondria of many organisms and even nuclear genes in some protozoa (e.g. ciliated protozoan *Tetrahymena* and *Paramecium*). The second category of variant codon assignments includes altered meanings in mitochondrial sense codons such as Met → Ile, Lys → Asn, Arg → Ser and Leu → Thr [62], while the only documented case of sense codon reassignment among nuclear genes is (CUG) Leu → Ser in *Candida cylindrica* [63]. It is believed that more than 10 million species belonging to Archaea, prokaryotic and eukaryotic kingdoms inhabit the Earth at present [64]. With the sequencing of whole genomes or at least part of them, interesting strategies for alternative assignments in the universal code table as well as various undiscovered re-coding "tricks" in translation are expected to come to light (see Fig. 4.4).

Various aspects and detailed accounts of alternative codon assignments can be found in the contemporary literature [62], to which the interested reader is directed for references and details which do not appear here. Extensive studies on this subject were also used as "recent evidence for the evolution of the genetic code" [65], which is hardly reconcilable with the previous considerations of code with an invariant structure preserved in all life taxa. Such reassignments might be seen as flexibility in the "frozen" structure of the code, since they do not affect the amino acid repertoire (i.e. resulting only in canonical amino acid "shifting" between different codons). The universality of the genetic code can only be questioned seriously in the case of the existence of such a species that regularly, in a codon-dependent manner (in the absence of any special context, i.e. without re-coding "tricks"), builds proteins with amino acids such as homoserine, homocysteine, α-aminobutyric acid, norleucine, phosphoserine, norvaline, etc. There are indeed no such examples in the known life kingdoms [37]. There are no codon reassignments that would introduce new amino acids, but only codon reassignments in different organisms resulting in the changing of one canonical amino acid to another, but always in the frame of the same, standard amino acid repertoire. In addition, the appearance of such codon reassignments is more species specific than universal and most of them are not disruptive at the level of the protein structure. In the case of *C. cylindrica*, the (CUG)Leu → Ser change should have a negative impact on the organism's fitness [63]. This reassignment is "reserved" for only for a tiny portion of genes involved in the stress response and pathogenicity; the CUG triplet

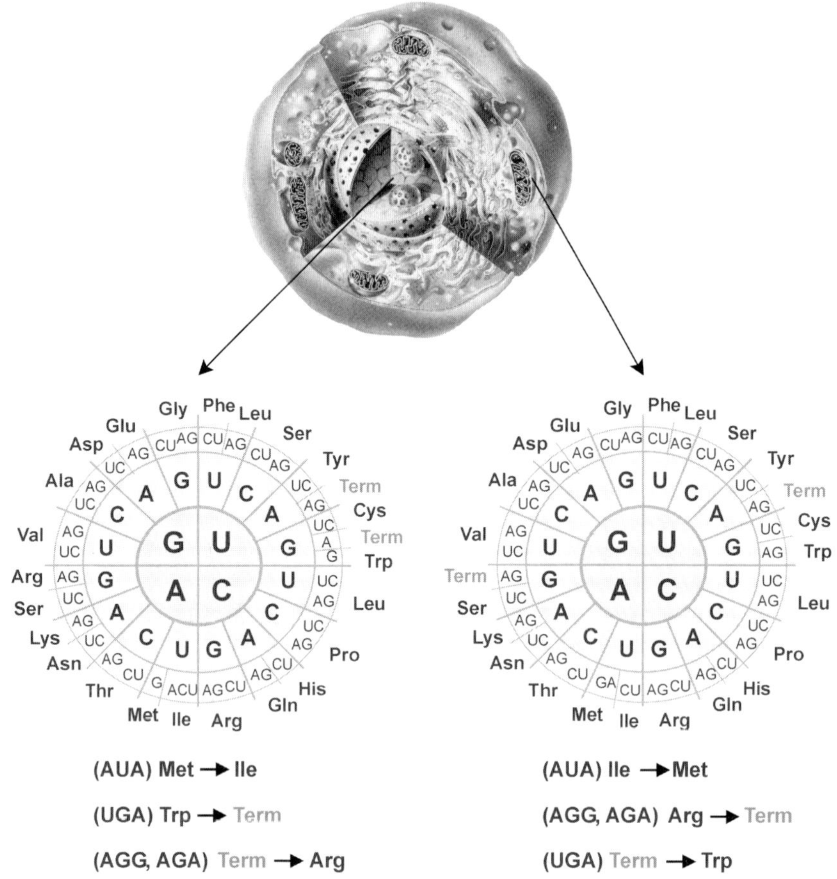

Fig. 4.4. The existence and cohabitation of mitochondrial and nuclear genetic codes in mammalian cells was discovered during human mitochondrial genome sequencing. Most proteins in animal mitochondria, including some translational components, are imported from the cytosol (i.e. encoded by nuclear genes), whereas there are only 10–20 "home-made" proteins [2]. The existence of these two arrangements results in the following differences in translation of mitochondrial relative to nuclear genes: (i) UGA is not a termination signal, but codes for Trp (therefore the anticodon of mitochondrial $tRNA^{Trp}$ recognizes both UGG and UGA, as if obeying Crick's "wobble rules"), (ii) internal Met is encoded by both AUG and AUA, while the initiator Met is cognate to AUG, AUA, AUU and AUC, and (iii) AGA and AGG are not assigned to Arg, but specify chain termination [1]. Thus, there are four termination and four Arg-coding triplets in the mitochondrial genetic code. Note that the amino acid repertoire is the same in both code variations. Plant mitochondria use an arrangement of the universal code [65].

does not appear in most of the cellular mRNA. Although co-translational incorporation of amino acids such as fMet, Sec and Pyl is a result of local changes in codon meaning or preprogrammed context-dependent re-coding phenomena (see Section 3.10), in contemporary literature they are often incorrectly assigned as natural variations in the code structure [64].

4.7.1
Nucleoside Modifications and Codon Reassignments

The study of nucleoside modification in the context of a natural departure from the universal code, in addition to serving as an excellent model for codon reassignment theories (Section 4.8.2), might also be a gold mine of useful information that can be applied in code engineering. It is well known that nucleoside modifications as a universal feature of tRNAs are not only necessary for fine tuning the tRNA structure, but also for the accuracy, strength and efficiency of the codon–anticodon interactions (see Section 3.8; Fig. 3.10). Therefore, ribosomal decoding might be influenced not only by tRNA structure and abundance, but also by modified nucleosides [66]. For example, AUA → Met reassignment in eubacteria might be achieved by control of anticodon base modification as follows. Bacteria use $tRNA^{Ile}$ with the *CAU anticodon [*C stands for (2-lysyl)cytidine] for translating AUA codon as cognate for Ile. By inhibition of the cytidine modification (i.e. with the CAU anticodon) $tRNA^{Ile}$ becomes charged with Met instead with Ile [67, 68]. This means that, by elimination of the C-modifying enzyme from bacteria, a strain having a novel codon reassignment (equal to those that already exists in mitochondria) can be designed. Another example includes the presence of pseuduridine (Ψ) in the anticodon of $tRNA^{Asn}$ of echinoderm mitochondria; $tRNA^{Asn}$ with the GΨU anticodon recognizes the AAA (Lys) coding triplet. These findings highlight the fact decoding flexibility (wobble rules) exists in the third codon position (first anticodon position); this feature might also be responsible for ambiguous decoding where two or more different tRNAs compete for the same coding triplet (e.g. "polysemous codon" [69]). Finally, modifications of bases that are direct neighbors of "wobble" nucleotides (e.g. N37) might have an influence on decoding since they might affect the interaction between the first anticodon and third codon base [70]. More details about the role of modified nucleosides in codon–anticodon interactions are available in [71].

4.8
Codon Reassignment Concepts Possibly Relevant to Code Engineering

4.8.1
Genome Size, Composition, Complexity and Codon Reassignments

The amino acid composition of proteins is correlated with the characteristic GC content of the genomic DNA of an organism, which is related to its phylogeny. Indeed, the neutral theory of protein evolution [72] expects that a significant proportion of the amino acids present in a protein should change by random genetic drift and therefore be influenced by the number of codons assigned to the different amino acids. The differential level of GC content in the different components of the genome in a given organism can be understood as the consequence of selective constraints that have been exerted to eliminate functionally deleterious mutants

[73]. Correspondingly, each organism has it own set of biases for use of 61 codons for the amino acid and tRNA population, which is in direct correlation with the codon composition of total mRNA. The frequency of appearance of particular synonymous codons (i.e. codon bias or codon usage which is correlated to the abundance of particular tRNA species in the cytosol) can differ significantly between genes, cell compartments, cells and organisms. Such a bias is also a measure of tRNA isoacceptor distributions for synonymous codons in particular genes and species [65]. Potential errors in tRNA recognition by AARS are avoided by positive and negative identity elements (see Section 3.6), whose repertoire obviously has limits. This is evident by the fact that endless variations in tRNA recognition elements are not possible; they would, sooner or later, lead to unacceptable levels of tRNA misacylations. For that reason, Schimmel and Ribas de Poplana proposed a concept that attributes the fixed state of the extant genetic code to the limited amount of productive intrinsic combinations among identity elements that are responsible for specific tRNA:AARS interactions [74].

Mitochondria which encode few proteins, importing most of what they need from the nucleus, are thus are not under strong selection for translational accuracy; this resulted in genome economization. This furthermore inevitably led to constrained codon usage acting as an efficient selective pressure for codon reassignments [70]. For example, most of the codon variations found in various animal mitochondria (11 different codons) are due to the dramatic reduction of the genome size and, in particular, the number of tRNA genes [74]. This reduction in the tRNA gene pool led to the loss of many of the conserved identity elements, and to a "relaxation" of the recognition constraints between AARS and tRNA. Indeed, the changed identity for $tRNA^{Arg}$, $tRNA^{Ile}$, $tRNA^{Met}$, $tRNA^{Lys}$ and $tRNA^{Ser}$ correlates remarkably with the variations (i.e. codon reassignments) in the mitochondrial genetic code [65]. Such "melting" of the genetic code is also conceivable in *Mycoplasma*, which is regarded as a degenerate form of Gram-positive bacteria. In fact, the "genome minimization" hypothesis proposes that both mitochondria and *Mycoplasma* are under extreme selection to reduce their genome size, which should confer an advantage in their replication [75]. For example, the genome of *M. capricolum* contains only 30 tRNA genes for 29 tRNA species with reduced extents of nucleoside modifications. In this manner, the genomes of *Mycoplasma* and mitochondria economize (simplify) their content by discarding redundant tRNAs and enzymes for nucleoside modifications (most probably as a result of AT pressure), leaving enough space for the appearance of new tRNA species capable of acquiring novel codon meanings [76].

Such codon reassignments within large, complex, relatively modern genomes would be much more difficult. For example, Jukes and Osawa [77] considered the possibility for tRNA mutation in *E. coli* that would make Gln cognate for the AAA codon together with Lys. Reading of this codon by both corresponding tRNAs would result in a mixture containing both amino acids at AAA sites of protein products. If 1% of such Lys $\leftrightarrow$ Gln replacements were disruptive (*E. coli* has 1562 genes with at least one AAA codon per copy) there would be 15 defective cellular proteins, enough to eliminate such an allele by negative selection. Similarly, there are 417 UGA termination signals distributed among the genes in the *E. coli* genome. In a hypothetical scenario of their general reassignment to a particular amino acid, proteome-wide disturbance accompanied with dramatic losses in cellular viability is the only possible outcome [78]. In this context, the emergence of special, sophisticated, context-dependent and species-specific mechanisms that keep internal stop codons separated from terminal ones during, for example, translation of Sec (see Section 3.10) is easy to understand. However, there are *E. coli* suppressor strains that are capable of suppressing particular stop codons with efficiencies of even 50% [79]. In some strains the error frequency during translation can be increased by a few orders of magnitude without any effects on exponential growth. Such ambiguity is not only limited to prokaryotes; eukaryotes have basal levels of ambiguity similar to prokaryotes. Obviously there are tolerable levels of nonsense suppressions without any apparent phenotypic feature [67]. The ambiguity is also used by parasitic species, e.g. when animal and plant viruses use ambiguous stop codons to adjust the level of stop-read-through for the translation of their essential gene products. This strategy can be "borrowed" from nature into the laboratory as well. For example, routine use of *amber* suppressor strains like *E. coli* DH5α is widespread since they enable (through restricted translational read-through) control of the number of plasmid copies per cell without any negative interference to its viability.

4.8.2
Stop Codon Takeover, Codon Capture and Codon Ambiguity

The departures from the universal code are often thought as a "relics of a primitive genetic code that possibly preceded the emergence of the so-called universal code" [80], whereas other thinkers in the field regard them to be "of recent origin and derived from the universal code" [65]. Based on the observation that the majority of the described variations in the universal code are in fact reassignments of standard termination codons, Lehman and Jukes [81] proposed code evolution via the takeover of stop codons ("stop codon takeover" hypothesis). Their crucial statement is that "all codons are essentially chain termination or stop codons until tRNA adaptors evolve having the ability to bond tightly to them". According to their hypothesis, new sense coding triplets have been derived from codons serving previously as exclusive nonsense signals (i.e. they did not have their "own" adaptor or RF). In particular, mutation in different tRNA adaptors (or in their isoacceptors) leads to recognition of a stop codon, while an additional change in the identity set leads to its acylation with a novel amino acid which is subsequently translated into proteins. The tRNA with new function should emerge by duplication of genetic material, which is also one of the fundamental mechanisms in protein evolution [64]. When parental sequences duplicate, the copies are confronted with three possible outcomes: (i) both copies retain native activity, (ii) one of the copies is inactivated by accumulation of a mutation and (ii) one copy might be further functionally diversified. The emergence of a new function with an associated novel codon assignment should have little effect on the phenotype since such an encoding

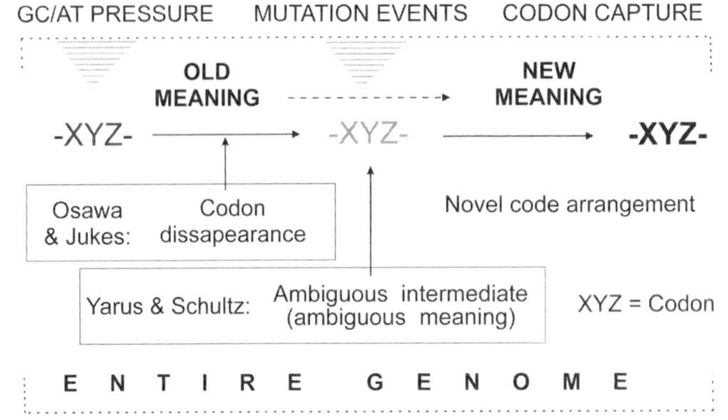

Fig. 4.5. The Osawa–Jukes "codons capture" [65] versus the Schultz–Yarus "ambiguous intermediate" theory [82] of codon reassignment. It is generally accepted that codon reassignments occur over relatively long evolutionary periods. In the "codon capture" theory the first step in the change of the genetic code is assumed to be the complete disappearance of a codon from a genome. The deleted codon might reappear with a new function upon, for example, specific mutation on tRNA. The "ambiguous intermediate" theory is similar, but does not envisage the full disappearance of a codon from genes before its reassignment. This theory proposes that the reassignment of a codon takes place via an intermediate stage during which the codon is recognized by two tRNAs assigned to different amino acids (or a tRNA and a release factor). Such a novel meaning might be fixed in the code only if there is a positive benefit arising from the reassignment.

should result in chain lengthening to a tolerable extent [81]. Evolutionarily advanced proteins tend to be longer and more complex in their amino acid composition, e.g. globins or highly specialized enzymes and functional enzymatic machinery like a ribosome or proteasome.

The codon capture (this term refers to a change in the meaning of a codon) theory of Osawa and Jukes [78] (Fig. 4.5) is in fact an extension and further sophistication of the Lehman–Jukes concept. Two crucial steps have been postulated. (i) An amino acid (or termination) codon temporally vanishes from the coding frames of translated sequences as a consequence of GC or AT mutational pressure. This loss of codons makes corresponding tRNAs functionless and they are eliminated. (ii) When future mutational pressure reverses and sequence composition (e.g. AT content) changes again, the emerging codons that lack cognate tRNA are in fact termination signals (i.e. inhibit translation). Their capture is thus advantageous; the emergence of tRNA that translates them with a specific assignment occurs via (i) a change in the anticodon of particular adapter, (ii) a change in identity sets that leads to charging with novel amino acid (without anticodon change) or (iii) a change in codon–anticodon pairing (e.g. by introduction of nucleotide modification). Such mutated tRNA adaptor originates most probably via gene duplication. This process of simultaneously evolving gene sequences, codons, anticodons and amino acids is accompanied by a series of nondisruptive, i.e. neutral, changes. According to codon capture theory, it would be deleterious for an organism if a codon was assigned to two amino acids simultaneously [77]. Thus, this theory requires a series of strict contingencies to prevent ambiguous codons from arising (to avoid disruptive effects on proteins) and these requisites must be fulfilled in order to add novel amino acids (or to reassign one from the existing repertoire) to the code [65].

In contrast, the Schultz–Yarus "ambiguous intermediate" [82] hypothesis (Fig. 4.5) does not require the complete disappearance of a codon from the genome before reassignment take place. This theory postulates that a mutant tRNA with expanded decoding properties starts decoding a codon belonging to a noncognate amino acid codon family, making it ambiguous. In other words, the codon undergoes a transitional stage in which it is ambiguously decoded. Such ambiguity may be one of aminoacylation identity, anticodon misreading or competition between two nonambiguous tRNAs. Thereby, an ambiguous intermediate is the preferred route by which a primordial code evolved to the present "universal" form. This leads to heterogeneity in the encoded proteins that, according to this theory and experimental evidence (see Section 5.1.4.1), can be tolerated by an organism. At the core of this proposal is the possibility that mutated tRNA could recognize a normally noncognate codon and insert its amino acid in competition with the cognate amino acid. If this new specificity confers an advantage, propagation of an organism with ambiguous codons can be favored by natural selection, resulting in an organism having an established genetic code with a new arrangement. This theory was recently reformulated by Santos and coworkers [83] by introducing a novel mechanism for codon reassignment through codon ambiguity, as follows:

> Mutant tRNAs with expanded decoding properties, or mutations in the translational machinery, which create decoding ambiguity, in conjunction with biased GC pressure, impose a negative selective pressure on particular codons, decreasing their usage to very low levels or forcing them to disappear from the genome. Some overall positive advantage must be present in order for this process to occur. These codons can be gradually reassigned through structural change of the translational machinery and loss of the ancestral cognate tRNAs.

There are also recent attempts to reconcile both theories in a single unifying model, which consider them as "opposite faces of the same coin" [62].

References

1 Watson, J., Gann, A., Baker, T., Levine, M., Losick, R. and Bell, S. (2004). *Molecular Biology of the Gene.* Benjamin Cummings, Amsterdam.

2 Stryer, L. (2001). *Biochemistry.* Freeman, New York.

3 Ardell, D. H. and Sella, G. (2001). On the evolution of redundancy in genetic codes. *Journal of Molecular Evolution* **53**, 269–281.

4 Woese, C. R., Dugre, D. H., Dugre, S. A., Kondo, M. and Saxinger, W. C. (1966). On fundamental nature and evolution of genetic code. *Cold Spring Harbor Symposia on Quantitative Biology* **31**, 723–736.

5 Di Giulio, M. and Medugno, M. (1999). Physicochemical optimization in the genetic code origin as the number of codified amino acid increases. *Journal of Molecular Evolution* **49**, 1–10.
6 Wheatley, D. N., Inglis, M. S. and Malone, P. C. (1986). The concept of the intracellular amino acid pool and its relevance in the regulation of protein metabolism, with particular reference to mammalian cells. *Current Topics in Cellular Regulation* **28**, 107–182.
7 Michal, G. (1999). *Biochemical Pathways: An Atlas of biochemistry and Molecular Biology.* Wiley, New York.
8 Zempleni, J. and Daniel, H. (2003). *Molecular Nutrition.* CABI, Cambridge, MA.
9 Meisenberg, G. and Simmonis, W. H. (1998). *Principles of Medical Biochemistry.* Mosby, St Louis, MO.
10 Neidhardt, F. C. and Umbarger, E. H. (1996). Chemical composition of Escherichia coli. In *Escherichia coli and Salmonella: Cellular and Molecular Biology* (Neidhardt, F. C. et al. eds), pp. 13–16. 2nd Ed. ASM Press, Washington D.C.
11 Lerner, J. A. (1978). *Review of Amino Acid Transport Processes in Animal Cells and Tissues.* University of Maine Press, Orono, ME.
12 Fillenz, M. (1995). Physiological release of excitatory amino acids. *Behavioral Brain Research* **71**, 51–67.
13 Fischer, W. N., Loo, D. D., Koch, W., Ludewig, U., Boorer, K. J., Tegeder, M., Rentsch, D., Wright, E. M. and Frommer, W. B. (2002). Low and high affinity amino acid H^+-cotransporters for cellular import of neutral and charged amino acids. *Plant Journal* **29**, 717–731.
14 Boudko, D. Y., Kohn, A. B., Meleshkevitch, E. A., Dasher, M. K., Seron, T. J., Stevens, B. R. and Harvey, W. R. (2005). Ancestry and progeny of nutrient amino acid transporters. *Proceedings of the National Academy of Sciences of the USA* **102**, 1360–1365.
15 Ritar, A. J., Dunstan, G. A., Crear, B. J. and Brown, M. R. (2003). Biochemical composition during growth and starvation of early larval stages of cultured spiny lobster (*Jasus edwardsii*) phyllosoma. *Comparative Biochemistry and Physiology A* **136**, 353–370.
16 Reimer, R. J., Chaudhry, F. A., Gray, A. T. and Edwards, R. H. (2000). Amino acid transport System A resembles System N in sequence but differs in mechanism. *Proceedings of the National Academy of Sciences of the USA* **97**, 7715–7720.
17 Kauzmann, W. (1957). Physical chemistry of proteins. *Annual Review of Physical Chemistry* **8**, 413–438.
18 Taverna, D. M. and Goldstein, R. A. (2002). Why are proteins marginally stable? *Proteins – Structure Function and Genetics* **46**, 105–109.
19 Trinquier, G. and Sanejouand, Y. H. (1998). Which effective property of amino acids is best preserved by the genetic code? *Protein Engineering* **11**, 153–169.
20 Radzicka, A. and Wolfenden, R. (1988). Comparing the polarities of the amino acids – side-chain distribution coefficients between the vapor-phase, cyclohexane, 1-octanol, and neutral aqueous solution. *Biochemistry* **27**, 1664–1670.
21 Rose, G., Geselowitz, A., Lesser, G., Lee, R. and Zehfus, M. (1985). Hydrophobicity of amino acid residues in globular proteins. *Science* **229**, 834–838.
22 Kyte, J. and Doolite, R. (1982). A simple method for displaying the hydropathic character of a protein. *Journal of Molecular Biology* **157**, 105–132.
23 Cornette, J., Cease, K. B., Margalit, H., Spouge, J. L., Berzofsky, J. A. and DeLisi, C. (1987). Hydrophobicity scales and computational techniques for detecting amphipathic structures in proteins. *Journal of Molecular Biology* **195**, 659–685.
24 Henikoff, S. and Henikoff, J. G. (1992). Amino acid substitution matrices from protein blocks. *Proceedings of the National Academy*

of Sciences of the USA **89**, 10915–10919.

25 Fauchere, J. L. and Pliska, V. (1983). Hydrophobic parameters π of amino acid side chains from the partitioning of *N*-acetyl amino acid amides. *European Journal of Medicinal Chemistry* **18**, 369–375.

26 Wolfenden, R. and Radzicka, A. (1986). How hydrophilic is tryptophan. *Trends in Biochemical Sciences* **11**, 69–70.

27 Taylor, W. R. (1986). The classification of amino acid conservation. *Journal of Theoretical Biology* **119**, 205–218.

28 Budisa, N., Huber, R., Golbik, R., Minks, C., Weyher, E. and Moroder, L. (1998). Atomic mutations in annexin V – thermodynamic studies of isomorphous protein variants. *European Journal of Biochemistry* **253**, 1–9.

29 Pal, P. P. and Budisa, N. (2004). Designing novel spectral classes of proteins with tryptophan-expanded genetic code. *Biological Chemistry* **385**, 893–904.

30 Bremer, H. and Dennis, P. P. (1996). Modulation of chemical composition and other parameters of the all by growth rate. In *Escherichia coli and Salmonella: Cellular and Molecular Biology* (Neidhardt, F. C. et al. eds), pp. 1553–1569. 2nd Ed. ASM Press, Washington D.C.

31 Wada, K., Wada, Y., Ishibashi, F., Gojobori, T. and Ikemura, T. (1992). Codon usage tabulated from the GenBank genetic sequence data. *Nucleic Acid Research* **20**, 2111S–2118S.

32 Chambers, J. A. A. (1993). Buffers, chelating agents and denaturants. In *Biochemistry Labfax* (Chambers, J. A. A. and Rickwood, D., eds), pp. 1–36. Blackwell Scientific, Oxford.

33 Charton, M. and Charton, B. I. (1982). The structural dependence of amino-acid hydrophobicity parameters. *Journal of Theoretical Biology* **99**, 629–644.

34 Dougherty, D. A. (1996). Cation-interactions in chemistry and biology: a new view of benzene, Phe, Tyr, and Trp. *Science* **271**, 163–168.

35 Forterre, P. (1997). Archaea: what can we learn from their sequences? *Current Opinion in Genetics and Development* **7**, 764–770.

36 Woese, C. (1998). The universal ancestor. *Proceedings of the National Academy of Sciences of the USA* **95**, 6854–6859.

37 Budisa, N., Moroder, L. and Huber, R. (1999). Structure and evolution of the genetic code viewed from the perspective of the experimentally expanded amino acid repertoire *in vivo*. *Cellular and Molecular Life Sciences* **55**, 1626–1635.

38 Crick, F. H. C. (1968). Origin of genetic code. *Journal of Molecular Biology* **38**, 367–379.

39 Delaye, L., Becerra, A. and Lazcano, A. (2004). The nature of the last common ancestor. In *The Genetic Code and the Origin of Life* (Ribas de Pouplana, L., ed), pp. 221–249. Landes Bioscience, Georgetown, TX.

40 Weber, A. L. and Miller, S. L. (1981). Reasons for the occurrence of the 20 coded protein amino acids. *Journal of Molecular Evolution* **17**, 273–284.

41 Galaev, I. Y. and Mattiasson, B. (1999). "Smart" polymers and what they could do in biotechnology and medicine. *Trends in Biotechnology* **17**, 335–340.

42 Govindarajan, S. and Goldstein, R. A. (1995). Optimal local propensities for model proteins. *Proteins – Structure Function and Genetics* **22**, 413–418.

43 Klimov, D. K. and Thirumalai, D. (1996). Factors governing the foldability of proteins. *Proteins – Structure Function and Genetics* **26**, 411–441.

44 Taverna, D. M. and Goldstein, R. A. (2002). Why are proteins so robust to site mutations? *Journal of Molecular Biology* **315**, 479–484.

45 Janin, J. (1979). Surface and inside volumes in globular proteins. *Nature* **277**, 491–492.

46 Budisa, N. (2004). Prolegomena to future efforts on genetic code

engineering by expanding its amino acid repertoire. *Angewandte Chemie International Edition* **43**, 3387–3428.

47 Rose, G. D. and Wolfenden, R. (1993). Hydrogen-bonding, hydrophobicity, packing, and protein-folding. *Annual Review of Biophysics and Biomolecular Structure* **22**, 381–415.

48 Taylor, F. J. R. and Coates, D. (1989). The code within codes. *BioSystems* **22**, 177–187.

49 Sonneborn, T. M. (1965). *Degeneracy of the Genetic Code: Extent, Nature and Genetic Implications*. Academic Press, New York.

50 Freeland, S. J. and Hurst, L. D. (1998). The genetic code is one in a million. *Journal of Molecular Evolution* **47**, 238–248.

51 Schimmel, P., Giege, R., Moras, D. and Yokoyama, S. (1993). An operational RNA code for amino acids and possible relationship to genetic code. *Proceedings of the National Academy of Sciences of the USA* **90**, 8763–8768.

52 Ribas de Pouplana, L. R. and Schimmel, P. (2001). Operational RNA code for amino acids in relation to genetic code in evolution. *Journal of Biological Chemistry* **276**, 6881–6884.

53 Bae, J. H., Alefelder, S., Kaiser, J. T., Friedrich, R., Moroder, L., Huber, R. and Budisa, N. (2001). Incorporation of beta-selenolo[3,2-*b*]pyrrolyl-alanine into proteins for phase determination in protein X-ray crystallography. *Journal of Molecular Biology* **309**, 925–936.

54 Knight, R. D., Freeland, S. J. and Landweber, L. F. (2004). Adaptive evolution of the genetic code. In *The Genetic Code and the Origin of Life* (Ribas de Pouplana, L., ed), pp. 75–91. Landes Bioscience, Georgetown, TX.

55 Maeshiro, T. and Kimura, M. (1998). The role of robustness and changeability on the origin and evolution of genetic codes. *Proceedings of the National Academy of Sciences of the USA* **95**, 5088–5093.

56 Judson, O. P. and Haydon, D. (1999). The genetic code: what is it good for? An analysis of the effects of selection pressures on genetic codes. *Journal of Molecular Evolution* **49**, 539–550.

57 Dufton, M. J. (1997). Genetic code synonym quotas and amino acid complexity: Cutting the cost of proteins? *Journal of Theoretical Biology* **187**, 165–173.

58 Mirny, L. A. and Shakhnovich, E. I. (1999). Universally conserved positions in protein folds: reading evolutionary signals about stability, folding kinetics and function. *Journal of Molecular Biology* **291**, 177–196.

59 Goethe, W. (1954). *Die Schriften zur Naturwissenschaft* 9. Herman Böhlaus, Weimar.

60 Monod, J. (1971). *Chance and Necessity*. Vintage Books, New York.

61 Barrell, B. G., Bankier, A. T. and Drouin, J. A. (1979). A different code in human mitochondria. *Nature* **282**, 189–194.

62 Santos, M. A. S. and Tuite, M. F. (2004). Extant variations in the genetic code. In *The Genetic Code and the Origin of Life* (Ribas de Pouplana, L., ed), pp. 183–220. Landes Biosciences, Georgetown, TX.

63 Tuite, M. F. and Santos, M. A. S. (1996). Codon reassignment in *Candida* species: an evolutionary conundrum. *Biochimie* **78**, 993–999.

64 Osawa, S. (1995). *Evolution of the Genetic Code*. Oxford University Press, Oxford.

65 Osawa, S., Jukes, T. H., Watanabe, K. and Muto, A. (1992). Recent evidence for evolution of the genetic code. *Microbiological Reviews* **56**, 229–264.

66 Muramatsu, T., Nishikawa, K., Nemoto, F., Kuchino, Y., Nishimura, S., Miyazawa, T. and Yokoyama, S. (1998). Codon and amino-acid specificities of a transfer RNA are both converted by a single post-transcriptional modification. *Nature* **336**, 179–181.

67 Schultz, D. W. and Yarus, M. (1996). On malleability in the genetic code. *Journal of Molecular Evolution* **42**, 597–601.

68 Muramatsu, T., Yokoyama, S.,

Horie, N., Matsuda, A., Ueda, T., Yamaizumi, Z., Kuchino, Y., Nishimura, S. and Miyazawa, T. (1988). A novel lysine substituted nucleoside in the first position of the anticodon of minor isoleucine tRNA from *Escherichia coli*. *Journal of Biological Chemistry* **263**, 9261–9267.

69 Suzuki, T., Ueda, T. and Watanabe, K. (1997). The "polysemous" codon – a codon with multiple amino acid assignment caused by dual specificity of tRNA identity. *EMBO Journal* **16**, 1122–1134.

70 Santos, M. A. S., Moura, G., Massey, S. E. and Tuite, M. F. (2004). Driving change: the evolution of alternative genetic codes. *Trends in Genetics* **20**, 95–102.

71 Yokoyama, S. and Nishimura, S. (1995). Modified nucleosides and codon recognition. In *tRNA: Structure, Biosynthesis, and Function* (Söll, D. and RajBhandary, U., eds), pp. 207–223. American Society of Microbiology, Washington, DC.

72 Kimura, M. (1983). *The Neutral Theory of Molecular Evolution*. Cambridge University Press, Cambridge.

73 Jukes, T. H. and King, J. L. (1969). Non-Darwinian evolution. *Science* **164**, 788–798.

74 Ribas de Pouplana, L. R. and Schimmel, P. (2004). Aminoacylations of tRNAs: Record-keepers for the genetic code. In *Protein Synthesis and Ribosome Structure* (Nierhaus, K. and Wilson, D. N., eds), pp. 169–184. Wiley-VCH, Weinheim.

75 Koonin, E. V. (2000). How many genes can make a cell: the minimal-gene-set concept. *Annual Review of Genomics and Human Genetics* **1**, 99–116.

76 Jose Castresana, J., Feldmaier-Fuchs, G. and Pääbo, S. (1998). Codon reassignment and amino acid composition in hemichordate mitochondria. *Proceedings of the National Academy of Sciences of the USA* **95**, 3703–3707.

77 Jukes, T. H. and Osawa, S. (1997). Further comments on codon reassignment. *Journal of Molecular Evolution* **45**, 1–3.

78 Jukes, T. H. and Osawa, S. (1989). Codon reassignment (codon capture) in evolution. *Journal of Molecular Evolution* **28**, 271–278.

79 Wang, L. and Schultz, P. G. (2002). Expanding the genetic code. *Chemical Communications*, 1–11.

80 Grivell, L. A. (1986). Deciphering divergent codes. *Nature* **324**, 109–110.

81 Lehman, N. and Jukes, T. H. (1998). Genetic code development by stop codon takeover. *Journal of Theoretical Biology* **135**, 203–214.

82 Schultz, D. W. and Yarus, M. (1994). Transfer-RNA mutation and the malleability of the genetic code. *Journal of Molecular Biology* **235**, 1377–1380.

83 Massey, S. E., Moura, G., Beltra, P., Almeida, R., Garey, J. R., Tuite, M. F. and Santos, M. A. S. (2003). Comparative evolutionary genomics unveils the molecular mechanism of reassignment of the CTG codon in *Candida* spp. *Genome Research* **13**, 544–557.

5
Reprograming the Cellular Translation Machinery

5.1
Enzyme Specificity of Aminoacyl-tRNA Synthetases (AARS) and Code Interpretation

5.1.1
Living Cells as Platforms for Amino Acid Repertoire Expansion

It is well known from the theory of evolution that mutations in living organisms occur spontaneously regardless of their fitness, i.e. they are random and the environment (selection) acts as a filter for those that are beneficial [1, 2]. The transmission of this strategy from nature into the laboratory provides material for anthropogenic selection of the phenotype of interest. Rapidly evolving organisms such as viruses, bacteria, protozoa or some organelles or fast-growing mammalian cells in culture are usually selected for work in laboratory conditions on very short human timescales. Engineering of the genetic code benefits from cell types such as auxotrophs, which are, in fact, metabolically engineered microorganisms or eukaryotic cells [3]. Cells with manipulated translational components, presumably AARS, offer an almost ideal platform to change the selection of the amino acids for protein biosynthesis, i.e. to expand its amino acid repertoire. The concept of the interpretation of the genetic code (elaborated in Section 3.5) postulates that AARS are crucial enzymes in the interpretation of the genetic code. Therefore, manipulation of their substrate specificity in the context of the living cell represents most the obvious way to expand the scope of ribosome-mediated protein synthesis (Fig. 5.1).

In general, to obtain new amino acids as building blocks for proteins, selective pressure should be specifically designed for such a compound of interest. Successful experiments include (i) uptake/import of the noncanonical amino acid, (ii) its intracellular accumulation at levels high enough for efficient substrate turnover (activation and tRNA acylation) by AARS, (iii) metabolic and chemical stability of the imported noncanonical amino acid, (iv) tRNA charging (acylation) that must be achieved at a demonstrable rate, and (v) translation of the noncanonical amino acid into a nascent polypeptide chain.

Engineering the Genetic Code. Nediljko Budisa

ISBN: 3-527-31243-9

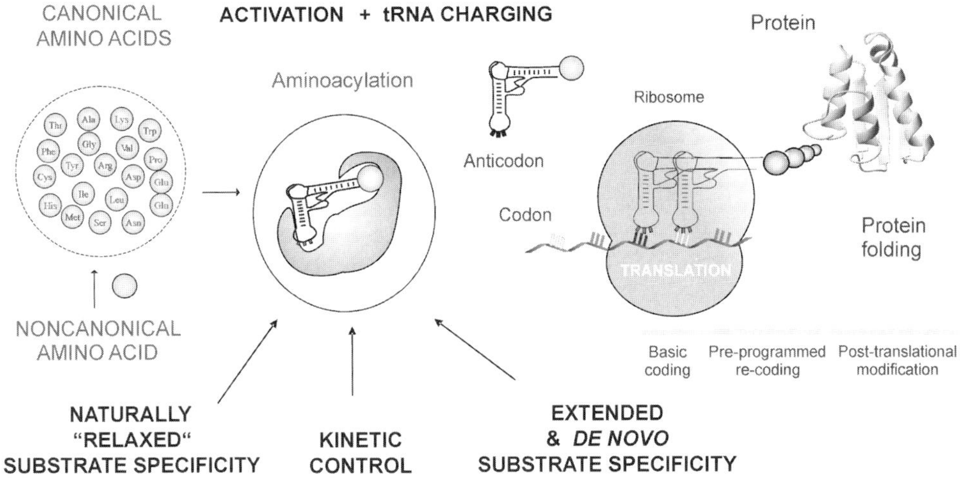

Fig. 5.1. The most straightforward way to reprogram the protein translational machinery is to manipulate the substrate specificity of AARS. The naturally "relaxed" substrate specificity of AARS refers to the absence of absolute substrate specificity of AARS, i.e. the "catalytic promiscuity" [4] of these enzymes. These enzymes have not evolved editing mechanisms and functions to discriminate against various noncanonical amino acids, especially those of anthropogenic origin. The stringent substrate specificity of natural AARS can also be bypassed by efficient elevation of the intracellular levels of AARS (i.e. by kinetic control; Section 5.1.4.3), this is especially suitable for those analogs that are poor substrates in activation reactions. A step further in this direction is to use designed AARS with extended substrate specificity (Section 5.1.4.4), or attenuated or changed editing functions (Section 5.1.4.5). An ultimate goal is certainly to engineer AARS with novel amino acid substrate specificity for *in vitro* and *in vivo* use (Section 5.1.5).

5.1.2 Uptake, Toxicity and Metabolic Fate of Noncanonical Amino Acids

5.1.2.1 General Considerations

Early fermentation experiments with noncanonical amino acid analogs revealed that the incorporation of an analog does not result in the synthesis of a radically different molecular species, but that the substitution of a canonical for a noncanonical amino acid(s) is the only change [5]. In addition, they have unambiguously shown that such analogs are transferred from the internal pool and contribute to protein synthesis without any necessity for different selectivity toward the noncanonical amino acid [6]. Furthermore, replacement of the residues of a canonical amino acid is random – there is an equal chance of replacement of any of the residues, irrespective of their position in the protein sequence. It was thought that the efficient incorporation of these substances into cellular proteins is possible due to their apparent resistance to intracellular degradation. However, such analogs appeared to be both incorporated into proteins and metabolized as well [7]. The incorporation capacity of such compounds is certainly influenced, among other things, by their (i) intracellular accumulation/removal, (ii) capacity to be transformed by other metabolic pathways and (iii) degree of participation in metabolic reactions (e.g. as substrates for amino acid oxidases and transaminases). For example, the phenylalanine analog β-2-thienylalanine inhibits the oxidation of Phe to Tyr; ethionine (Eth) and SeMet are metabolically converted to adenosyl-ethionine or adenosyl-SeMet; 4- and 5-methyltryptophan are found to block the synthesis of Trp, while a variety of analogs can participate in transamination reactions [8, 9].

Although living cells inspect many substances which can and do interfere with their metabolism, amino acid transport systems are quite promiscuous, and are not able to distinguish between chemically and sterically similar amino acids (and other small molecules) [10]. Therefore, cells pay the penalty of being penetrated by harmful substances like various amino acid analogs whose pools are formed in a same manner as those of their endogenous counterparts [11]. Due to such a lack of specificity of the cellular uptake mechanisms, a particular cell would treat D- and L-forms in a similar, if not, identical manner and allow for the incursion of all kinds of small molecules forming intracellular pools of β-imino, β- and γ-amino acids, cyclized forms, etc. Once such substances are present in the cytosol they can influence a variety of metabolic, synthetic and other physiological functions. Examples include (i) influence in amino acid biosynthesis, (ii) interactions with catabolic enzymes, (iii) action as mechanism-based irreversible enzyme inhibitors, (iv) interference with amino acid transport and storage [12], (v) they can serve as a substrates for translation into protein sequences (i.e. enter the genetic code) and, finally, (vi) noncanonical amino acids incorporated into proteins might be targets for various enzymatic or nonenzymatic (e.g. oxidations, hydrolyses) posttranslational modifications (see Section 3.11).

The synthesis or uptake of amino acids has to be regulated, not only to enable protein translation, but also to provide a proper balance of metabolites in the cell. Under normal physiological conditions a smaller amount of toxic amino acids can be tolerated without gross effects on cellular viability. Indeed, in the translation process, isosteric noncanonical amino acids are often suitable substrates for polypeptide synthesis, like norleucine (Nle) that can fully substitute for Met in bacterial proteins [13]. However, Nle is toxic since its inhibits methylation reactions, i.e. complete replacement of Met with Nle will inhibit cellular growth due to the inhibition of DNA methylation and replication, resulting in host restriction and degradation of chromosomal DNA, and subsequently in the cell death [14]. On the other hand, *Escherichia coli* can biosynthesize and accumulate elevated amounts of Nle under conditions that depress the enzymes of the Leu biosynthetic pathways and most probably Nle is always present in small quantities under normal growth conditions [15]. The result is that Met intracellular pools effectively compete with the Nle pools and heterologous proteins expressed in such *E. coli* host cells occasionally contain Nle residues at the positions normally occupied by Met. Living cells obviously have to find a way to push the balance between these pools in the direction that leads to the prevention of a larger extent of Met $\rightarrow$ Nle replacements through proteome.

Such studies on canonical/noncanonical amino acid pairs in the context of cellular metabolism reveal the usual metabolic fate of each particular canonical amino acid: they are "multitasking" substrates perfectly integrated into the fine network of cellular biosynthetic, metabolic and energetic functions [16]. Their synthesis or uptake has to be regulated, not only to enable protein translation, but also to provide a proper balance of metabolites in the cell.

5.1.2.2 **Amino Acid Transport**

The issue of cellular transport of noncanonical amino acids and the ways it influences cell uptake systems was also the subject of intensive research in early studies [6]. For example, Janecek and Rickenberg found that Phe surrogate 2-thienylalanine (Section 5.2.1.3), although a substrate for protein synthesis, inhibits permease action, i.e. its active transport in the cytosol [17]. It was also observed that 5- and 6-methyltryptophan are not incorporated into proteins since their intracellular penetration is not efficient in tested Trp auxotrophic strains [18]. Bacterial strains resistant to the presence of Nle, norvaline (Nva) or Eth as well as fluorinated amino acids have been isolated and characterized; they normally overproduce and excrete the corresponding canonical amino acid which than competes with the analog for entry in the cell [5]. It is also possible that particular endogenous AARS change their specificity and even the existence of alternative enzyme copies in the genome cannot be excluded as a protective mechanism (Sections 5.2.1.5 and 5.2.1.10).

The basic prerequisite for noncanonical analog/surrogate incorporation into proteins is that it must be accumulated in the cell at concentrations high enough to give adequate aminoacylation. This important issue gained attention again in the recent work [19]. For example, Gruskin, Conticello and coworkers [20, 21] have shown that *trans*-4-hydroxyproline, although a poor substrate for ProRS, can be incorporated into recombinant proteins if present at sufficient intracellular levels. This was achieved by manipulation of *E. coli* proline transport systems to effect intracellular accumulation via fermentation in media with hyperosmotic sodium chloride concentrations (around 600 mM). It is well known that Pro is actively accumulated intracellularly in response to hyperosmotic shock in *E. coli* and other prokaryotes [22]. Low-affinity Pro transporters (products of *putP*, *proP* and *proU* genes) [23] are upregulated in hyperosmotic expression cultures. Their hyperosmotic "relaxation" also enables ample intracellular uptake of Pro analogs and surrogates such as *trans*-4-hydroxyproline and their subsequent translation into proteins.

In general, in each experimental setup for noncanonical amino acid incorporation, the possibilities for manipulation with amino acid transport systems that lead to higher intracellular accumulation of the analog/surrogate of interest should be borne in mind. This might be applied for other organisms as well. For example, yeasts such as *Saccharomyces*, *Pichia* or *Candida* alter metabolism (and even codon assignment [24]) in response to environmental stresses [25], and conditions may be found in these systems to promote the cellular uptake, accumulation and incorporation of novel amino acid analogs.

5.1.2.3 Metabolic Conversions and Toxicity of Analogs and Surrogates

Most of noncanonical amino acids are known to posses growth-inhibitory properties, particularly towards microorganisms. A general characteristic of almost all amino acid analogs and surrogates is that their toxic effects can be specifically reversed by the canonical amino acid which is antagonized by the noncanonical counterpart. This toxicity is often the result of competition for an active transport system or permease or by analog/surrogate conversion into a toxic substance by a relatively complex metabolic route. This is because the amino acids are also utilized as metabolic fuels for the formation of a variety of metabolites. The particularly well-documented toxicity of *m*-fluorotyrosine and *p*-fluorophenylalanine (Sections 5.2.1.2 and 5.2.1.3; Fig. 5.11) is due to the formation of fluoroacetate generated via the dominant tyrosine metabolic pathway [26]. The toxin fluoroacetate is the most ubiquitous of the small class of organofluorine compounds, and has been identified in more than 40 tropical and subtropical plant species, and is also produced by some microorganisms when incubated in media containing fluoride [12]. Such enzymatic conversion of fluoride into C-F was recently documented as well: fluoroacetate and 4-fluorothreonine are secondary metabolites of the Actionmycete bacterium *Streptomyces catteya* [27]. On the other hand, the irreversible oxidation of Met to Met sulfone is an example of a nonenzymatic change which is often deleterious for living cells (Section 5.2.1.6; Fig. 5.14).

The toxic effects of amino acid analogs toward bacteria can be revealed thorough their bacteriostatic (e.g. Nle, some fluorinated amino acids) and bactericidal (e.g. canavanine) effects. While Nle and most of the synthetic and natural amino acids (more that 700 amino acid natural products have been identified until now) [28] are normally growth inhibitors which are more or less tolerated by living cells, incorporation of the Arg analog canavanine (Section 5.2.1.9; Fig. 5.17) leads to immediate cell death [29]. It is reasonable to expect that all amino acids with extremely reactive/unstable side-chains activity also will do so. In addition, tolerance toward different noncanonical amino acids is dependent of many factors, such as the bacterial strain or cell line used, chemical nature or reactivity of the amino acid side-chain and surrounding milieu or phase of the cellular growth. Changes associated with the cell cycle phase, for example, might also result in analog metabolic activation, degradation or even conversion into potent cytotoxins. Some eukaryotic cells such as plants certainly develop sophisticated protective mechanisms against toxic analogs they produce by preventing their entry in metabolism or contact with enzymes via localization into specific vacuoles, for example. Other mechanisms might include selection or emergence of the strain having a permease with a lower affinity for the analog/surrogate or availability of a facile degradative mechanism that converts the noncanonical analog into a harmless compound [10]. For example, Pro analogs dehydroproline and thiaproline (Section 5.2.1.5; Fig. 5.13) are substrates for intracellular oxidative degradation mediated by some endogenous *E. coli* enzymes such as Pro dehydrogenase or pyrroline reductase [21].

Furthermore, it should be kept in mind that toxic effects might also be caused by analog/surrogate modification after its charging of cognate tRNAs. Protein synthesis can also be interrupted by blocking chain initiation or chain elongation, by analog with selenaproline [30]. However, full insertion of the novel amino acid into the cell proteome would be in many cases fatal for the host cells, i.e. protein translation with analogs and surrogates might be responsible for the generation of abnormal, biologically impaired proteins. Replacements in and around the enzyme active site or that compromise protein structural integrity are especially destructive. This might have a profound effect on intracellular physiology, especially enzyme activity, protein turnover, protein–protein interactions, signaling pathways, etc., and might induce cessation of growth and death of the cell [10].

5.1.3
Constrains and Levels in Code Engineering

The extent of cellular substitution of canonical amino acids with noncanonical ones, especially *in vivo*, is constrained by general limits imposed by this specific experimental milieu. These limits (summarized below in Fig. 5.2) are metabolic constraints (related to the metabolic toxicity of most of the amino acids outside the canonical 20 as prescribed by the genetic code), proofreading mechanisms in genetic message transmission that are provide the fidelity of protein translation (quality control [31]), and, finally, protein folding rules which are closely related to the structure of the genetic code and physicochemical properties of the amino acids [3] (see Section 4.5). To challenge these constraints and find means to bypass, change or modulate AARS activities is one of the most fundamental problems of genetic code engineering.

After bypassing these three principal barriers, a particular noncanonical amino acid is regularly translated into a protein sequence and attains "full membership" in the expanded genetic code. Proteins expressed in this way that contain noncanonical or artificial amino acids are termed "alloproteins" [32] or, in the context of rational protein design, they can be regarded as "tailor-made" [33], since they are created with tailored functions capable of performing specific roles in user-defined environments. The expansion of the amino acid repertoire performed in this way reveals that the postulated universality of the genetic code should not be limited to its first (restricted) part, but also to its second part (relaxed or second code) which can be experimentally assessed [34]. See Fig. 5.2. These considerations are elaborated in more detail in the Section 5.2.

5.1.4
Auxotrophism and Natural AARS with Manipulated Functions

A special class of mutants of bacterial or eukaryotic cells deprived of their own amino acid supply (auxotrophs) can be bred and selected using the methods of classical genetics. As per the definition, auxotrophic mutant strains of microorganisms represent such a population that will proliferate only when the medium is supplemented with some specific substance not required by wild-type organisms [36]. In other words, the auxotroph is a naturally or genetically generated strain which cannot grow on "minimal" medium (e.g. mineral salts and glucose) without

First (obligatory) coding level:
20 canonical amino acids
(standard amino acid repertoire)

Second (facultative) coding level:
noncanonical amino acids
(expanded amino acid repertoire)

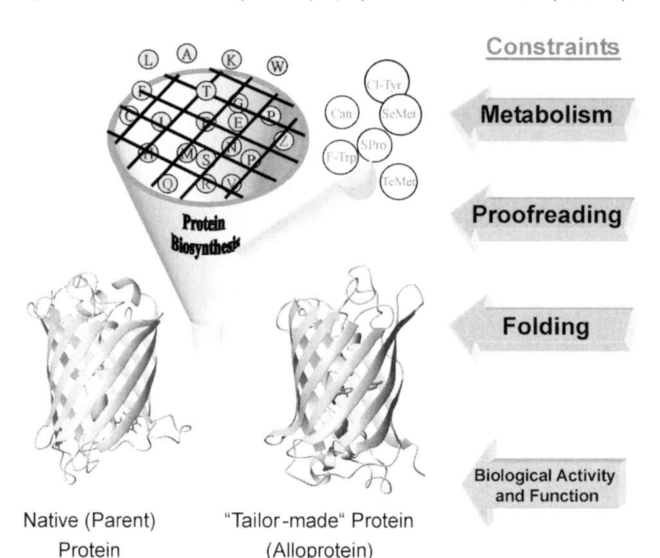

Fig. 5.2. The recruitment of canonical and noncanonical amino acids for protein translation. Canonical amino acids are not metabolically toxic, neither are they subject to any cellular proofreading functions (Section 4.2). On the other hand, the entry of noncanonical amino acids into protein translation is allowed only after bypassing three principal barriers: metabolic toxicity (Section 5.1.2), proofreading mechanisms (see Section 3.7) and they should allow folding of the target protein (see Section 4.5). In this way it can be postulated that the universality of the code is not limited to its "restricted" ("first") part (reserved for only 20 amino acids), but consists also of a "relaxed" or "second" part that can be experimentally assessed [34]. Proteins generated by translation of noncanonical amino acids can be taxonomically assigned as "tailor-made" [33] or "alloproteins" [32]. (Adapted from Minks [35] and Budisa [3].)

the addition of one or more specific supplements (e.g. a specific amino acid). Metabolically engineered cells where one or more amino acid biosynthetic pathways are knocked out or switched off offer an almost ideal platform to affect the selection of the amino acids for protein biosynthesis.

Another widespread natural phenomenon among enzymes, i.e. catalytic promiscuity (defined as the capacity to catalyze a secondary reaction at an active site that is specialized to catalyze primary reactions [4]), can be also employed for this purpose. In fact, the simplest way to expand the amino acid repertoire *in vivo* is to exploit the feature that the AARS proofreading machinery was not evolved to exclude or to "edit" the specific features of some sterically or chemically similar noncanonical amino acids [16] (Fig. 5.2). This allows for their efficient activation and charging onto cognate tRNAs, enabling rather large numbers of noncanonical amino acids that resemble canonical counterparts in terms of shape, size and chemical properties to be successfully incorporated *in vivo* into proteins.

Finally, the cellular uptake mechanisms also exhibit quite strong substrate promiscuity, i.e. they are incapable of distinguishing between structurally and chemically similar amino acids under defined conditions (see Sections 4.2 and 5.1.2). The combination of all these features in the context of controlled fermentation provides a clever approach for amino acid repertoire expansion without the requirement for mutations or other changes in the genes of host cells. Indeed, almost quantitative replacement of many canonical residues in proteins can be achieved by subjecting the auxotrophic host cells to an efficient selective pressure without changing its translational components [37].

5.1.4.1 **Proteome-wide Replacements: "Unnatural Microorganisms"**

By the 1950s it was well known that amino acid analogs are not only inhibitors of bacterial growth, but also might serve as substitutes for canonical amino acids in protein synthesis [6]. The most important and far-reaching finding from that time was the discovery of quantitative SeMet incorporation into proteins, reported by Cowie and Cohen in 1957 [38]. The incorporation experiment is based on use of the Met auxotrophic *E. coli* ML304d strain under defined fermentation conditions. Since the culture growth was quantitatively dependent on the external Met supply, it was possible to calibrate a defined synthetic medium where Met was replaced by SeMet. Although in such cultures, *E. coli* ML304d cells grow slowly in comparison with cultures supplied with Met, the growth curve was exponential. This is possible because the SeMet, when offered as a sole source of replacement for Met, completely and uniformly substituted for Met in all cellular proteins. Since cellular viability is not dramatically harmed by Met → SeMet replacements through the whole bacterial proteome, this first "unnatural" organism had similar properties to its Met-containing parental cells. In the following decade, Anker and coworkers [39, 40] also demonstrated that some leucine auxotrophs of *E. coli* are able to grow to a certain extent in synthetic media containing trifluoroleucine (TFL) instead Leu, although it was not possible to incorporate it into eukaryotic cells [41]. On the other hand, these early reports of the successful adaptation of *E. coli* auxotrophs for grow on TFL as a sole source for Leu contradict recent observations that this substance is a strong inhibitor of bacterial growth [42].

Since the landmark experiment of Cowie and Cohen [38] as the first documented demonstration of proteome-wide SeMet insertions, this amino acid still remains one of the few analogs with such a great substitution capacity. Moreover, the work of Cowie and Cohen is an excellent example of experimentally imposed changes in the interpretation of the genetic code (i.e. Met → SeMet reassignment for the AUG coding triplet) by use of metabolically engineered cells (Met auxotrophs). In the following decades, the focus was to study the mechanisms of SeMet utilization in different cells as well as to study its role in aminoacylation reactions and protein synthesis. The recent renaissance of SeMet incorporation into proteins was possible only after Hendrickson rediscovered Cowie and Cohen's approach as an important tool for structural biology [43].

Wong reported in 1983 a selection of microorganisms capable of intrinsically changing their preference for amino acids. In this experiment, a second example of an "unnatural" microorganism was provided, where the *Bacillus subtilis* QB928 Trp auxotroph was serially mutated and cultured for many generations in defined minimal medium containing 4-fluorotryptophan [44] [(4-f)Trp; Section 5.2.1.1; Fig. 5.8]. The resulting *B. subtilis* mutant HR15, which shows a higher preference for (4-f)Trp than for canonical Trp, has been described as the "first free-living organism in the past couple of billion years to has learnt to thrive on a genetic code that departs from the universal code". Although an accompanying genotype analysis was not presented and experiments that would reproduce these exciting findings have not been performed yet, the analyses of ribosomal proteins isolated from *B. subtilis* HR15 revealed almost quantitative labeling. Such analyses are quite straightforward since (4-f)Trp exhibits no characteristic fluorescence at room temperature, enabling in this way the identification of labeled species via "silent" (i.e. suppressed) fluorescence [45]. The importance of Wong's experiments is, however, that he demonstrated the power of genetic selection and the possibility to impose experimental pressure on the membership mutation of the genetic code, i.e. to make a microorganism with a changed genetic code vocabulary. The choice of (4-f)Trp as a model amino acid for such experiments has the advantage that natural TrpRS could not distinguish between Trp and (4-f)Trp since they have almost the same affinity for this enzyme in activation and aminoacylation reactions [10].

Most recently, Ellington and Bacher [46] reported two experimental attempts to design an "unnatural" organism following Wong's approach. In their first report they described attempts to evolve *E. coli* variants which would be able to survive on (4-f)Trp via a series of fermentations. The working hypothesis was that a relatively small number of proteins might be adversely affected, and some cells might mutate and emerge through many generations created in a simple serial transfer experiment in a way that (4-f)Trp could be accommodated in the proteome. Although mutant strains generated after 250 h of continuous evolution and 14 serial transfers were able to grow on 99.97% of (4-f)Trp in the synthetic medium, they still maintained an absolute requirement for Trp. Ellington and Bacher concluded "that the incorporation of unnatural amino acids into organismal proteomes may be possible but that extensive evolution might be required to reoptimize proteins and metabolism to accommodate such analogs". In their most recent contribution, Bacher and coworkers switched to bacteriophage Qβ as a model system for the generation of chemically ambiguous proteomes [47]. The Qβ phage is a relatively simple model system, defined as a quasi-species since its genome is a weighted average of a large number of different individual sequences. The replication of this phage was examined in the *E. coli* Trp auxotroph in the presence of a series of monofluorinated Trp analogs. Even though the fitness of the phage after 25 serial passages increased substantially, an absolute requirement for traces of L-Trp still remained.

Is it possible to select an organism capable of being submitted for selective pressure such that it becomes dependent on a noncanonical amino acid? Coupling the codon reassignment with the survival of the host cells in order to design "microbial strains with a clear-cut requirement for an additional amino acid that should be instrumental for widening the genetic code experimentally" was the approach used in Marliere's laboratory [48]. In particular, it was expected that mutagenesis/selection cycles should result in *E. coli* mutant strains capable of installing translational pathways specific for additional amino acids *in vivo*. In the first reported experiment, a defective *E. coli* mutant for thymidylate synthase as an essential enzyme for growth and viability was constructed. After the essential Arg codon (at position 126) in thymidylate synthase had been mutated to a Leu codon, cells were set to grow in the presence of azaleucine. The activity of this enzyme could be restored only when azaleucine (whose structure mimics Leu, but serve as a functional substitute for Arg) was incorporated at position 126 of the enzyme. In other words, carefully designed selective pressure led to a phenotypic change that causes Arg compensation with the antibiotic amino acid azaleucine. In other experiments, Döring and coworkers [49] succeeded selecting stable bacterial strains with coding triplets that can be read ambiguously. The captured coding signal was the isoleucine AUA rare codon and cysteine was used as the "codon captor" due to the rarity of its codons in genomes. Most recently, Marilere and coworkers constructed an *E. coli* strain with a defective editing function of ValRS (Thr222Pro) capable of overall amino acid replacements by almost 24% valine with α-aminobutyric acid (Abu) in the cellular proteome [50].

It is obvious that the simplest way to create an organism with permanent changes in the interpretation of the genetic code (codon capture) would be to engineer suitable cells or even organelles with a "minimal genome" [51] that affords enough space for the generation of new AARS:tRNA species. Such cells should more easily acquire novel codon meanings capable of inserting the desired noncanonical amino acid in the context of the whole proteome. On the other hand, in substitution experiments in highly evolved cells simple "nonfunctional" aliphatic amino acids like Ala, Leu, Ile and Val that indeed have relatively limited metabolic choices might be more suitable targets for proteome-wide substitutions since their replacement should be less harmful. Thus, they are most probably the best suitable candidates for future noncanonical amino acid translations in the total protein output of the particular cells. For example, Döring and Marliere [49] constructed stable bacterial strains with ambiguous reading of the rare coding triplet AUA by overexpression of CysRS:tRNA^{Cys}. Experimentally designed (and imposed) selective pressure yielded almost full reassignment (Ile $\rightarrow$ Cys) of the AUA codon in *E. coli*. Such proteome-wide substitutions were possible since Ile side-chains are not identified as catalytic residues in any enzyme of the *E. coli* proteome.

It is also conceivable to create mutant *E. coli* strains where rare codons are deleted from the wild-type genome by replacing them with degenerate codons specifying the same amino acids. They can then be reintroduced back into the genome but with novel assignment (i.e. codon capture) that would lead to amino acid repertoire extension which provide an evolutionary advantage for such an organism (see Section 4.8).

5.1.4.2 Substitutions at the Level of Single Proteins – Selective Pressure Incorporation (SPI)

In spite of the fact that noncanonical amino acids are overwhelmingly either bacteriostatic or even bactericidal substances (with SeMet as an exception to this rule), they might be substrates for the protein translation machinery. This was observed in early experiments with amino acid analogs, where it was clearly demonstrated that metabolically toxic analogs (fluorinated phenylalanines, Nle and canavanine) might serve as substrates in protein synthesis [6, 7]. However, when such toxic analogs are supplied together with their canonical counterparts in the synthetic growth media only lower incorporation levels in all cellular proteins were achieved. For example, the Met analog Nle leads to only 38% replacement of total cellular Met in proteins accompanied by greatly impaired viability of *E. coli* ML304d [52]. More recent experiments have indeed shown that it is possible to replace in this way about 50% of the total cellular proteins since one further division of bacterial cells takes place after Met was exhausted in the intracellular pool and extracellular medium [14, 15].

To achieve full substitution of single target proteins, the translation capacity should be resolved from metabolic toxicity. In this respect the autotrophic mutants of bacteria provided an excellent vehicle to circumvent the problem of analog toxicity during translation. These cells are suitable for the design of successful replacement experiments because preferential incorporation of a canonical amino acid over a noncanonical one can be achieved. A particularly effective approach was to starve host cells of a specific canonical amino acid and to induce the synthesis of specific protein, with concomitant addition of the analog to the culture medium. Skyes and coworkers [53] succeeded in this way to achieve in alkaline phosphatase a 10% replacement of all Tyr residues. Bacteriophage-coded proteins labeled with fluorinated amino acids have been prepared in a similar manner by simultaneous initiation of bacteriophage infection and amino acid analog addition [54]. However, the auxotrophic approach for complete labeling of target proteins could be fully generalized to a single target protein level only after the introduction of the recombinant DNA techniques. Cohen and coworkers [55] found a means to reassemble isolated DNA fragments and genes in new arrangements, and to propagate these fragments in living cells. With this method at hand, highly efficient expression systems capable of generating heterologous proteins in large amounts became available. Such a controlled extrachromosomally encoded protein expression system also provides an excellent platform for amino acid repertoire expansion. Namely, the possibility for stringent control of heterologous gene expression allows for the full exercise of selective pressure on the translation apparatus to produce a single substituted target protein in the context of the host cell with an unchanged proteome. In this way the desired noncanonical amino acid incorporations can be performed at single target proteins without harmful global effects on the expression host. The basic requirements include (i) selection of a proper cell and expression system, (ii) control of fermentation conditions (i.e. environment), and (iii) selective pressure for the amino acid replacement [i.e. sense codon(s) reassignment] at the level of the single protein [34]. This approach is termed the SPI method [37]

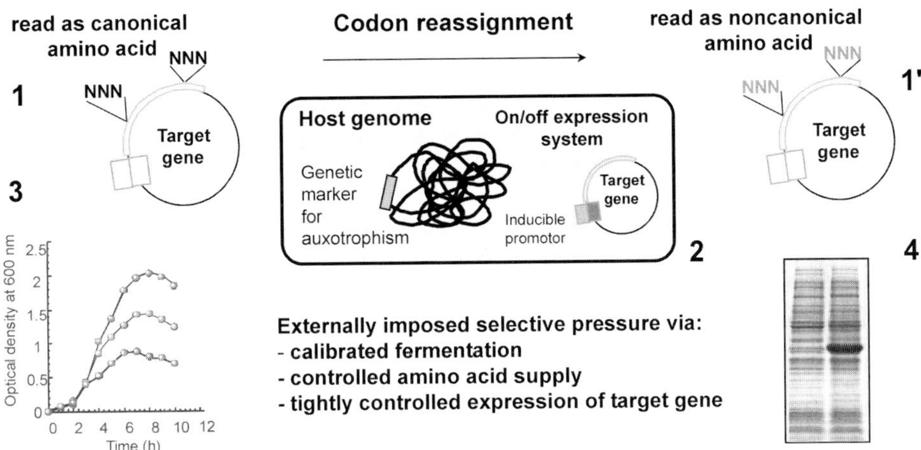

Fig. 5.3. Biosynthetic incorporation of noncanonical amino acids at the level of a single protein under experimentally imposed selective pressure (SPI method). Naturally assigned (1) sense coding triplets (NNN) of the target gene are reassigned (1′) during fermentation and expression experiments under the experimentally imposed selective pressure. Such changes in the interpretation of the genetic code can efficiently be achieved in the context of the intact proteome of the host cell (2) via calibrated fermentation (3) in defined minimal media combined with a controllable expression system (4). These components when combined together in a particular expression experiment results in a biosynthetic approach for large-scale production of proteins with a single or set of canonical → noncanonical amino acid substitutions. In this context, the SPI approach can be seen as an extension of extrachromosomal heterologous expression of the target gene that enables a general reassignment of the sense coding triplets.

since it confirms the principle that the choice of amino acids for protein synthesis could be conditioned by controlling environmental factors such as amino acid supply and fermentation parameters, i.e. by an experimentally imposed selective pressure, as shown in Fig. 5.3.

SPI has been practiced under various names by different research groups [20, 56] (e.g. auxotroph method, media shift, etc.) over the past decade, resulting in a large number of noncanonical amino acid analogs and surrogate incorporation into proteins using a variety of bacterial strains (Tab. 5.1). The experimental pressure on a selected expression host has been exercised as follows. The target gene activity is kept silent while growing the host cells to an appropriate density. After sufficient "healthy" cells are produced, the synthesis of the target protein is induced. From this point on, the host cells serve only as a "factory" to produce the desired recombinant protein. Fermentation in minimal media with growth-limiting concentrations of the native amino acid as natural substrate leads to its depletion from the medium. At this point, the noncanonical amino acid is added and the protein synthesis is induced. As a result of such pressure on the translation apparatus, the target protein containing exclusively noncanonical amino acid(s) is accumulated. This procedure leads to the arrest of cellular growth, but

Tab. 5.1. The list of the *Escherichia coli* auxotrophic strains used for incorporation experiments in Budisa's laboratory over the last decade.

Auxotrophic E. coli strain	*Genotype*	*Stable auxotrophic marker of interest*
B834 (DE3)	*ompT F hsdS$_B$ (r_B^- m_B^-) gal dcm metE*	Met deficiency
DSM 1563	*(F λ^s thr leu$^-$ arg$^-$ pro$^-$ his$^-$ thi$^-$ SmA^r ara$^-$ xyl$^-$ mtl$^-$ gal)*	Leu, Arg, Pro, His deficiency
DG 30	*(F Δ(gpt proA) 62 lacYl tsx-33 supE44 galK2 λ^- aspC13 rac$^-$ sbcB15 hisG4 rfbDl recB21 recC22 rpsL31 kdgK51 xyl-5 mtl$^-$ ilvE12 argE3 thi$^-$ tyrB507 hasdSl4 (hppT29))*	Phe deficiency
AT2471	*(Hfr (valS uxuBA) λ^- tyrA4 relAl spoTl thi$^-$)*	Tyr deficiency
ATCC31882	*(aroF aroG tyrR pheA pheAo tyrA trpE)*	Tyr deficiency
ATCC49979	*(WP2 uvrA)*	Trp deficiency
ATCC49980	*(WP2)*	Trp deficiency

These strains were selected for their capacity to serve as good hosts for target gene transformation, expression and protein preparation on a larger scale. In addition, they are stable auxotrophs, which enables tight control of fermentation conditions (i.e. control of amino acid supply) (from [35]).

in the light of a successful expression of target protein variants this fact is of less importance. Thus, the toxicity can be circumvented in quite a simple and straightforward way. Alternatively, the biosynthetic pathways of the host cells, acting as the endogenous supply, can be blocked by intracellular supplementation with proper inhibitors during incorporation experiments.

The power of the SPI method results from the use of sense DNA coding triplets and the endogenous AARS:tRNA pair, providing the possibilities for sense codon-directed (residue-specific) substitutions, expression and purification of variant proteins at the wild-type levels, and simple, easy and reproducible methodology. For these reasons, The SPI approach still holds the most promise for the large-scale synthesis of proteins with noncanonical amino acids for therapeutic, biomaterial or bioengineering applications. The Section 5.2.1 provides a detailed survey of amino acids incorporated into recombinant proteins by using this method.

5.1.4.3 **Kinetic Control – Enhanced System for Protein Translation**

Although the natural AARS have evolved their rather strict substrate specificity toward cognate canonical amino acids over billions of years, they have never encountered most of the synthetic analogs and surrogates in their evolutionary paths [57]. For that reason, some amino acid analogs are quite poor substrates for activation reactions mediated by AARS. The native genomic background activity of AARS within the host bacterium is sufficient only for trace levels of their incorporation, detectable in target proteins only with extraordinarily sensitive fluorescence- or

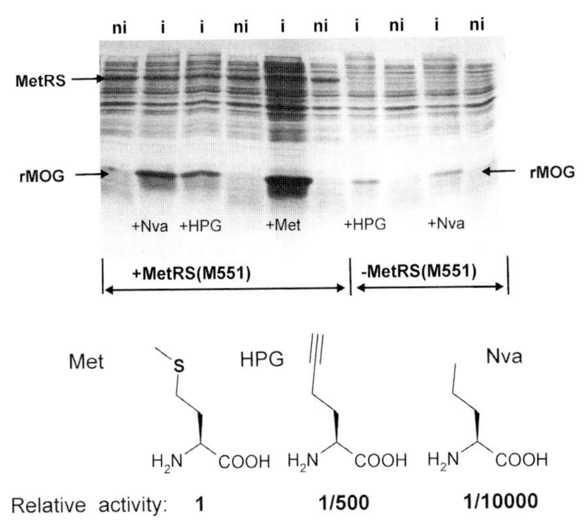

Fig. 5.4. The elevated intracellular amounts of MetRS and enhanced Met surrogate expression. The incorporation of homopropargylglycine (HPG, L-2-amino-5-hexynoic acid) and norvaline (Nva, L-2-amino heptanoic acid) into human recombinant myelin oligodendrocyte glycoprotein (rMOG) by using *E. coli* Met auxotroph B834 strain. Note enhanced the level of rMOG expression in the presence of elevated amounts of intracellular MetRS. "Relative activity" refers to the relative ratio k_{cat}/K_M in activation (pyrophosphate exchange) assay of these substrates with *E. coli* MetRS. ni = noninduced cellular lysates; i = induced lysates; M551 is truncated active MetRS fragment of 551 amino acids. The expression gel is generously provided by Dr. B. Mulinacci.

radioactivity-based assays. Classic kinetics teach us that low substrate turnover associated with poor substrate quality can be efficiently circumvented by increasing the amounts of catalysts in the reaction test tube. This strategy can be transmitted into the intracellular milieu of the living cells as well (Fig. 5.4), where incorporation of weakly recognized amino acid analogs and surrogates into target proteins can be enhanced by co-expression of wild-type AARS under the control of an orthogonal promoter system.

In this way, traditional auxotrophism-based methods can be improved via kinetic control of intracellular aminoaclytion reactions. Even AARS with well-documented editing activity can be kinetically manipulated in this way (but only if the desired noncanonical amino acid is not a substrate for editing reactions). For example, the Leu analogs TFL and hexfluoroleucine, although quite poor substrates in comparison with Leu in the ATP–pyrophosphate (PPi) exchange activation assay (Tab. 5.2), might be incorporated into the relatively small coiled-coil protein HA1 by using this approach [58]. Similarly, Conticello and coworkers succeeded in efficiently incorporating hydroxyl-proline analogs (Section 5.2.1.5; Fig. 5.13) into elastin in the presence of intracellularly co-expressed wild-type ProRS [21].

In addition, it is well known that the fidelity of the tRNA aminoacylation reaction is determined not only by AARS catalytic specificity, but also by specific intra-

Tab. 5.2. Kinetic parameters for activation of Leu, trifluoroelucine and hexafluoroleucine by *E. coli* LeuRS (reproduced with permission from [2]).

Substrate	k_{cat} (s^{-1})	K_M (μM)	Rel. k_{cat}/K_M
Leucine	3.8 ± 0.3	17.8 ± 5.1	1
TFL	0.49 ± 0.05	561 ± 121	1/242
Hexafluoroleucine	0.11 ± 0.01	1979 ± 804	1/4100

cellular concentrations of both tRNA and AARS in cytosol [59, 60]. It was even demonstrated that variations of cytosolic concentrations of AARS in bacterial expression hosts allow a single RNA message to be read in different ways. In particular, Tirrell and coworkers [2] have shown that the identity of the noncanonical amino acid (2*S*,3*R*)-4,4,4-trifluorovaline can be changed in protein translation by its assignment either to isoleucine or to valine codons, depending on intracellular levels of the isoleucyl- or valyl-tRNA synthetase in the bacterial expression host. This landmark experiment is an excellent example that shows how kinetic knowledge about relative rates of competing aminoacylation reactions could be used to affect the fidelity of protein synthesis in the context of expanded the amino acid repertoire (see also Section 6.2.1 and Fig. 6.1.).

The elevated intracellular concentrations of MetRS also proved to be a crucial factor for efficient translation of various Met analogs and surrogates in response to the AUG coding triplets in the mRNA sequence of target proteins. For example, by testing various terminally saturated Met analogs, Tirrell and coworkers found that *trans*-crotylglycine (Section 5.2.1.6; Fig. 5.15) is translationally active while *cis*-crotylglycine is almost "silent" in spite of its modest activation capacity [56]. However, the efficient incorporation of *cis*-crotylglycine (Section 5.2.1.6; Fig. 5.15) (over 90%) into model proteins has been achieved by endogenous coexpression of recombinant MetRS [61]. These examples clearly demonstrate an obvious correlation between activation and translation capacity, at least for some AARS (Met, Leu, Val and Ile, Section 5.2.1) coding overwhelmingly simple aliphatic amino acids. Thus, at least part of the noncanonical amino acids that do not support efficient protein synthesis in conventional expression conditions can indeed be efficiently incorporated into proteins after the microbial host is equipped with elevated intracellular amounts of AARS [62].

5.1.4.4 **Extension of the Existing Specificities of AARS**

The efficiency of analog/surrogate incorporation into native proteins in place of canonical amino acids often depends on the degree of structural and chemical similarity between these molecular species. As the structural and chemical differences between canonical and noncanonical amino acids increase, the facility of incorporation decreases associated with reduced protein yields and the appearance of products induced by amino acid starvation. This is not surprising since the activation of the larger-sized and chemically different amino acid substrates is normally ex-

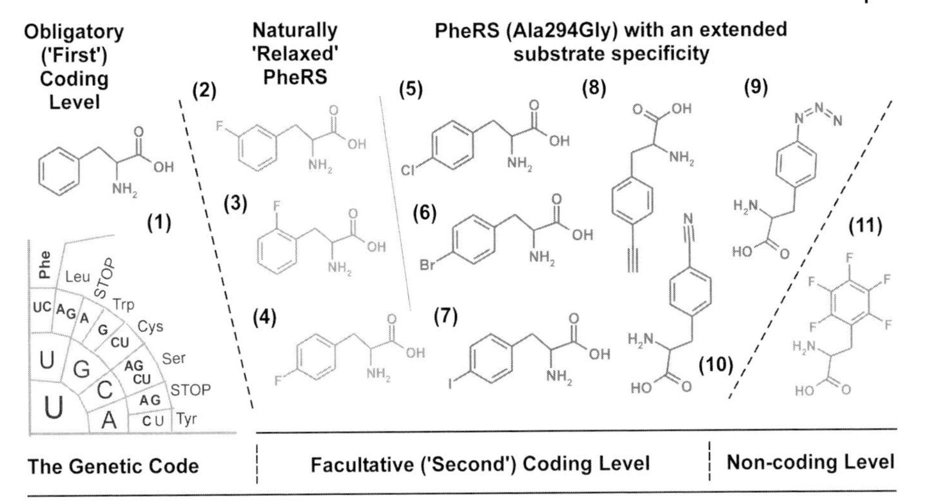

Fig. 5.5. The repertoire for Phe-coding triplets can be greatly expanded by kinetic control of intracellular PheRS, extension of its substrate specificity or a combination of both approaches. The number of Phe analogs (e.g. fluorinated amino acids) and surrogate (e.g. thyenylalanine) amino acids capable of entering the genetic code ("second" coding level) using PheRS catalytic promiscuity is relatively small when compared to suppression-based methodologies. The range of the desired chemical diversity is greatly expanded by using functionally active PheRS mutants in protein synthesis. These amino acids can be introduced as Tyr analogs and surrogates as well as using TyrRS mutants with altered substrate specificity. Amino acids: (**1**) Phe, (**2**) *m*-fluorophenylalanine, (**3**) *o*-fluorophenylalanine, (**4**) *p*-fluorophenylalanine, (**5**) *p*-chlorophenylalanine, (**6**) *p*-bromophenylalanine, (**7**) *p*-iodophenylalanine, (**8**) *p*-ethynylphenylalanine, (**9**) *p*-azidophenylalanine, (**10**) *p*-cyanophenylalanine and (**11**) 2,2′,3,3′,4-pentafluorophenylalanine.

cluded to insignificant levels by steric hindrance of their binding. For example, almost absolute specificity of TyrRS toward Tyr is possible since discrimination against Phe is achieved exclusively through the differences in initial binding energies [63]. Therefore, in cases when the catalytic promiscuity of the wild-type AARS is not sufficient enough to expand the amino acid repertoire, topological redesign of the binding pocket of an existing enzyme offers an interesting route to extend the existing specificities of natural AARS.

Kast and Hannecke [64] reported an engineered mutant of PheRS with a mutation in the binding pocket (Ala294Gly) which charges $tRNA^{Phe}$ with *p*-chlorophenylalanine (Fig. 5.5, **5**). Later, Ibba and coworkers [65, 66] demonstrated that it was possible to introduce amino acids like *p*-chlorophenylalanine and *p*-bromophenylalanine (Fig. 5.5, **6**) into recombinant luciferase and dihydrofolate reductase by using bacterial hosts extrachromosomally supplemented with this PheRS variant that exhibits relaxed substrate specificity. Similarly, by using a PheRS mutant with broadened substrate specificity (PheRS-αAla294Gly), the *E. coli* expression host supplemented with this enzyme was able to incorporate a whole set of Phe analogs and surrogates into the model protein dihydrofolate reductase [67], such as *p*-iodophenylalanine (Fig. 5.5, **7**), *p*-cyanophenylalanine (Fig. 5.5, **7**), *p*-ethynylphenylalanine (Fig. 5.5, **10**), *p*-azidophenylalanine (Fig. 5.5, **9**), and 2-, 3- and 4-azaphenylalanine (Section 5.2.1.3; Fig. 5.11). Bentin and coworkers reported a high-level of incorporation of a photoreactive bicyclic amino acid benzofuranylalanine into a model protein [68]. Based on the structural data available for *Thermus thermophilus* PheRS, a novel mutant of *E. coli* PheRS(Thr251Gly) with as enlarged binding pocket was computationally predicted and subsequently experimentally designed [69]. This mutant allowed successful incorporation into proteins *in vivo* of *p*-acetylphenylalanine carrying aryl ketone functionality [70]. Similarly, yeast TyrRS engineering based on the available three-dimensional structure of the *Bacillus stearothermophilus* enzyme yielded the mutant Tyr43Gly-TyrRS which was able to utilize several 3-substituted Tyr analogs as substrates for *in vivo* aminoacylation [71]. The ring-expanded Pro analog (2*S*)-piperidine-2-carboxylic acid (Section 5.2.1.5; Fig. 5.13) was incorporated into recombinant elastin only in the presence of ProRS with the Cys443Gly mutation in the binding pocket of the enzyme [21]. Although such a mutation resulted in minor attenuation of aminoacylation activity, the increased volume of the active site of the mutant enzyme allowed for accommodation and subsequent kinetic turnover of sterically bulky (2*S*)-piperidine-2-carboxylic acid.

It should always be kept in mind that it is notoriously difficult to predict amino acid substitutions that would lead to an extended specificity even when ample structural information is available. Nevertheless, a computational approach for predicting the relative energies of binding of different noncanonical amino acids into the binding site of AARS on the basis of crystal structures or homology-derived models might represent a useful approach for virtual screening of amino acid analogs. For example, Goddard and coworkers [69] reported that the calculated binding energies of some Phe analogs to PheRS correlate well with their translation activities in *E. coli*. Another recently reported example includes a computing procedure termed the "clash opportunity progressive computational method" for designing mutants of AARS with a higher preference for noncanonical amino acids [72].

5.1.4.5 AARS with an Attenuated Editing Function

Aminoacyl-tRNA synthetases for branched-chain amino acids such as ThrRS, ValRS, LeuRS and IleRS evolved additional domains or insertions in their structures, where each amino acid (cognate, noncognate, noncanonical and biogenic) is checked though a "double sieve" [73]. For this purpose, a single (catalytic) active site is supplemented with an additional editing (hydrolytic) site that dissociates noncognate amino acids. Manipulation (deletion or attenuation) of the editing function might also be used in order to expand the amino acid repertoire. This approach was first successfully demonstrated with ValRS from *E. coli*. This enzyme activates $tRNA^{Val}$ not only with Val, but also with Cys, Thr and α-aminobutyrate (Abu, Section 5.2.1.8; Fig. 5.16) [74]. The strain carrying ValRS with impaired editing function was selected and used for proteome-wide incorporation of Abu by Marliere and coworkers [50].

Recently, Mursinna and Martinis [75] reported a rational design that led to blocking of amino acid editing in LeuRS from *E. coli* (Fig. 5.6). In this enzyme a single

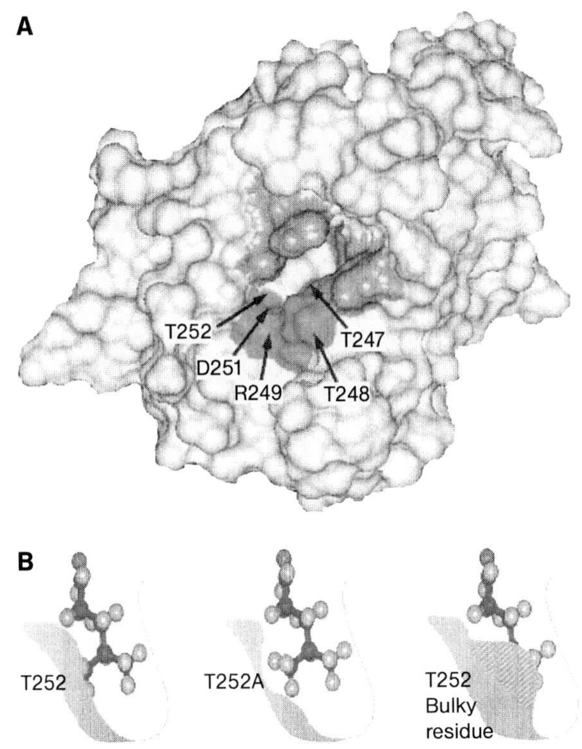

Fig. 5.6. Manipulating the editing function of LeuRS. (A) Surface representation with the cavity proposed to serve as an editing site of *Thermus thermophilus* LeuRS (reproduced from [75] with permission of American Chemical Society. Copyright 2002). Among other marked residues that participate in the cavity building, the highly conserved residue Thr252 is supposed to play a crucial role in LeuRS editing by preventing the cognate amino acid Leu from binding and hydrolysis. Noncognate and noncanonical amino acids such as Ile, Met and Nva bind in the editing site, which results in their hydrolysis from $tRNA^{Leu}$ isoacceptors. This is postulated on the basis of mutagenesis experiments on residue Thr252 (B). The mutation Thr252Ala yielded mutant LeuRS which hydrolyzes (i.e. edits) its "own" cognate amino acid, since the pocket volume of this cavity increases allowing Leu to bind. On the other hand, bulky side-chains (e.g. Thr252Phe) would significantly reduce the volume of the editing binding pocket and might block amino acid binding. For example, the Thr252Tyr mutant generates stable $tRNA^{Leu}$ misacylated with isoleucine, homoallylglycine, 2-butynylglycine and homopropargylglycine [76].

Thr at position 252 (in the amino acid-binding pocket of the editing active site) plays a critical role in the editing reaction; its mutation with bulky amino acids like Phe or Tyr results in the loss of the editing reaction. Such LeuRS, inactivated in its editing function, has the potential to provide an efficient and facile enzymatic synthetic route to generate tRNAs loaded with noncanonical amino acids. Tang and Tirrel [76] isolated and characterized three LeuRS mutants (Thr252Tyr, Thr252Leu and Thr252Phe) with attenuated editing activity. The LeuRS Thr252Tyr mutant was endogenously overexpressed together with target proteins, which allows for a successful incorporation of the noncanonical amino acids Nva and allylglycine. In addition, noncanonical amino acids described as Met surrogates were incorporated at the Leu positions of the model protein as well, e.g. homoallyglycine, homopropargylglycine and 2-butynylglycine (Section 5.2.1.6; Fig. 5.15) [76].

5.1.5 Beyond Auxotrophism: Towards AARS with *De Novo* Substrate Specificity

The structural diversity of natural proteins used to promote relatively simple chemical reactions and transformation is impressive. Enzyme substrate recognition is usually coupled with catalysis and takes place via multitude of intermolecular interactions of cooperatively built substructures, leading to accurate positioning of the catalytic amino acid residues in the active site, which explains the high level of specificity. On the other hand, a general lack of understanding of the structure–function relationship of enzymes generally hinders the rational design (only a few successful examples of designed novel substrates specificity are known) and makes "nonrational" approaches such as directed evolution extremely useful tools for developing new substrate specificities [77]. This is not surprising, since the informational content of protein sequences is notoriously nonuniform, with many residues being highly tolerant to a wide range of substitutions, whereas others cannot be altered without dramatic consequences for folding or function. Directed evolution, which is a sort of "molecular breeding" in laboratory, is based on the generation of a library of different mutants from which the enzyme(s) with the desired catalytic function(s) have to be selected [78]. Coupling the target reaction to survival in the selection step usually requires the development of complex, not-trivial intelligent assays (e.g. use of fluorescence, intracellular toxicity, antibiotic resistance and specific antibodies) [79].

In this context, customizing AARS towards exclusive recognition of desired noncanonical amino acids is a far more ambitious goal than extensions of existing specificities since it requires *de novo* generation of substrate specificity. Such experiments are intended to a generate mutant orthogonal AARS that in an ideal case should depart from its original substrate specificity and activate a noncanonical/noncognate amino acid as well as its "own" (orthogonal) tRNA which should be designed as well (Section 5.6.1). Such orthogonal AARS should not be acylated by any endogenous host tRNAs, canonical amino acids or other noncanonical amino acids naturally present in cytoplasm [e.g. homocysteine (Hcy), ornithine], but exclusively with desired noncanonical amino acid [80]. Pioneering steps in this direction were taken by Schultz and coworkers who were surprisingly successful in generating *Methanococcus jannaschii* TyrRS mutants with substrate specificities quite different from natural ones [81]. Based on the crystal structure of the homologous TyrRS from *B. stearothermophilus* [63], five residues (Tyr32, Asn123, Asp158, Ile159 and Leu162) in the amino acid-binding pocket of *M. jannaschii* TyrRS, that are within 6.5 Å of the *para* position of the aryl ring of bound Tyr, were mutated to Ala and subsequently randomized to all possible canonical amino acids (Fig. 5.7A). After library generation (containing usually up to 1.6×10^9 mutants [79]), a positive selection procedure based on suppression of an amber termination codon in the chloramphenicol acetyltransferase (CAT) was applied. The cells which survived in the presence of chloramphenicol and desired noncanonical amino acid (positive selection) were then grown in the same conditions with the absence of noncanonical amino acid and those that did not survive (since they had mutants with exclusive specificity for the target noncanonical amino acid) were isolated from a replica plate supplemented with canonical amino acid (negative selection) (Fig. 5.7B). This approach was used for the first time to select TyrRS mutants that selectively recognize and charge related orthogonal suppressor tRNA with the Tyr analog *O*-methyl-Tyr (*O*-Me-Tyr) (Fig. 5.7, **20**) [82]. The enzyme was not toxic for *E. coli* host cells, whereas the in-frame amber codon of the target mRNA was translated into the target protein sequence with *O*-Me-Tyr with sufficient suppression efficiency. This initial selection procedure was further improved by the recruitment of the well-known and well-characterized intracellular protein toxin barnase, a highly active ribonuclease from *Bacillus amyloliquefaciens* [83] often used in screening processes of directed evolution strategies [84]. For the purpose of the negative selection procedure, the specific nonsense codons were introduced into the barnase gene whose efficient suppression prevents its lethal effects, i.e. allows host cells to survive [85, 86].

The screening and selection procedure can be simplified to a great deal if an antibiotic were combined with the amplifiable fluorescence reporter. A T7/GFPuv reporter system coupled with T7 polymerase that contains amber nonsense codons at permissive sites was developed [87]. The efficient translation of T7 polymerase enables expression of GFP, allowing visual inspection of the cells based on amino acid-dependent fluorescence. The cells containing this system can, in combination with negative selection procedures (based either on chloramphenicol resistance or barnase expression suppression), be selected by fluorescence-activated cell sorting (FACS). Cells that fall within a positive window of fluorescence are isolated, propagated and submitted to the next shuffling cycle. Phage-display technology was also used for the directed evolution of AARS substrate specificity as well [88]. In the first step, the gene for AARS is cloned into a phage-display vector with the phage enzyme with an in-frame stop codon. Then a library of the mutants is constructed using either random mutagenesis or DNA shuffling [78] methods. Phages containing an active AARS capable of exclusively recognizing noncanonical amino acids are selected by specific monoclonal antibodies. The noncanonical amino acids of interest are in fact haptens against which these antibodies are directed [85].

These selection methods allowed the evolution of a series of novel *M. jannaschii* TyrRS mutants with preferential substrate specificity for more than 30 different Tyr analogs and substrates [86]. Moreover, this enzyme can be evolved to recognize specifically diverse amino acid side-chains that are structurally and chemically diverse from natural aromatic Tyr side-chains to a large extent (shown in Fig. 5.7C), and that include even neutral or charged aliphatic side chains and even sugars. Analysis of such substances reveals amino acids substituted with different functional moieties and groups to be very useful for selective protein labeling, e.g. metoxy [82] (Fig. 5.7, **20**), amino (Fig. 5.11, **23**), [89] alkene (Fig. 5.8, **5**) [90], keto (Fig. 5.7, **11**), [91] azido (Fig. 5.7, **16**), [92] photocrosslinking [93] (Fig. 5.7, **9**), naphtyl groups [94], etc. (for more detailed surveys, see [80, 81, 85, 86]).

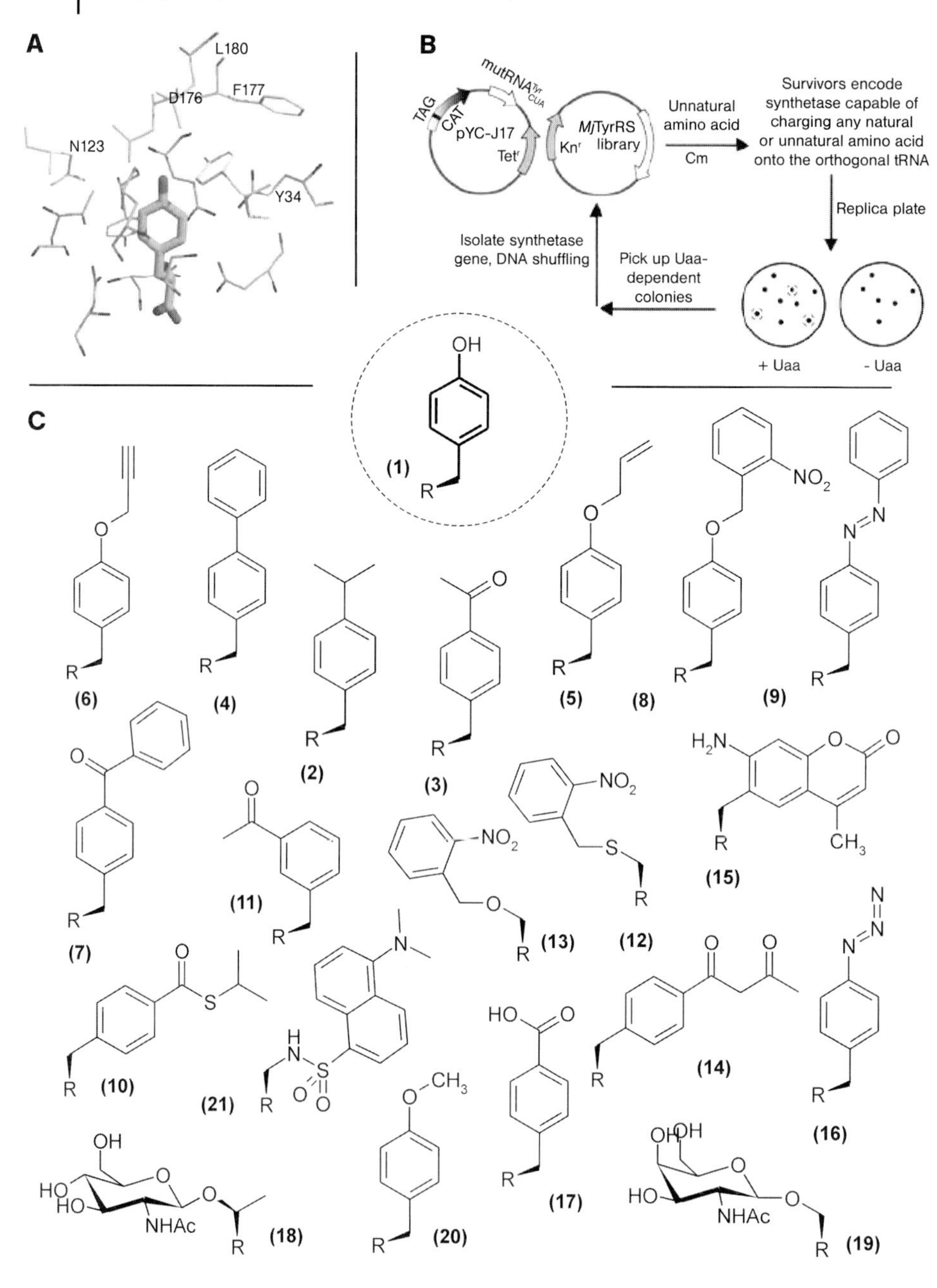
A
L180
D176
F177
N123
Y34
B
mutRNA$^{Tyr}_{CUA}$
TAG
CAT
pYC-J17
Tet^r
MjTyrRS
Kn^r library
Unnatural
amino acid
Cm
Survivors encode synthetase capable of charging any natural or unnatural amino acid onto the orthogonal tRNA
Replica plate
Isolate synthetase gene, DNA shuffling
Pick up Uaa-dependent colonies
+ Uaa
- Uaa
C
(1) (2) (3) (4) (5) (6) (7) (8) (9) (10) (11) (12) (13) (14) (15) (16) (17) (18) (19) (20) (21)

AARS mutants with enhanced preference for noncanonical amino acids were reported by other groups as well. Namely, AARS without editing activity like TyrRS are especially attractive targets for substrate specificity engineering. Nishimura and coworkers [95] identified mutant Phe130Ser of TyrRS from *E. coli* with elevated affinity for 2-azatyrosine (Fig. 5.11, **25**), i.e. mutant TyrRS recognizes azatyrosine 1.5-fold better than Tyr. This mutant was selected by screening *E. coli* cells transformed with a library of the mutant TyrRS gene (*tyrS*) harboring various plasmids. Further progress in this direction was reported by Yokoyama and coworkers [71]. Using the crystal structure of *B. stearothermophilus* TyrRS (in a similar manner Schultz group), they engineered *E. coli* TyrRS by introducing Y37V and Q195C mutations into the enzyme, to preferentially recognize L-3-iodotyrosine (Fig. 5.11, **22**) rather than L-Tyr. The TyrRS(V37C195) mutant activates L-3-iodotyrosine 10-fold more efficiently than L-Tyr (Tab. 5.3). It was used together with an *E. coli* and *B. stearothermophilus* suppressor $tRNA^{Tyr}$ in eukaryotic *in vitro* translation systems (e.g. wheat germ) or even in mammalian cells (Chinese hamster ovarian cells and human embryonic kidney 293 cells) for protein labeling with 3-iodotyrosine [96].

5.1.5.1 Kinetic Issues of Orthogonal AARS: Catalytic Efficiency and Selectivity

The *E. coli* TyrRS(V37C195) activates 3-iodotyrosine 10-fold more efficiently ($k_{cat}/K_M = 3.3 \times 10^3$ M^{-1} s^{-1}) than Tyr itself (the $k_{cat}/K_M = 3.2 \times 10^2$ M^{-1} s^{-1}) [71]. Although *M. jannaschii* TyrRS specific for *O*-Me-Tyr has a k_{cat}/K_M value towards this substrate about 10^2 times higher than that of Tyr [82], this is still a relatively low level of discrimination when compared with natural systems. Indeed, the Tyr versus Phe discrimination in TyrRS of *E. coli* and *B. stearothermophilus* was experimentally determined to give a value of about $1\text{–}2 \times 10^5$ for the preferential activation of Tyr (Tab. 5.3; an extensive discussion of TyrRS selectivity is given in

Fig. 5.7. Active-site engineering and positive/negative selection for novel catalytic preferences of engineered *M. jannashii* TyrRS mutants for desired amino acid substrates. (A) Portion of the active site of TyrRS from *B. stearothermophillus* (which is homologous to the *M. jannashii* TyrRS) with marked residues which are first systematically randomized in order to gain preferential specificity for *O*-methyl-Tyr (**20**) substrate. (Reproduced from [82] with permission.) (B) Positive/negative selection principle – in a positive selection step AARS mutant that activate desirable noncanonical amino acid is identified and selected; with the negative selection screen those mutants that still activate canonical amino acids are eliminated. (Reproduced from [86] with permission.) (C) The AARS evolved and selected by these procedures no longer activate the canonical amino acid Tyr (**1**), and their substrate specificity is not only extended to Tyr analogs and surrogates, but also toward structurally diverse side-chains such as sugars. Amino acids: (**1**) Tyr, (**2**) *p*-isopropyl-phenylalanine, (**3**) *p*-acetyl-phenylalanine, (**4**) *p*-phenyl-phenylalanine, (**5**) *O*-allyl-tyrosine, (**6**) *O*-(2-propynyl)-tyrosine, (**7**) *p*-benzoyl-phenylalanine, (**8**) *o*-nitrobenzyl-tyrosine, (**9**) *p*-phenylazo-phenylalanine, (**10**) *p*-ethylthiocarbonyl-phenylalanine, (**11**) *m*-acetyl-phenylalanine, (**12**) *o*-nitrobenzyl-cysteine, (**13**) *o*-nitrobenzyl-serine, (**14**) *p*-(3-oxobutanoyl)-phenylalanine, (**15**) 7-aminocoumarine-alanine, (**16**) *p*-azido-phenylalanine, (**17**) *p*-carboxyl-phenylalanine, (**18**) α-*N*-acetylglucosamine-*O*-threonine, (**19**) β-*N*-acetylglucosamine-*O*-serine, (**20**) *O*-methyl-tyrosine, (**21**) *dansyl*-alanine (Cm – chloramphenicol; Uaa – unnatural amino acid).

Tab. 5.3. Amino acid activation efficiency of Tyr by *E. coli* TyrRS compared with the experimental data obtained for activation of (3-I)Tyr by the *E. coli* TyrRS mutant V37C195A and *M. jannaschii* TyrRS mutant specific for *O*-Me-Tyr. These catalytic parameters are derived from a ATP–PPi isotopic exchange reaction (all amino acids presented are in the L-chiral stereoisomeric form).

Origin of TyrRS	k_{cat} (s^{-1})	K_M (μM)	k_{cat}/K_M ($M^{-1}s^{-1}$)	*Selectivity*
E. coli [63] (natural evolution)	21	2.8	7.5×10^6	Tyr versus Phe: 100000
Orthogonal *Mj*TyRS (Schultz laboratory) [82]	14.0 ± 0.001	443 ± 93	3.2×10^4	*O*-Me-Tyr versus Tyr: 100
*Ec*TyrRS (V37C195) (Yokoyama laboratory) [71]	0.43 ± 0.08	130 ± 20	3.3×10^3	(3-I)Tyr versus Tyr: 10

*Mj*TyRS is derived from *M. jannaschii*, while *Ec*TyrRS (V37C195) is from *E. coli* wild-type enzyme; both mutant enzymes are proved to be orthogonal in the context of either *in vivo* or *in vitro* translations. Note the unusual high turnover number (k_{cat}) in spite of the relatively high K_m value (almost 0.5 mM) of the *Mj*TyrRS mutant enzyme.

Fersht [73]). It is well known that natural enzymes are 10^2–10^6 times more effective than related catalytic antibodies [79] and the designed AARS are obviously not an exception from this rule. In fact, wild-type TyrRS enzymes either from *Bacillus*, *Escherichia*, *Saccharomyces* or *Methanoccocus* activate their cognate substrate Tyr at least 10^2 times better than TyrRS mutants activate "cognate" *O*-Me-Tyr or 3-iodotyrosine (Tab. 5.3). In other words, this generation of designed enzymes was always a few orders of magnitudes catalytically less efficient than endogenous (natural) TyrRS enzymes. Not surprisingly, an expression system using such TyrRS mutant enzymes requires higher intracellular noncanonical amino acid concentrations (since TyrRS mutant enzymes have higher K_M values for selected substrates) and calibrated minimal media (as in routine auxotroph approaches) with higher analog/surrogate concentrations (usually 1–5 mM). Catalytic parameters are also available for the "natural" 21st synthetase PylRS, whose apparent K_M towards Pyl is much better (53 μM) than those of "man-made" enzymes. In contrast, the turnover number of PylRS is much lower ($k_{cat} \sim 0.1\ s^{-1}$). The reason for the unusually high k_{cat} of the "man-made" synthetases presented in Tab. 5.3 (whose quite high K_M values indicate bad substrates) lies most probably in the experiment setup: extremely high (i.e. nonphysiological) substrate and enzyme concentrations in the reaction mix usually allows for estimation of higher k_{cat} values.

All these examples clearly illustrate how far we still are from the *de novo* design of enzymes that rival natural ones, despite the robust screens and selection procedures for AARS substrate specificity evolution. Therefore, contemporary works on AARS substrate specificity changes can be seen as the first inroads towards the novel generation of more effective mutant AARS with efficiently altered substrate specificity. However, it should be always kept in mind that the problem of "bad" substrates can also be efficiently circumvented by kinetic control of natural endogenous AARS (Section 5.1.4.3). For example, the activation capacity of LeuRS towards TFL is about 2×10^2 reduced when compared with the natural substrate Leu (Tab. 5.2). Nevertheless, kinetic control in combination with fermentation and expression control (i.e. selective pressure) was sufficient to obtain translation with TFL with high efficiency, fidelity and almost wild-type level yields [97].

Catalytic parameters of mutant TyrRS enzymes presented in Tab. 5.3 refer to activation kinetics, i.e. aminoacyl-adenylate formation from the amino acid and ATP by an AARS. This reversible reaction is normally assayed by the isotopic exchange of [^{32}P]pyrophosphate into ATP upon addition of an enzyme and amino acid into the reaction tube [6, 98]. Thereby, it should be always kept in mind that aminoacyl-adenylate formation is, in general, the least accurate step in an aminoacylation reaction. There is no data about attempts to measure directly the transfer of activated amino acids to orthogonal tRNAs. This is not surprising, since most of the aminoacylation assays are mainly based on the use of radioactivity (usually ^{14}C or ^{3}H isotopes), whereas radioactively labeled noncanonical amino acids are scarce. Aminoacylation parameters for noncanonical amino acids would provide a more realistic picture of the effectiveness of AARS with altered specificity in tRNA charging in comparison with native ones. Current and future works should be also directed to not only expand the pool the genetically encoded amino acids at the proof-of-principle levels, but also to improve the performance and wide applicability of these systems.

Recent crystallographic studies of native and mutant synthetases confirm the well-known fact that structural plasticity evident in wild-type AARS plays a major role in the establishment of modified specificities as well [99]. Indeed, the specificity of the aminoacylation reaction (termed as discrimination or D-factor) with regard to the 20 canonical amino acids varies considerably among the different AARS *in vitro* [100]. The highest D-factor as measure for specificity between 28 000 and >50 0000 is found for TyrRS, while the lowest values between 130 and 1700 were observed for LysRS [101]. The existence of "intrinsically relaxed" AARS such as LysRS is a feature that can also be exploited for further chemical diversification of the expanded code, with such amino acids having structural and chemical features that are quite different from the standard 20. In other words, they can be used as an excellent starting point for the design of novel generations of orthogonal AARS with considerably improved specificity and catalytic performance.

5.2 Reassigning Coding and Noncoding Units

5.2.1 Sense Codon Reassignment: Most Commonly Used Substitutions

At the conceptual level of the genetic code and genetic message translation, the traditional approach based on use of the auxotrophic strains results in sense codon reassignments seen as a canonical → noncanonical amino acid substitution in the resulting protein sequence. The first substitution experiments were performed at the level of the whole proteome (see Section 2.2). Later, when extrachromosomal plasmid-directed protein expression systems became available, it was possible to generate amino acid replacements in single target proteins (Section 5.1.4.2). Expansion of the amino acid repertoire, especially *in vivo*, is constrained by general limits imposed by this specific experimental milieu. These (summarized in Fig. 5.2) are metabolic constraints, proofreading mechanisms (i.e. quality control) in the genetic message transmission and requirements for proper protein folding. The expansion of the amino acid repertoire as presented in Fig. 5.2 reveals that the postulated universality of the genetic code should not be limited to its first (restricted) part, but also to its second part (relaxed or second code) which can be experimentally assessed. Three levels in the structure of the genetic code can be postulated: (i) the first coding level includes canonical amino acids whose entry in the genetic code is obligatory, (ii) the second coding level includes noncanonical amino acids whose entry is allowed in the code (i.e. entry is facultative) and (iii) the third or noncoding level, consisting of amino acid substrates for which experimental attempts for incorporation have so far been unsuccessful (i.e. entry is forbidden). In other words, for each amino acid in the standard genetic code table it should be possible to establish or define a whole library of translationally active substances capable of being incorporated in proteins. The final goal would be to build up an extended structure of the genetic code which would have an additional (i.e. second) coding level with a defined library of translationally active analogs and surrogates for each canonical amino acid. Alternatively, the goal can be defined as one of the main tasks of code engineering, i.e. to "move" translationally inactive amino acids from the noncoding level to coding levels. In following sections give a brief overview of experimental reassignments of coding triplets (as well as triplet families) naturally assigned to 11 amino acids. These experiments are designed so that target amino acids are substituted normally at the level of the single protein, rarely being proteome-wide.

5.2.1.1 **Tryptophan**

Trp as a unique amino acid is an especially attractive substitution target in protein engineering and design studies for several reasons. First, the diverse and rich indole chemistry offers numerous analogs/surrogates (Fig. 5.8) to be tested for translation activity. Many of them are also substrates for "naturally relaxed" TrpRS, a feature that was extensively used in recent years to expand the coding capacity of the Trp UGG codon as shown in Fig. 5.8. Second, Trp has a relatively low abundance in proteins (1.2%) [102] and a single triplet in the genetic code (UGG); thus, the substitution of Trp side-chains in globular proteins provides an almost site-directed mode for analog incorporation. Third, it possesses special biophysical properties which allow for its participation in numerous interactions in proteins (π–π stacking, hydrogen bonding and cation–π interactions). Finally, Trp is the main source of UV absorbance and fluorescence of proteins [103]. The evolutionarily optimized amino acid frequency in the sequences of most natural proteins is

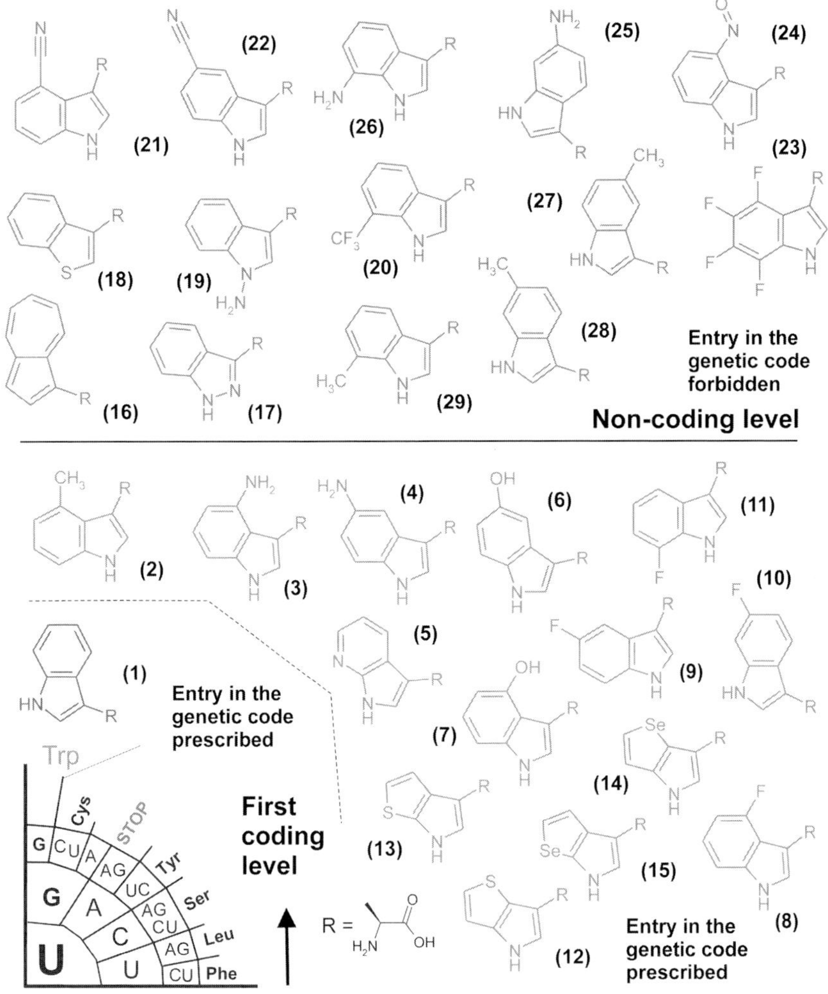

Fig. 5.8. An expanded amino acid repertoire for the UGG coding triplet. The "first" or obligatory coding level marks the evolutionary established assignment of Trp for the UGG coding triplet during protein biosynthesis. The translation of various noncanonical Trp-like amino acid analogs and surrogates as a response to the UGG coding triplet into target protein sequences enables building of a second or facultative coding level in the genetic code. The noncoding level is represented here with mostly methylated Trp analogs for which the translational activity was not previously experimentally confirmed. This is of course only a tiny fraction of all possible translationally inactive Trp analogs and surrogates. The common chemical names of Trp (**1**) analogs and surrogates are: 4-methyltryptophan (**2**) 4-aminotryptophan (**3**), 5-aminotryptophan (**4**), 7-azatryptophan (**5**), 4-hydroxytryptophan (**6**), 5-hydroxytryptophan (**7**), 4-fluorotryptophan (**8**), 5-fluorotryptophan (**9**), 6-fluorotryptophan (**10**), 7-fluorotryptophan (**11**), L-β-(thieno[3,2-*b*]pyrrolyl)alanine (**12**), L-β-(thieno[2,3-*b*]pyrrolyl)alanine (**13**), β-selenolo[3,2-*b*]pyrrolyl-L-alanine (**14**), β-selenolo[2,3-*b*]pyrrolyl-L-alanine (**15**), azulene (**16**), 2-azatryptophan (**17**), benzothieno-phenylalanine (**18**), 1-, 6- and 7-aminotryptophan (**19**, **25** and **26**), 7-trifluoromethyl-tryptophan (**20**), 4-cyanotryptophan (**21**), 5-cyanotryptophan (**22**), 4,5,6,7-tetrafluoro-tryptophan (**23**), 4-nitrozotryptophan (**24**), 5-methyltryptophan (**27**), 6-methyltryptophan (**28**), and 7-methyltryptophan (**29**). (Reproduced from [105] with permission.)

relatively rare for amino acids such as Trp, Cys and Met. However, these amino acids often play crucial roles in protein functionality, serving as ideal intrinsic probes for studying the structure, dynamics and function (for more details, see [104–108]).

Trp (together with Met) is not strictly polar or apolar, but combines both properties, as reflected in its "promiscuous" topological distribution in proteins. In fact, there are controversies regarding the hydrophobic nature of Trp in the context of the canonical amino acid repertoire [109] (see Section 4.3; Fig. 4.2). Namely, more than 40 published hydrophobicity scales in the literature with many cases of contradictions about the hydrophobic/hydrophilic nature of Trp are available [110, 111]. It is well known that the aromatic amino acids Phe, Tyr and Trp harbor a combination of two properties, i.e. hydrophobicity and polarity (ability to bind ions through cation–π interactions), that are often considered to be mutually exclusive [112]. Such a combination of properties makes them suitable candidates for both the interior of a protein or for interactions with cell membranes. This is best exemplified in the case of Trp187 in annexin-V where one crystal form is closely packed in the hydrophobic niche of domain III. Upon addition of Ca^{2+} in the presence of membranes, Trp187 undergoes large local conformational changes which lead to its insertion into the membranes [113]. Trp side-chains are also known to play a major role in protein stability, folding and assembly [114]. It participates in other essential biological activities as well, such as receptor–ligand interactions, enzyme–substrate binding and antigen–antibody recognition. These essential functions, both in the structure of many proteins as well as their means of interacting with other molecules, make Trp conservation throughout many protein families inevitable [115].

The TrpRS is thought to have a relatively short evolutionary history since Trp is believed to be the latest addition to the genetic code [116]. This enzyme from class I (subclass Ic) catalyzes activation of Trp by ATP and its transfer to $tRNA^{Trp}$, ensuring correct translation of the UGG coding triplet. The active enzyme form in both eukaryotes and prokaryotes is a dimer [117, 118]; the second subunit is required to recognize the CCA anticodon of $tRNA^{Trp}$, whereas indole nitrogen donates a hydrogen bond to the carboxylate of Asp132 in the binding pocket [119], which is essential for enzyme catalytic efficiency [120] (Fig. 5.10). In general, the imino function of indole obviously has unique significance in the recognition of Trp and its analogs by TrpRS and the translation apparatus. This was demonstrated by synthesis of analogs where the imino group was replaced with other heteroatoms (carbon, sulfur or oxygen, e.g. compound **18** from Fig. 5.8), which resulted in translationally inactive amino acids [121]. For example, the Trp-like blue amino acid azulene (Fig. 5.8, **16**) is not a substrate for the native translational machinery even when is chemically charged onto a tRNA for *in vitro* reactions [122]. This is an important guideline for the synthesis of possibly useful surrogates or analogs of Trp. This suggests that the related synthetic Trp analog would probably not be recognized and activated by the TrpRS in the protein translation step. Even changes in positions adjacent to the imino group were expected to be disturbing, since 2-azatryptophan (Fig. 5.8, **17**) was also not a substrate for protein synthesis [123].

Conversely, 7-azatryptophan (Fig. 5.8, **5**) is recognized by cellular TrpRS and incorporated into proteins [124]. Thus, the benzene ring of the indole moiety is apparently a much more permissive target for chemical transformations that might produce analogs and surrogates containing useful ring substitutions. The effects of fluorinated Trp analogs on the catalytic efficiency of TrpRS from *B. subtilis* was also studied in the frame of classical steady-state kinetics [125].

The difficulties of cellular uptake of methylated Trp analogs (Fig. 5.8, **28**, **29**) reported earlier [18] were confirmed in recent experiments which indicate that only 4-methyltryptophan (Fig. 5.8, **2**) is a substrate for native *E. coli* translational machinery (Fig. 5.8) [126]. In addition to the problems of cellular uptake, catalytic parameters toward TrpRS of these substances are also unknown and might well represent a barrier for their translational activity. It was also recently found that is possible to introduce only (4-NH_2)Trp into *Aequorea* GFP proteins, but not (5-NH_2)Trp as well as Trp hydroxylated analogs, although all these substances were substrates of TrpRS [33]. The main reason for their inability to serve as substrates for protein synthesis might be their poor recognition by TrpRS. On the other hand, even if these amino acids can be loaded onto $tRNA^{Trp}$ it might well be that certain still unknown editing mechanisms during ribosome synthesis or cotranslational folding prevent their incorporation in *av*GFPs (see Section 6.2.3, Fig. 6.3).

The history of the expanded amino acid repertoire for the Trp-coding triplet UGG started with reports from 1955 by Kidder and Dewey [127], 1956 by Pardee and coworkers [128] and 1957 by Brawerman and Yčas [129] about the incorporation of 7-azatryptophan and 2-azatryptophan (Fig. 5.8, **5**, **17**) into total protein output of *E. coli* as well as Trp analog incorporation into eukaryotic proteins. Schlesinger reported alkaline phosphatase as the first single enzyme to be labeled with these amino acids in the context of intact cells [124]. Pratt and Ho incorporated for the first time various fluorotryptophan analogs (Fig. 5.8, **8–10**) into proteins as tools suitable for ^{19}F-NMR analysis [130]. In the last decade, Szabo, Ross and their coworkers have further extended this repertoire by introducing 5-hydroxyryptophan (Fig. 5.8, **6**) as a useful intrinsic fluorescent probe [107]. Selenophen- and thienyl-containing Trp analogs have proved to be useful as pharmacologically active substances and markers for X-ray crystallography of proteins (Fig. 5.8, **12–15**) [121, 123]. One of the most recent advances represents incorporation of aminotryptophan analogs (Fig. 5.8, **3**, **4**) that offer the possibility to design protein-based sensors [131] (see Section 7.6.1, 7.6.2 and Fig. 7.8). The optical properties of 4-hydroxytrytophan (Fig. 5.8, **7**) are similar to those of Tyr [131], whereas 7-fluorotryptophan (Fig. 5.8, **11**) is found to be a silent fluorophor [105, 132] in the same manner as 4-fluorotryptophan [45] (Fig. 5.8, **8**).

It is not difficult to foresee how much additional chemical diversity in proteins could be achieved with such engineered TrpRS at hand. Thus, attempts to create an orthogonal pair that would use the seldom used UGG codon would be the most straightforward way to achieve both position-specific and the multiple-site mode of replacements due to the rare appearance of UGG in protein sequences either in eukaryotic or prokaryotic cells. An orthogonal pair for *opal* suppression with *B. subtilis* $tRNA^{Trp}{}_{UCA}$ charged with 5-hydroxytrytophan for target protein expression in mammalian cells (e.g. human 293T cell line) was developed [86]. First, Zhang and coworkers have shown that *B. subtilis* $tRNA^{Trp}$ and its suppressor derivative are orthogonal in yeast and mammalian cells. Second, by using essentially the methods described in Section 5.1.5 (selection procedure in yeast, Section 5.6.4), they modified the active site of *B. subtilis* TrpRS (which is also naturally orthogonal in mammalian cells) so that it preferentially activates and charges 5-hydroxytrytophan (Fig. 5.8, **6**) [133]. This opens the possibility for the design of TrpRS-based enzymes with stronger preferences for more complex Trp analogs and surrogates.

5.2.1.2 Tyrosine

The aromatic amino acid Tyr is normally found buried in the hydrophobic cores of proteins, where it is usually involved in various stacking interactions with other aromatic side-chains. Since it contains a reactive hydroxyl group at the *para* position on the benzene ring, it is suitable for additional interactions like hydrogen bonding or various interactions with nonprotein atoms. In proteins of advanced metazoan cells Tyr side-chains are, together with Ser and Thr, targets for posttranslational phosphorylation reactions – crucial events in signal transduction processes [134].

The first experimental attempts to successfully replace Tyr with its analogs in single proteins were reported by Munier and Sarrazin in 1963 [135]. Their demonstration of the incorporation of (3-f)Tyr (Fig. 5.11, **19**) in β-galactosidase marks the beginning of the history of expanded amino acid repertoire for the Tyr-coding triplets UAU and UAC. A few decades later, Sykes and coworkers [53, 136] used (3-f)Tyr to measure ^{19}F-NMR spectra of the labeled *lac* repressor and alkaline phosphatase from *E. coli* as tools suitable for protein ^{19}F-NMR analysis. Ring and Huber [137, 138] probed the role of catalytic Tyr in the active center of β-galactosidase by the global replacement of all Tyr side-chains with (3-f)Tyr. More recently, Brooks and coworkers succeeded in expanding the repertoire of translationally active Tyr analogs by incorporating 2-fluorotyrosine (Fig. 5.11, **18**) and 2,3-difluorotyrosine into ketosteroid isomerase from *Pseudomonas testosteroni* [139]. The engineering efforts on *E. coli* and *M. jannashii* TyrRS yielded a variety of mutants capable of charging cognate or orthogonal tRNAs with an impressive wide range of noncanonical amino acid substrates (Section 5.1.5; Fig. 5.7) [71, 86].

Interesting facts regarding the behavior of fluorinated Tyr residues in protein structures recently came to light – the crystal high-resolution crystal structures of rat glutathione transferase [141] and GFP from *A. victoria* [142] (Fig. 5.9) with fluorinated Tyr residues allowed not only mapping of novel, unusual interaction distances created by the presence of fluorine atoms, but also an interesting dynamic behavior. Fluorine atoms in the *ortho* position of tyrosyl side-chains (including chromophore) exhibit a single conformation, while two conformer states were observed in the *meta* position (Fig. 5.9). The co-existence of two rotameric species of *m*-fluorotyrosines in the crystal state with similar populations seems to be a general feature of the globally fluorinated proteins [140]. This raised an additional and interesting question about the dynamic behavior of these residues in the context of fluorine-containing protein structures.

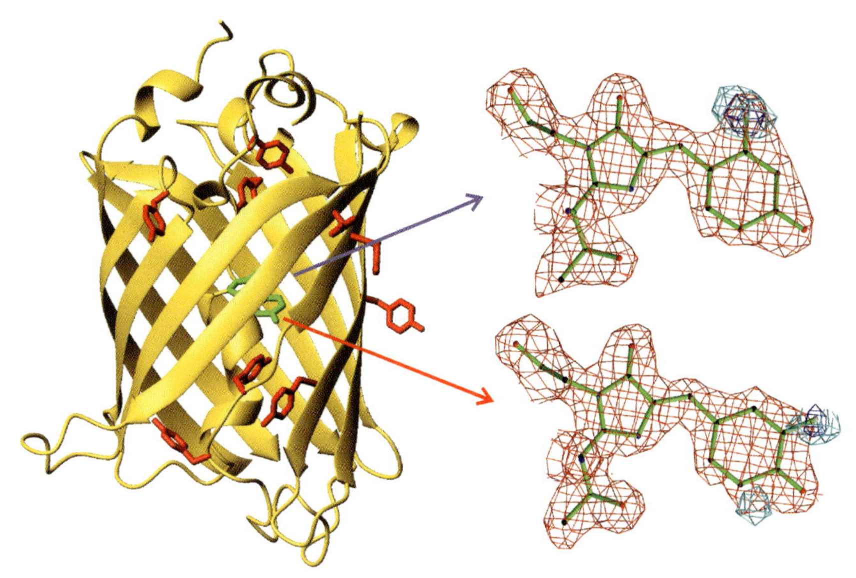

Fig. 5.9. Crystallographic evidence for Tyr fluorinations in GFP from *A. victoria*. This protein family contains 11 Tyr residues (marked red) with Tyr66 being an integral part of the chromophore (green). Upon global fluorination of the protein, the tyrosyl moiety of the chromophore is fluorinated as well. Analyses of the related protein crystals revealed experimental electron densities (red) which indicate two conformers of *meta*-fluorine (blue, lower part) and only one for *ortho*-fluorine (blue, upper part) in the Tyr moiety (for more information, see [140]).

Since Fersht performed his seminal kinetic experiments on *B. stearothermophillus* TyrRS, this protein has been considered as an archetypal enzyme for studying substrate specificity and energetic of catalysis [73]. Although highly selective, class I (subclass Ic) TyrRS does not exhibit editing, which was recognized as advantage for active site engineering in order to gain novel substrate specificities. The eubacterial and yeast enzymes even charge cognate $tRNA^{Tyr}$ with D-tyrosine and the resulting D-Tyr-$tRNA^{Tyr}$ is hydrolyzed by specific deacylase (in the absence of this enzyme some D-amino acids are toxic for cells) [119]. As shown in Fig. 5.10, the amino acid-binding domains of TrpRS and TyrRS from *B. stearothermophilus* are highly homologous, thus providing mechanistic support for a common origin of the two enzymes. Future experiments should show to what extent this feature can be used for engineering their substrate specificity.

5.2.1.3 **Phenylalanine**

The aromatic amino acid Phe is described in most of the hydrophobic scales as strictly hydrophobic, which is in agreement with its preference to be buried in the hydrophobic interiors of proteins [110, 111]. Phe is usually involved in stacking interactions with other aromatic side-chains. Since Phe residues are not reactive, they rarely play a role in protein recognition or catalytic functions [134].

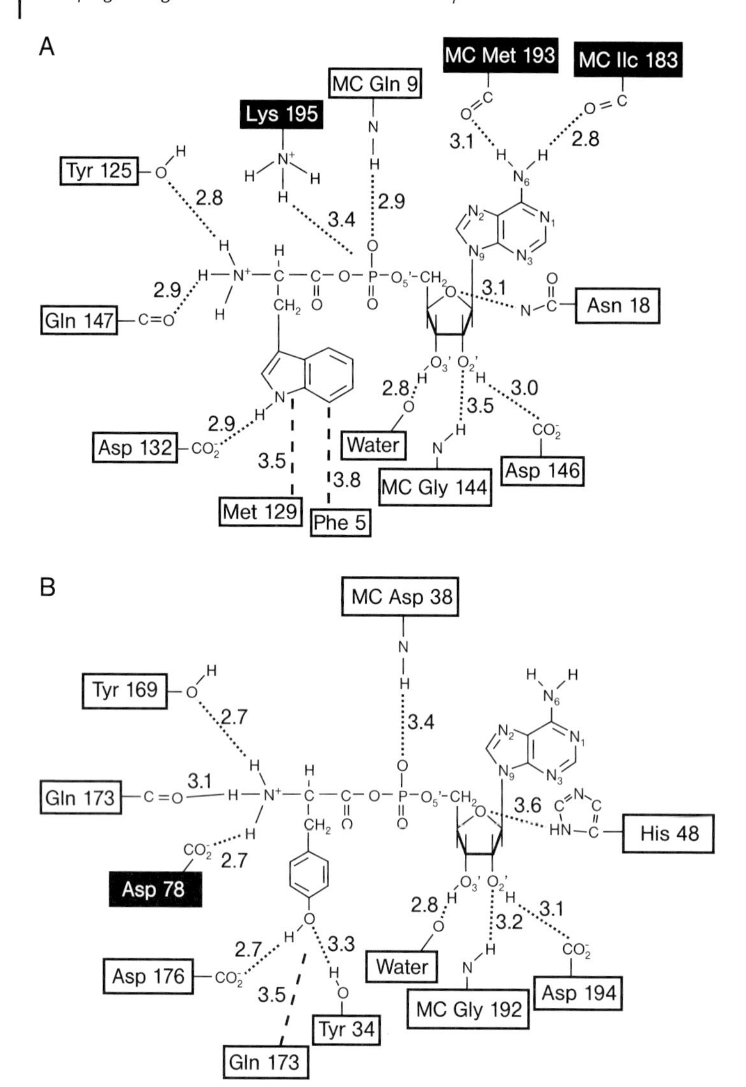

Fig. 5.10. The high homology between the amino acid-binding domains of TrpRS (A) and TyrRS (B) from *B. stearothermophilus*. Residues that differ among these two systems are marked in black. Residue Asp176 is crucial in TyrRS since its replacement by site-directed mutagenesis led to an inactive enzyme [143]. Residues Phe5, Met129 and Asp132 are in direct interaction with the indole side-chain of Trp adenylate. Residue Asp132 (which corresponds with Asp176 in TyrRS) is crucial for TrpRS catalytic efficiency since it interacts directly with indole nitrogen. It was speculated that these similarities suggest that conserved residues in TrpRS may be responsible for both determining tryptophan recognition and discrimination against tyrosine [144]. Nonetheless, such a close evolutionary relationship between these two class I (subclass Ic) enzymes might be used for substrate specificity engineering, especially in the light of considerable insights gained from the works of the Schultz and Yokoyama groups on TyrRS engineering. (Reprinted from Ref. [117], copyright 1995, with permission from Elsevier.)

Numerous enzyme studies from the early part of 20th century have shown that fluorinated substrate analogs such as *o*-, *m*- and *p*-fluorophenylalanine can substitute native substrates for certain enzymes [145]. These findings inspired Armstrong and Lewis in 1951 to design an experiment in which *o*- and *p*-fluorophenylalanine served as substitutes for Phe [146]. Their intention was to prove the hypothesis that "... an animal might be able to incorporate a structurally similar derivative into proteins and survive on a phenylalanine-free diet". However, they found that the growth of rats was inhibited. This inhibitory activity was later demonstrated for bacteria as well as mammalian cells in culture [9]. Indeed, fluorinated Phe analogs were often used in the early incorporation experiments [6] as well as in more recent years mainly for protein NMR spectroscopy. Fluorinated analogs of Phe in single proteins produce various effects. Yoshida found the stability of *B. subtilis* α-amylase remains unchanged upon partial replacement of Phe with *p*-fluorophenylalanine, but its activity was altered [147]. Similarly, *E. coli* alkaline phosphatase, crystalline lysozyme, ovalbumin and aldolase exhibited no detectable effects in terms of stability and activity upon replacement of Phe with *o*- and *p*-fluorophenylalanine (up to 56%) [7]. Conversely, Richmond found that *Bacillus cereus* exopenicillase changed its immunological properties upon substitution with *p*-fluorophenylalanine [5]. More recently, *p*-, *m*- and *o*-fluorophenylalanine (Fig. 5.11; **3**, **4** and **5**) were systematically incorporated into recombinant proteins or recombinantly produced materials, providing them with interesting optical and thermal properties as well as with altered kinetics [148, 149]. The surrogate of Phe, 2-thienylalanine, can be incorporated into β-galactosidase, but the activity of the protein is lost [18], while 3-thienylalanine (Fig. 5.11, **11**) was shown to be incorporated into a recombinant periodic protein that forms lamellar crystals comprising of regularly folded β-sheets [150]. Koide and coworkers [32] reported incorporation in recombinant human epidermal growth factor of Phe analogs having pyridyl groups that are protonated in acidic solutions such as 4-azaphenylalanine, 3-azaphenylalanine and 2-azaphenylalanine (Fig. 5.11, **13**, **14** and **15**).

The discovery of a bacterial PheRS variant with relaxed substrate specificity by Kast and Hennecke [64] paved the way for further expansion of the translationally active Phe surrogates and analogs. The halogenated analogs *p*-chloro- or *p*-bromophenylalanine (Fig. 5.11; **6** and **9**) were incorporated into full-length luciferase *in vitro*. Further redesign of the PheRS active site and its use for coexpression in *E. coli* resulted in replacement of Phe residues by *O*-acetlyltyrosine, *p*-amino-, *p*-iodo-, *p*-bromo-, *p*-azido-, *p*-cyano-, *p*-ethynyl-, *p*-azido- and 2,3-diaza-phenylalanine in dihydrofolate reductase (Figs. 5.5 and 5.11). These designed PheRS variants also allow efficient *in vivo* incorporation of aryl ketone functionality into proteins, which is normally rejected by the cellular translational machinery [70]. A rather large number of Phe-based amino acid substrates, such as *p*-bromo- and *p*-iodophenylalanine, were incorporated into proteins as well as a response to the amber stop codon using engineered TyrRS from *M. jannaschii* [86]. In this way, all these amino acids can have a "dual identity", serving as substrates for both systems (Fig. 5.11).

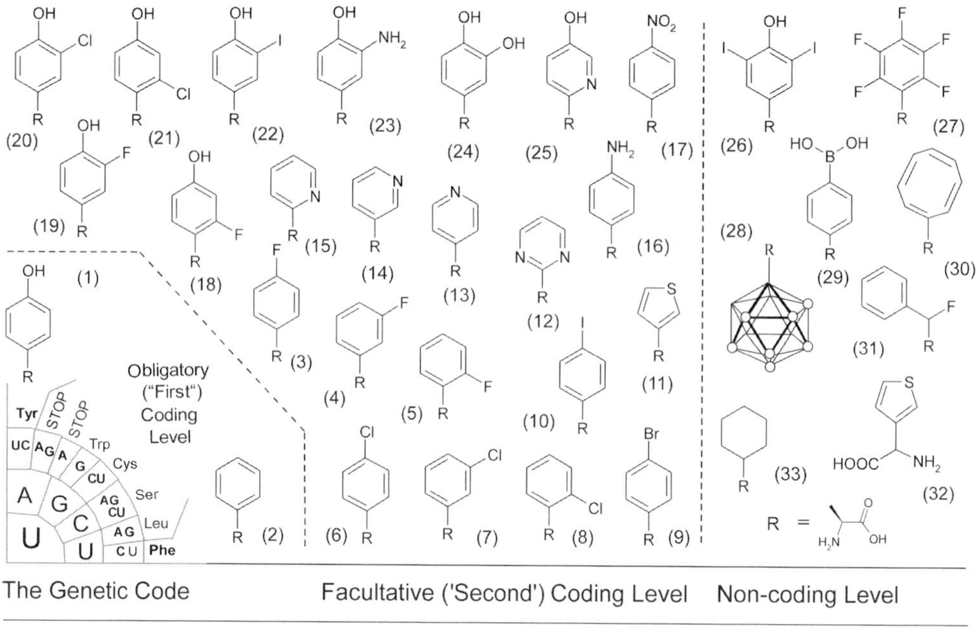

Fig. 5.11. The library of translationally active analogs and surrogates of Tyr and Phe. Their "second" coding levels include at least 24 analogs and surrogates with wide chemical and structural diversity (without counting those from Figs 5.5 and 5.7). Note that in terms of "identity", many of them can be assigned either as Tyr or Phe analogs, or surrogates, depending on the system where they are used (see also Section 6.2.1). For example, *p*-iodo- (**10**) and *p*-bromophenylalanine (**9**) can be incorporated into proteins either by use of *E. coli* cells whose PheRS exhibits relaxed substrate specificity or by use of plasmids harboring *M. jannashii* TyrRS with novel substrate specificity. Although analogs like *p*-aminophenylalanine (**16**) are introduced into model proteins as a response to the termination UAG coding unit by using suppressor tRNA, there is no doubt that they can be translated into proteins as a response to UAU or UAC sense coding triplets by using designed TyrRS and cognate tRNAs. The chemical names of the presented amino acids are as follows: tyrosine (**1**), phenylalanine (**2**), *o*-, *m*- and *p*-fluorophenylalanine (**3**–**5**), *o*-, *m*- and *p*-chlorophenylalanine (**6**–**8**) β-fluorophenylalanine (**31**), 2- and 3-thienylalanine (**11**), 3-thienylglycine (**32**), 2-, 3- and 4-azaphenylalanine (**13**–**15**), 2,6-diazaphenylalanine (**12**), *p*-nitrophenylalanine (**17**), *o*- and *m*-fluorotyrosine (**18**, **19**), *m*-iodotyrosine (**22**), *o*- and *m*-chlorotyrosine (**20**, **21**), 2-azatyrosine (**25**), 3,4-dihydroxytyrosine (DOPA, **24**), 3,5-diiodotyrosine (**26**), *m*-aminotyrosine (**23**), cyclooctatetraenylalanine (**30**), cyclohexylalanine (**33**) (4-boronic acid)phenylalanine (**29**), 2,2′,3,3′,4-pentafluorophenylalanine (**27**), and carboranylalanine (**28**).

5.2.1.4 Histidine

Many enzymes use general acid–base catalysis as a way to increase reaction rates. The amino acid His is optimized for this function because it has a pK_a (7.1) near physiological pH. This means that in the pH region where most enzymes function, His can act as a donor or an acceptor of protons – a fact that highlights the great importance of this residue in the chemistry of various kinds of enzymes [73]. In a number of proteins, His side-chains are able to adopt favorable conformations to form metal-binding sites, a feature that is exploited for engineering of metal-binding sites in recombinant proteins for various purposes [151]. For example, routine protein purification procedures profited from the possibility of the addition of the more His residues (e.g. at the N- or C-terminus) in the primary sequence of target proteins. Such proteins have a high affinity for the metal ions immobilized at the chromatographic matrix, a feature useful for their efficient separation from other proteins [152].

Although the biosynthesis of His is energetically very expensive (about 40 ATP per single His molecule) its intracellular concentration is similar to other canonical amino acids (25–75 μM) [36]. These intracellular concentrations correspond well with the L-His Michaelis constant for the class II enzyme HisRS ($K_M \sim$ 10–30 μM) [153]. The engineering of substrate specificity of HisRS will certainly contribute to the incorporation of additional various His analogs/surrogates into proteins. Taking into account the essential importance of His residues in many protein catalytic functions, this engineering will certainly result in a great expansion of the scope and versatility of the enzyme catalysis.

Isosteric replacements in the frame of some canonical amino acids such as Ser/Cys, Thr/Val, Asp/Asn or Glu/Gln represent an excellent tool to study the nature of the catalytic mechanism of residues directly involved in enzymatic action. However, such replacements are not possible for His side-chains since the imidazole group is quite unique and does not particularly substitute well with any other canonical amino acid. For that reason isosteric replacement of His with its analogs is an almost ideal tool to study its biological functions. For example, His analogs 1,2,4-triazole-3-alanine and 2-methyhistidine (Fig. 5.12; **2** and **4**) have been successfully incorporated into *E. coli* alkaline phosphatase, resulting in the enzyme being inactive since its dimerization was prevented [12]. 1,2,4-Triazole-3-alanine cause dramatic differences in pH activity profiles between the native and modified enzymes upon incorporation into aspartate transcarbamylase and a mutant phospholipase A2 [154]. In fact, the replacement of His by its noncanonical counterpart 1,2,4-triazole-3-alanine yields an enzyme with high activity at acidic pH, which is not surprising since the pK_a value of the 1,2,3-triazole conjugate acid is around 5 units lower than that of imidazole. A similar effect was achieved by the incorporation of 4-fluorohistidine and 2-fluorohistidine (Fig. 5.12; **6** and **7**) that reduce the pK_a of the imidazole ring by approximately 5 pH units [155]. Such a dramatic increase in the acidity of the imidazole ring dramatically alters the activity and properties of the fluorohistidine-containing proteins. This strong effect on pK_a possibly induces a loss in enzymatic activity in many proteins in the physiological milieu, making such His replacements with 2-fluorohistidine and 4-fluorohistidine responsible for the toxicity of these substances [12]. These considerations were fully confirmed in the studies of ribonuclease S with semisynthetically incorporated 4-fluorohistidine [156]. Recently, Ikeda and coworkers [157] succeeded in increasing the number of translationally active His analogs by incorporating 1,2,3-triazole-3-alanine (Fig. 5.12; **3**) into the chitin-binding domain (CBD). Interestingly, the amino analog of His 3-amino-1,2,4-triazole-1-alanine (Fig. 5.12; **8**), which is a natu-

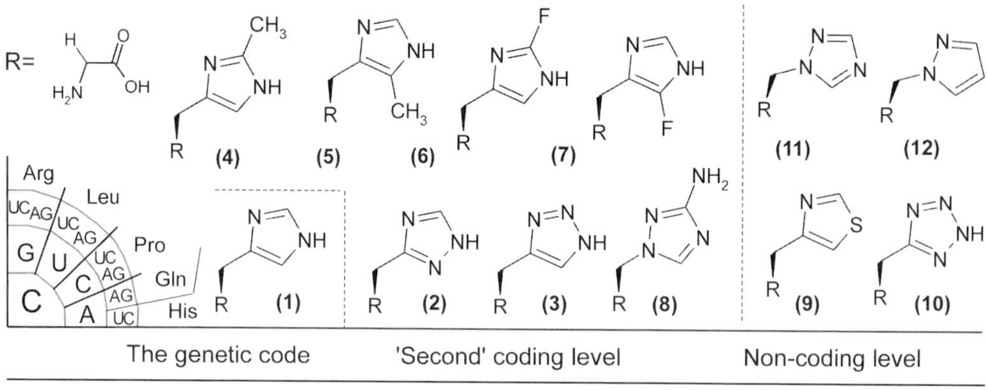

Fig. 5.12. Histidine analogs incorporated into protein using the traditional auxotrophism-based (SPI) biosynthetic approach. All analogs belonging to the "second" coding levels are substrates for wild-type HisRS, for which physiological levels are sufficient for $tRNA^{His}$ charging and subsequent analog translation into target protein sequences. Chemical names: His (**1**), 1,2,4-triazole-3-alanine (**2**), 1,2,3-triazole-3-alanine (**3**), 2-methylhistidine (**4**), 4-methylhistidine (**5**), 4-fluorohistidine (**6**), 2-fluororhistidine (**7**), 3-amino-1,2,4-triazole-3-alanine (**8**), 2-thiohistidine (**9**), 1,2,3,4-tetrazole-3-alanine (**10**), 1,2,4-triazole-1-alanine (**11**) and pyrazol-1-alanine (**12**).

ral product biosynthesized upon treatment of green plants with the herbicide amitrole (3-amino-1,2,4-triazole), was reported to be translationally active as well [10]. The explicit translational activity of other natural product His analogs including β-1,2,4-triazole-1-alanine (Fig. 5.12; **11**) and β-pyrazol-1-alanine (Fig. 5.12; **12**) has not yet been reported.

5.2.1.5 Proline

Proline is structurally unique imino acid (having -NH- instead -NH_2 groups) whose side-chain is connected to the protein backbone twice, forming a five-membered nitrogen-containing ring. Although Pro is assigned as an aliphatic and hydrophobic molecule, it is usually found on the protein surface, especially there where the polypeptide chain must change direction [110]. Therefore, Pro residues act as classical breakers of both the α-helical and β-sheet structures in water-soluble proteins and peptides. The insertion of turns in proteins, in general, or kinks, in particular structures such as α-helices, are possibly due to the inability of Pro residues to adopt a normal helical conformation. This property is due to an endocyclic imino group with a fixed ϕ angle which narrowly limits its accessible conformational space, making Pro unable to occupy many of the main-chain conformations easily adopted by all other amino acids [134]. Pro residues have critical importance for protein structure and function since the rate-limiting step in the slow folding reactions of many proteins involves *cis*/*trans* isomerization about peptidyl-prolyl-amide bonds [158]. Pro plays important roles in molecular recognition, particularly in intracellular signaling, and is often part of repetitive oligopeptide motifs [134]. In eukaryotic proteins, Pro residues are often posttranslationally hydroxylated, while in prokaryotes the enzyme modification is absent [20]. In peptide synthesis, pseudoprolines are well known from peptidomimetic studies as serine- or threonine-derivatized oxazolidines and cysteine-derived thiazolidines [159].

It should be kept in mind that together with His, Pro also has no similar counterparts among the remaining 19 canonical amino acid. Therefore, a fine dissection of its structural and functional roles in proteins can be performed only by its cotranslational substitution with various analogs and surrogates. Early reports assigned 3,4-dehydroproline (**9**), 4-thiaproline (γ-thiaproline) (**10**), 4-selenoproline (γ-selenoproline) (**14**), *cis*-4-hydroxyproline (**5**), *trans*-4-hydroxyproline (**6**), *trans*-4-fluoroproline (**3**) and azetidine-2-carboxylic acid (**13**) [54] (Fig. 5.13) as potential proline analogs in protein synthesis. Recent research confirmed that *cis*-4-fluoroproline, *trans*-4-fluoroproline, difluoroproline [160] (Fig. 5.13; **4**) and γ-thiaproline [161] (Fig. 5.13; **11**) can indeed be incorporated into recombinant proteins. In addition, the proline analog 3,4-dehydroproline was incorporated in a repetitive pe-

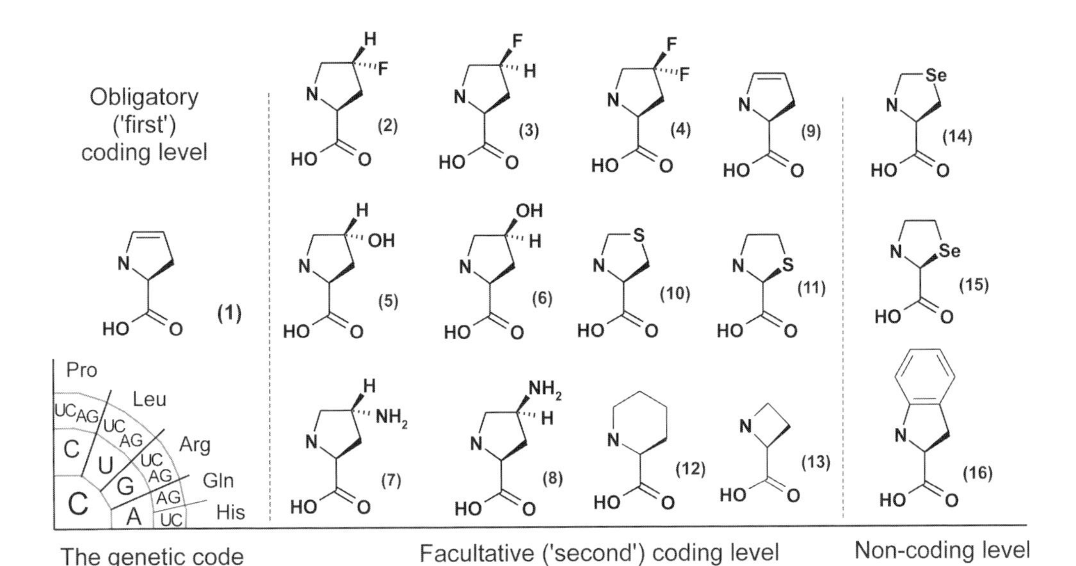

Fig. 5.13. The alternative coding level for the canonical imino acid proline (**1**) achieved by the traditional SPI approach. Analogs *cis*-fluoroproline (**2**), *trans*-fluoroproline (**3**), dehydroproline (**9**), β-thiaproline (**10**) and γ-thiaproline (**11**) can be easily translated into protein by exploiting the "promiscuity" of natural *E. coli* ProRS to misacylate $tRNA^{Pro}$. The following amino acids are activated by ProRS a few orders of magnitude less efficiently than Pro: *cis*-hydroxyproline (**5**), *trans*-hydroxyproline (**6**), *cis*-aminoproline (**7**), *trans*-aminoproline (**8**) and azetidine-2-carboxylic acid (**13**). Other presented analogs and surrogates: (2*S*)-piperidine-2-carboxylic acid (**12**), γ-selenoproline (**14**), β-selenoproline (**15**) and 3,4-phenylproline (**16**).

riodic protein with lamellar morphology [162]. The manipulation with amino acid uptake systems (Section 5.1.2.2) in combination with a simple elevation of intracellular levels of ProRS (Section 5.1.4.3) yielded translation of *trans*-4-hydroxyproline, *cis*-4-hydroxyproline, and azetidine-2-carboxylic acid into recombinant elastin [21] and Type I collagen [20] by using Pro auxotrophic strains with native genomic background activity of ProRS. Furthermore, the Pro surrogate piperidine-2-carboxylic acid (Fig. 5.13; **12**) acid was also successfully incorporated into elastin by engineering of the binding pocket of *E. coli* ProRS (Section 5.1.4.4). Incorporation was achieved as follows. First, *E. coli* Pro uptake systems were "relaxed" by higher NaCl concentrations in growth media (400–600 mM) which facilitate their efficient intracellular accumulation. Second, elevated amounts of native or mutated ProRS were expressed under the control of an orthogonal promoter system. The same procedure was applied to incorporate the Pro analog (2*S*)-piperidine-2-carboxylic acid (**12**), but in the presence of ProRS with a mutation (C443G) in the binding pocket of the enzyme. The proof-of-principle for translational activity of these imino acids was provided by previous *in vitro* experiments using chemically aminoacylated suppressor tRNAs [163]. Most recently, the Pro analogs *cis*-4-aminoproline and *trans*-4-aminoproline (Fig. 5.13; **7** and **8**) have been incorporated into a few recombinant proteins (Budisa, unpublished material).

While 3,4-phenylproline (**16**) is not expected to be activated by ProRS, both β-selenoproline (**15**) and γ-selenoproline (**14**) are also not translated into proteins [164]. Interestingly, although β-selenoproline is not activated by wild-type ProRS, γ-selenoproline is activated and charged onto $tRNA^{Pro}$, but induces premature chain termination upon its incorporation into the growing polypeptide chain on the ribosome [30], a phenomenon that certainly deserves to be more closely investigated and clarified (see also Section 6.2.3 and Fig. 6.3).

Analogs and surrogates of Pro azetidine-2-carboxylic acid, thiaproline and 3,4-dehydroproline strongly inhibit the growth of several bacterial species and induce developmental abnormalities in animals tested [10]. Azetidine-2-carboxylic acid and pipecolic acid can (along with 4-hydroxyproline, 4-ketoproline and sarcosine) act as Pro surrogates during the synthesis of actinomycin antibiotics by *Streptomyces antibioticus* [165]. Azetidine-2-carboxylic acid is also a natural product of the secondary metabolism of plants such as *Phaseolus*, *Canvalaria* and *Polygonatum*. Interestingly their cellular extracts are not able to activate this analog (and most of the analogs/surrogates from Fig. 5.13). On the other hand, cellular extracts of the plant *Asparagus* (which does not produce azetidine-2-carboxylic acid) are indeed capable of activating azetidine-2-carboxylic acid and most of the analogs/surrogates from Fig. 5.13. Not surprisingly, these analogs/surrogates are toxic for *Asparagus*, while *Phaseolus*, *Canvalaria* and *Polygonatum* are resistant to their toxic actions. These interesting early experiments (extensively reviewed in [10]) should be confirmed (or disproved) by cloning and characterization of ProRS enzymes from these organisms or by sequencing their genomes. Most probably, these differences in substrate specificity towards Pro analogs and surrogates are due to the presence of different forms of ProRS in different species, in a manner that was recently discovered for LysRS (Section 5.2.1.10).

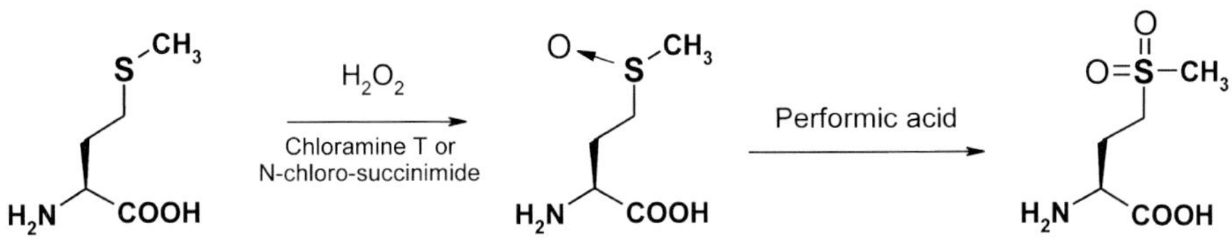

H_2O_2

Chloramine T or N-chloro-succinimide

Performic acid

L-Methionine L-Met-D,L-Sulfoxide L-Met-Sulfone

Fig. 5.14. Chemical oxidation of Met to the sulfoxide and sulfone forms. The oxidation of Met sulfoxide (which has an asymmetric sulfur atom and consequently two diastereomers) is a reversible reaction, but is irreversible once the reaction proceeds to the sulfone stage. The sulfoxide form can be chemically reduced to Met by a number of sulfhydryl or nonsulfhydryl reagents. In the context of intact protein structures these reactions are often associated with the loss of biological activity or deactivation of proteins because the reactive oxygen species (oxygen radicals) are ubiquitously present during storage and handling of the proteins in aerobic biological conditions [167].

5.2.1.6 **Methionine**

Met residues are relatively rare (1.5% of all residues in proteins of known structures) and usually located in positions inaccessible to the bulk solvent, with only 15% of all Met residues exposed to the surface [102, 166]. While Cys residues play crucial roles in the catalytic cycle of many enzymes and they can form disulfide bonds that contribute to protein structure, no such specific catalytic functions can be assigned to Met side-chains. Nevertheless, it is well known that Met metabolism plays an important role in many aspects of cellular physiology. For example, Met serves as the methyl donor in biological methylation reactions [upon reaction with ATP to produce *S*-(adenosyl)methionine] [134]. Met is an aliphatic and hydrophobic amino acid that contributes to protein structures with both hydrophobic interactions and hydrogen bonding.

A distinct property of Met side-chains is their sensitivity toward oxidation – the thioether moiety is weakly nucleophilic, not protonated at low pH; it can be selectively oxidized at physiological pH to become hydrophilic upon conversion to the sulfoxide form (Fig. 5.13). Solvent-accessible Met side-chains when exposed to oxidizing agents undergo reversible oxidation into hydrophilic Met sulfoxide (Fig. 5.14) leading to dramatic changes in surface hydrophobicity. *In vivo* oxidation of Met by an oxidative system is an important event for the change of protein conformation, i.e. Met is a "switchable" amino acid with different (hydrophillic ↔ hydrophobic) conformational preferences [168]. However, Met sulfoxide is normally subjected to reduction by the Met sulfoxide reductase system in living cells. Following the pioneering work of Weissbach and Brot [169], the physiological importance of Met oxidation is gaining more and more recognition since the presence of Met sulfoxide in particular proteins is correlated with some pathological conditions such as emphysema and cataracts. Nowadays, it is widely believed that reversible Met oxidations are in function of (i) reduction of the level of reactive oxygen species, (ii) repair of oxidized proteins and (iii) regulation of cellular functions [167].

Protein synthesis initiates with Met in the cytosol of eukaryotes, and formylmethionine (fMet) in prokaryotes and eukaryotic organelles (see Section 2.10). This initiator Met is often cotranslationally removed by a Met aminopeptidase [170]. Removal of Met is a physiologically critical process – in certain cases the inhibition of Met aminopeptidases arrests cell growth, thus making these enzymes an interesting target for tumor therapy [171]. Recent observations on the efficient blockage of N-terminal processing of Met by trifluoromethionine (TFM, Fig. 5.15; **17**) in GFP [42] makes TFM a potentially efficient inhibitor of Met aminopeptidases, i.e. a promising substance for tumor therapy. One might expect that future research should yield more Met analogs and surrogates (e.g. diaza-oxo-Met, Fig. 5.15; **26**) with interesting biomedicinal properties.

The analogs of Met, Nle, SeMet and Eth (Fig. 5.15; **2**, **4** and **9**), which were demonstrated very early to be incorporated *in vivo* into proteins, are activated *in vitro* by MetRS at a rate approaching that of Met [5]. Eth is one of the first amino acid analogs to be synthesized and its physiological effects were extensively studied [172]. Namely, Dyer in 1938 prepared this substance in order to study physiological effects in growing rats whose diet contained Eth as a substitute for Met [173]. Levine and Tarver in 1951 fed [^{14}C]Eth to rats and found it distributed in proteins from several tissues. Numerous works in the 1950s (reviewed in [5, 6]) established Eth as an inhibitor of protein synthesis and inducer of the accumulation of abnormal proteins in some tissues (e.g. the liver of rats fed with Eth). Yoshida [174] reported that incorporation of Eth into crystalline α-amylase did not influence its specific enzymatic activity, although about one-third of Met residues were substituted.

Early reports of the trace presence of Nle in proteins that generated the opinion that Nle was a canonical amino acid were not later confirmed [6]. Interestingly it is now well known that Nle is indeed present in traces in intracellular proteins since it can be generated metabolically [15] (Section 5.1.2.1). The study of the integration of radioactive [^{14}C]Nle into carcinoma cells in 1954 [9] was followed by the first report of its incorporation into a single, well-characterized protein (casein) in 1955 by Black and Kleiber [175].

Binding of Met and its analogs to *E. coli* MetRS is accompanied by a series of conformational changes within the active site [176], a scenario typical of other AARS as well. Rearrangements of Tyr15 and Trp253 enable the protein to form a hydrophobic pocket around the ligand side-chain. In the presence of Hcy, this hydrophobic pocket is not formed and cyclization of the Hcy adenylate takes place [177]. The side-chain of Tyr15 controls the size of the hydrophobic pocket, while His301 appears to participate in the specific recognition of the sulfur atom of Met [178]. A target of the editing activity of MetRS is the Met metabolic precursor Hcy [74] (Fig. 5.15; **23**). During editing, the side-chain SH group of Hcy reacts with its activated carboxyl group forming a cyclic thioester, Hcy thiolactone. It is well known that all Met-like amino acids smaller in size than Met itself are edited (with Nle on the border of this size-based exclusion) [179]. More recently, Jakubowski [180] had shown that *S*-nitrosomethionine can be transferred to $tRNA^{Met}$ by MetRS. Removal of the nitroso group from *S*-nitorosomethionine-$tRNA^{Met}$ results in Hcy-$tRNA^{Met}$. The incorporation of Hcy into proteins in place of Met side-

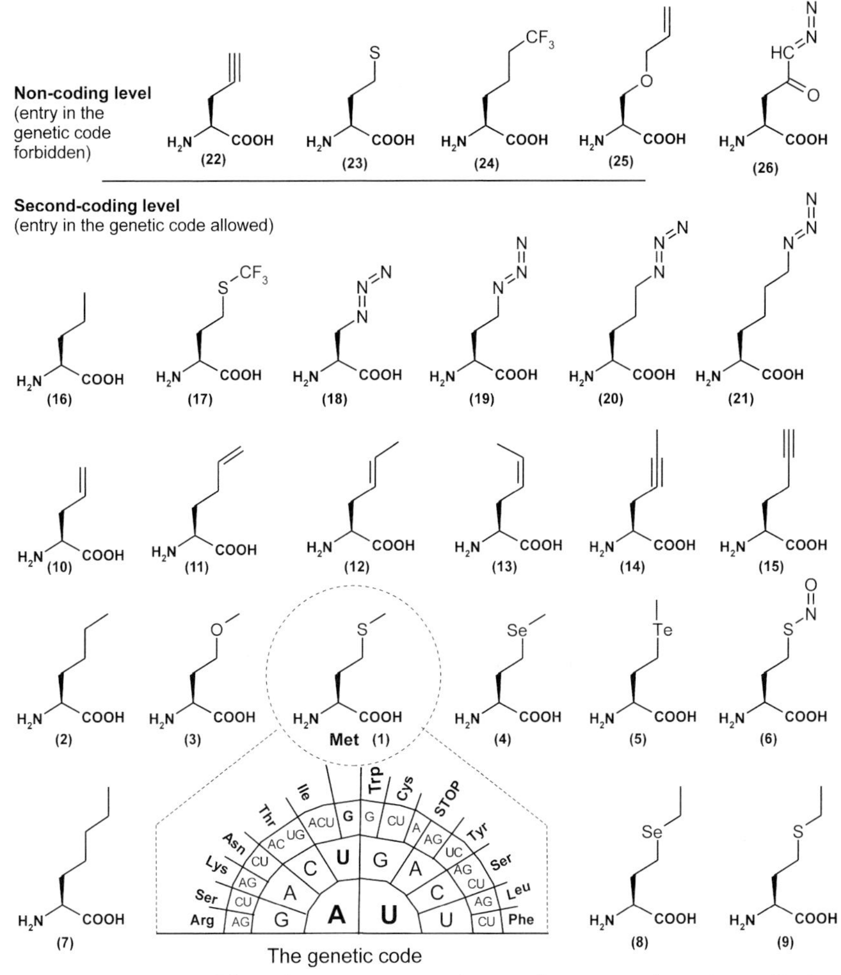

Fig. 5.15. Expansion of the coding capacities of AUG triplet which in the genetic code is assigned to the canonical amino acid Met (**1**). Experimental substitutions of Met in proteins yielded an entire library of Met analogs and surrogates able to not only serve as substrates for protein biosynthesis, but also build a second or facultative coding level for the AUG codon. Most of the translationally active amino acids presented here were incorporated into proteins by the groups of Budisa [3] and Tirrell [62]. Note that translationally active allylglycine (**10**), 2-butynylglycine (**14**) as well as norvaline (**16**), norleucine (**2**), homoallylglycine (**11**) and homopropargylglycine (**15**) might serve as substrates for engineered LeuRS as well. This novel feature in the translational process ("identity problem") is discussed in detail in Section 6.2.1. The common names of the other presented Met analogs and surrogates are: 2-aminoheptanoic acid (**7**), metoxinine (**3**), selenomethionine (**4**), telluromethionine (**5**), *S*-nitrosomethionine (**6**), ethionine (**9**), selenoethionine (**8**), *trans*-crotylglycine (**12**), *cis*-crotylglycine (**13**), trifluoromethionine (**17**), azidoalanine (**18**), azidohomoalaine (**19**), azidonorvaline (**20**), azidonorleucine (**21**), propargylglycine (**22**), homocysteine (**23**), trifluoronorleucine (**24**), *O*-allylserine (**25**) and diazo-oxo-norleucine (**26**).

chains upon addition of Hcy-tRNA^{Met} was demonstrated in rabbit reticulocyte and *E. coli*-based cell-free systems.

Thus, it should not be surprising that plasticity in substrate binding in wild-type MetRS can be exploited to incorporate cotranslationally relative large numbers of analogs and surrogates as shown in Fig. 5.15. It is reasonable to expect that telluromethionine (TeMet, Fig. 5.15; **5**) is a good substrate for MetRS as well. Conversely, the translation of TFM (Fig. 5.15; **17**) into phage lysozyme polypeptide sequence (demonstrated for the first time by Honek and coworkers [181]) was not as facile as with the analogs mentioned above. It is generally observed that trifluoroamino acids are much more toxic for the living cells than their di- or mono-fluoro counterparts. Using lysozyme from bacteriophage λ as an model protein, Vaughan and coworkers [182] succeeded in obtaining an almost 100% replacement of Met residues by 5′,5′-difluoromethionine (DFM), while substitution levels achieved by TFM were always about 30%. There is no doubt that the poor activation of TFM by MetRS is the chief reason for its poor translation activity. Thus, its efficient incorporation requires sufficient intracellular concentrations of the analog and MetRS in order to drive better misacylation of the tRNA^{Met}. Such intracellular enhancement of MetRS by coexpression experiments in *E. coli* Met auxotroph strains and the model protein dihydrofolate reductase allowed for further extension of the number of Met-like surrogates, known to be quite poor substrates for MetRS [61, 183]: homoallylglycine (**11**), homopropargylglycine (**15**), *cis*-crotylglycine (**13**), *trans*-crotylglycine (**12**), allyglycine (**10**), 2-butynylglycine (**14**), 2-aminoheptanoic acid (**7**), Nva (**16**) and azidohomoalanine (Fig. 5.15). Even substrates such as azidoalanine (**18**), azidonorvaline (**20**) and azidonorleucine (**21**) (Fig. 5.15), whose activation was not measurable, can be detected in recombinant proteins by using sensitive fluorescent assays [184]. Such "naturally broad" substrate specificity of AARS enables not only rather large numbers of chemically versatile substances to be translated into protein sequences as a response to relatively rare AUG codons, but also a relatively simple, straightforward and efficient experimental setup for amino acid repertoire expansion in the context of the SPI method (Fig. 5.3).

Met contains a unique thiol ether moiety whose sulfur atom can undergo facile substitution with methylene, oxygen, selenium and even tellurium [185]. However, the chalcogen analogs of Met, methoxinine (Fig. 5.15, **3**) or SeMet, are not represented in the genetic code repertoire. It is well known that the chalcogen elements oxygen, sulfur and selenium are essential constituents of functional groups of amino acids, which in related proteins play unique chemical and structural roles [186]. Interestingly, nature achieved such diversity in chalcogen elements by assigning coding units to Ser, Cys and by introducing an in-frame UGA coding for the special canonical amino acid Sec (Fig. 5.16, **2**, **3** and **4**). Tellurocysteine (TeCys; Fig. 5.16, **5**) has not yet been identified as a naturally occurring amino acid, while the efforts to incorporate synthetic TeCys into proteins biosynthetically failed [187].

5.2.1.7 Leucine

Nature assigned the strictly hydrophobic amino acid Leu (which is usually buried in the hydrophobic cores of proteins) to six coding triplets in the genetic code. It

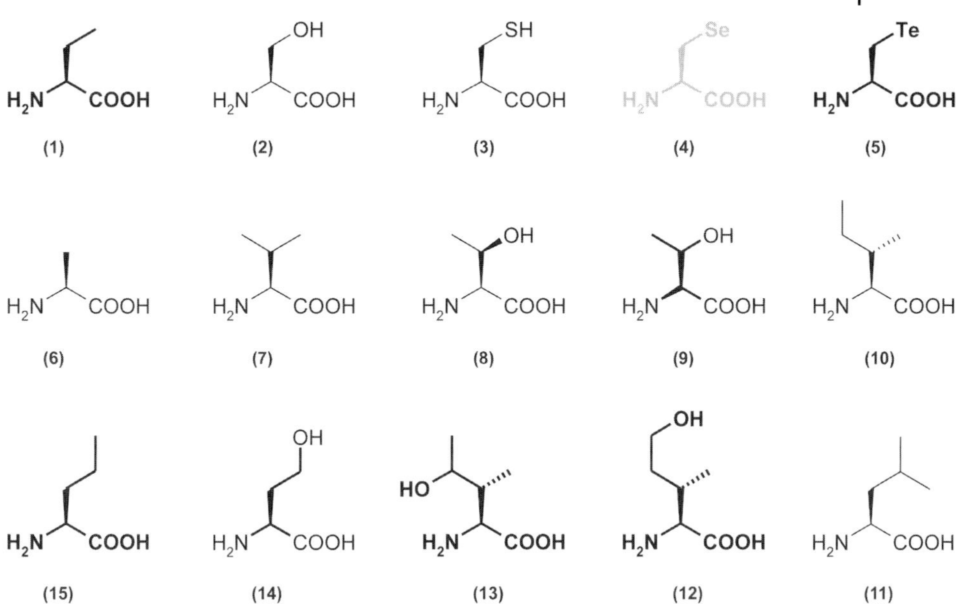

Fig. 5.16. Survey of some simple aliphatic and chalcogen-containing canonical and noncanonical amino acids. In the chalcogen series (first row), α-aminobutyrate (**1**) is not genetically encoded, but is active in reprogrammed translation, serine (**2**) and cysteine (**3**) have "membership" in the genetic code, whereas Sec (**4**) occasionally appears in a small fraction of cellular proteins as result of naturally preprogrammed decoding (see Chapter 3 and Section 3.10). The synthetic amino acid tellurocysteine (TeCys, **5**) is neither genetically encoded nor active in protein translation. Canonical amino acids alanine (**6**), valine (**7**), leucine (**11**) and isoleucine (**10**) have numerous analogs and surrogates such as norvaline (**15**) that might be incorporated into proteins. The hydroxy analogs of Val, Leu and Ile (third row) such as γ-hydroxy- (**13**) and σ-hydroxy-isoleucine (**12**) are efficiently prevented from mischarging onto cognate tRNAs by the editing functions of ValRS, LeuRS and IleRS [191, 192] The Thr (2*R*,3*S*) diastereomer L-*allo*-Thr (2*R*,3*S*) (**9**) as well as hydroxynorvaline (**14**) are competitive inhibitors of Thr in the aminoacylation reaction with ThrRS [194].

shows a higher preference for building α-helices (e.g. Leu zippers) than β-sheets. Since the Leu side-chain is chemically inert, it is thus rarely directly involved in protein catalytic functions. However, Leu side-chains might play a role in substrate recognition where hydrophobic amino acids are involved in binding/recognition of hydrophobic ligands [134]. Already in the 1960s, Anker and coworkers in their pioneering works demonstrated that 5′,5′,5′-TFL can be incorporated into bacterial proteins [39], but not in eukaryotic ones [41]. These early results were recently reproduced in Tirrel's laboratory where, in addition, the hexafluoroleucine was incorporated into peptides expressed *in vivo* [97]. Leu residues of *Lactobacillus casei* dihydrofolate reductase expressed in *E. coli* were successfully substituted with (2*S*,4*S*)-5-fluoroleucine in order to study conformational changes of the protein upon ligand binding with ^{19}F-NMR spectroscopy [188]. The translational activity of the noncanonical plant amino acid *threo*-3-hydroxyleucine was reported by Hortin and Boime [189], while Apostol and coworkers [190] found that Nva can be incorporated in the place of Leu in recombinant proteins. It is also well documented that analogs of Leu and Ile carrying hydroxy groups in the γ or δ positions (Fig. 5.16) are hydrolytically proofread by *E. coli* and yeast LeuRS either in a pre- or post-transfer modus [191, 192]. The T252Y mutant of *E. coli* LeuRS with attenuated editing activity (Section 5.1.4.5) enabled replacement of Leu residues with Nva, Nle, allylglycine, homoallylglycine, homopropargylglycine and 2-butynylalanine in recombinant proteins [76]. Some of these amino acids have a "dual identity" (see Section 6.2.1), like Nva, 2-butynylglycine, homoallyglycine and allylglycine (Fig. 5.15), since they can serve as substrates for both MetRS and LeuRS with attenuated editing activity. The generation of the *E. coli* $tRNA^{Leu}$:LeuRS orthogonal pair capable of selectively incorporating *O*-Me-Tyr, α-aminocaprylic acid and the photo-caged amino acid *o*-nitrobenzyl-Cys into caspase 3 in response to the amber nonsense codon was also reported [193].

5.2.1.8 **Valine and Isoleucine**

Val and Ile residues are also chemically nonreactive, and thus seldom involved in protein function. Exceptionally, they might be involved in binding/recognition of hydrophobic ligands such as lipids. An additional property often overlooked in the literature is that Val, like Ile and Thr, is branched at the β-carbon, thus generating more bulkiness near to the protein backbone. This in also restricts the conformations that the main-chain can adopt – a phenomenon absent in Leu side-chains. In fact, these amino acids would much more easily and even preferably lie within β-sheets rather than adopt an α-helix conformation [134]. The systems for Leu, Val and Ile selection and tRNAs aminoacylation evolved additional domains in their synthetases for a "double-sieve" discrimination of cognate from noncognate, noncanonical and biogenic amino acids [195, 196]. Engineered ValRS, IleRS or LeuRS with an inactivated or attenuated editing function should have a great potential to fill up the second coding levels of Leu, Val and Ile with a large number of analogs, surrogates or similar amino acids.

The noncanonical amino acid L-Abu (Fig. 5.16, **1**) is often suspected by some researches in the field of code evolution to be present in the repertoire of the primitive code (see Section 6.3.2 and Fig. 6.4). Marliere and coworkers demonstrated how it can gain access to the genetic code again [50]. This is achieved by editing attenuation of ValRS and thorough invasion of the Val-coding pathway in selected *E. coli* mutant strains. This experiment was based on the observation that Cys can be erroneously attached to $tRNA^{Val}$ by ValRS [49]. Mutagenesis and selection of bacteria with impaired editing allowed the selection of ValRS mutants with a defective editing activity that exhibit an increased efficiency for Cys, Thr and L-Abu attaching onto $tRNA^{Val}$ (Sections 5.1.4.1 and 5.1.4.5). The Val analogs 2-amino-3-chlorobutyroneic acid, cyclobutaneglycine, penicillamine and *allo*-isoleucine were reported to be used in order to probe the process of membrane transversion of secretory proteins through the endoplasmic reticulum [54]. However, the extent of their incorporation is still unclear. For example, Porter and coworkers [197] reported that "cyclobutaneglycine is activated by ValRS and forms cyclobutaneglycyl-tRNA^{Val}, and presumably is incorporated into proteins in lieu of valine". Therefore, additional experiments are necessary to resolve these ambiguities.

Some Ile analogs were also used in pioneering experiments, such as 4-thiaisoleucine, *O*-ethylthreonine, *O*-methyltreonine, 4-fluoroisoleucine and *allo*-isoleucine [5, 54, 189]. Novel isoleucine analogs incorporated more recently into a few model target proteins include furanomycin [198] and trifluoroisoleucine [199]. Although the chemical structure of furanomycin is not similar to that of Ile, it can be charged onto *E. coli* tRNA^{Ile} by IleRS as efficiently as with cognate amino acid [200]. Kohno and coworkers [198] also demonstrated the *in vitro* protein biosynthesis of β-lactamase in the presence of furanomycin as a substitute for Ile. Since furanomycin is a natural antibiotic product, its translational activity as well as the possibility for species-specific selective inhibition of AARS raises an interesting question about AARS as targets for new anti-infectives [201].

5.2.1.9 **Arginine and Canavanine**

Both Lys and Arg are composed of a longer hydrophobic backbone and a positively charged terminal group. This feature is reflected in the observation that part of the side-chain is buried, while only the charged end is exposed to solvent [166]. For that reason they can be classified as amphipathic, although Arg and Lys side-chains generally prefer protein surfaces, and in most hydrophobic scales are characterized as strictly polar residues [111]. Both participate in the process of building salt-bridges as a positively charged molecule, where it pairs with a negatively charged amino acid (usually Asp). In addition, Arg often plays the role of a critical residue in the protein active site or binding pockets, which is not surprising since its complex guanidinium group has a geometry and charge distribution that is ideal to form multiple hydrogen bonds as well as for binding negatively charged phosphate groups.

Canavanine (Fig. 5.17, **11**), an Arg analog from raw seeds of the plant *Canavalia ensiformis*, strongly affects cell development and has growth-inhibiting properties [5]. Its probable natural function is chemical defense, i.e. to serve as an allelochemic agent that deters the feeding activity of plant-eating insects and other herbivores [202]. For that reason, its antimetabolic, anticancer, antibacterial, antifungal and antiviral properties were intensively studied [203]. Chemical, physical and immunological studies from the 1950s and 1960s established that incorporation of canavanine into proteins provides the biochemical basis for canavanine's antimetabolic properties [29]. Canavanine is an efficient substrate for ArgRS ($K_M = 45$ μM and $K_i = 50$ μM for yeast enzyme [204]); thus it is incorporated into proteins in place of Arg, and the resulting canavanyl proteins are functionally impaired and structurally aberrant since canavanyl residues disrupt local protein conformation and tertiary structure, as demonstrated in substituted vitellogenin from an insect *Locusta migratoria* [206]. Canavanine as a potent antimetabolite of Arg affects regulatory and catalytic reactions of Arg metabolism, uptake and Arg participation in other cellular processes. For example, it is now well known that overproduction of nitrite oxide (by enzymatic conversion of L-Arg to L-citrulline; Fig. 5.17, **13**) by an

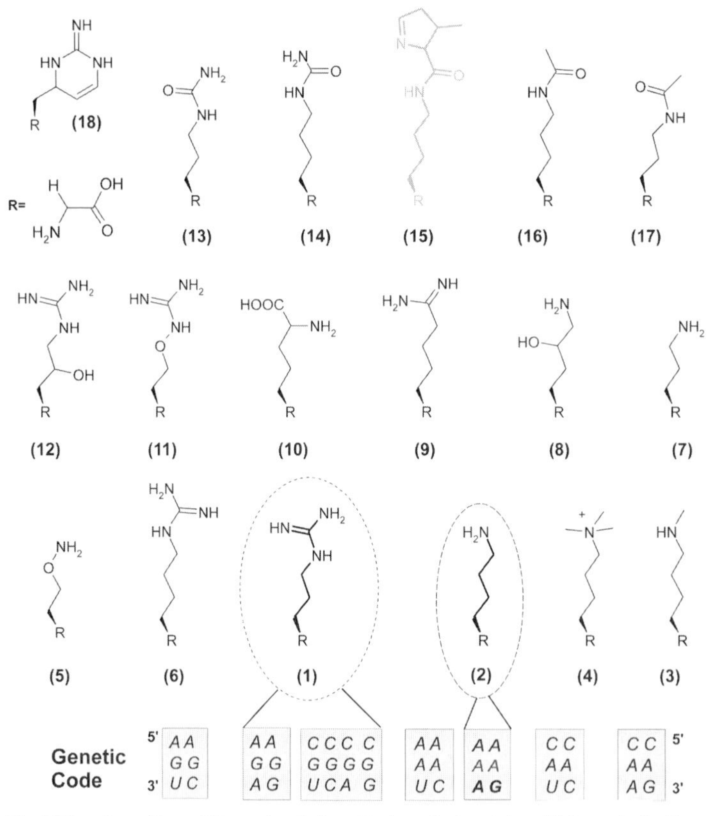

Fig. 5.17. A portion of the natural diversity in the chemistry and structures of Arg (**1**) and Lys (**2**) analogs and surrogates. These natural amino acids are identified either as secondary metabolism products or metabolic intermediates in various organisms, mainly plants (for more a comprehensive review, see [205]). Natural amino acids like ornithine (**7**), citrulline (**13**) and homocitrulline (**14**) are widely distributed, whereas others are isolated from particular species, e.g. N^6-methxyllysine (**3**) from *Sedum acre*, laminine (**4**) from *Reseda luteola*, homoarginine (**6**) and tetrahydro-lathryne (**18**) from plant *Lathyrus* sp., 5-hydroxylysine (**8**) from plant *Medicago* sp., indospicine (**9**) from *Indigofera spicata*, canaline (**5**) and canavanine (**11**) from *Canvalia ensiformis*, 2,6-diaminopimeic acid (**10**) from *Pinus* pollen, and N^{ω}-acetyl-lysine (**16**) and N-acetyl-ornithine (**17**) from *Beta vulgaris*. From this vast reservoir of α-amino acids, nature recruited only Lys (by assigning two coding triplets to it) and Arg (six codons) as canonical substrates for regular protein biosynthesis. A special canonical amino acid pyrrolysine (Pyl, **15**) is exceptionally translated in a few proteins of the Archaea family *Methanosarcinidae* via read-through of the in-frame UAG termination codon, as a result of naturally preprogrammed decoding of their mRNAs (see Section 3.10).

inducible nitric oxide synthase is responsible for some pathological states associated with vascular smooth muscle vasodilatation mechanisms [207]. In this context, the discovery that L-canavanine is capable of selective inhibition of inducible nitric oxide synthase, [208] was not surprising. Other interesting compounds with similar properties are nitro-L-Arg as well as various guanidino methylated analogs [209].

A large number of Arg analogs and surrogates are commercially available (e.g. *N*-ω-nitro-D-arginine) and quite a large pool of natural analogs of this amino acid has been identified (Fig. 5.17). Homoarginine (Fig. 5.17, **6**) is a substrate for activation, but not for the charging reaction. Common metabolic intermediates such as ornithine (Fig. 5.17, **7**), citrulline (Fig. 5.17; **13**) and homocitrulline (Fig. 5.17; **14**) together with nitroarginine, sulfaguanidine lysine, arginosuccinate and α-amino-γ-quanidinobutyric acid are expectedly not substrates for the $tRNA^{Arg}$ aminoacylation reaction mediated by native ArgRS [204].

5.2.1.10 Lysine and Lysyl-tRNA Synthetases – Enzymes with Many Talents

Lys residues are frequently placed in the active sites of proteins; the positively charged primary amine on its side-chain often participates in hydrogen bond formation with negatively charged molecules [134]. Lys residues in proteins play an important role either as targets for chemical derivatization or targets for numerous posttranslation modifications (see Section 3.11). Lys shares many common physicochemical features as well as protein distribution patterns with Arg residues (as described in previous section). Lys natural product chemistry offers numerous interesting hydroxylated, cyclic, acetylated, extended, reduced, etc., analogs and surrogates like mimosine, canaline (**5**), laminine (**4**), methxyllysine (**3**), hydroxylysine (**8**), indospicine (**9**), 2,6-diaminopimeic acid (**10**), acetyllysine (**16**) or lythrine derivatives (**18**) (shown in Fig. 5.17). However, there has still not been any systematic research on their systematic global or position-specific incorporation into proteins.

In the pioneering studies on noncanonical amino acid incorporation into proteins, a few analogs and surrogates of Lys with the potential to be translated into proteins were reported: methxyllysine (Fig. 5.17; **3**), 5-hydroxylysine (Fig. 5.17; **8**), 4-oxalysine (Fig. 5.18; **2**), 4-thialysine (*S*-2-aminoethylcysteine, Fig. 5.18; **3**), 4-selenolysine (Fig. 5.18; **4**), *trans*-4-dehydrolysine (Fig. 5.18; **8**) and 2,6-diamino-4-hexynoic acid (Fig. 5.18; **9**) (for reviews, see [5, 6, 54, 189]). Particularly well documented are attempts to incorporate 4,5-dehydrolysine and thialysine into collagen [54]. A remarkable structural similarity between thialysine and Lys is also reflected in their kinetic properties: thialysine is strong competitive inhibitor of Lys in the activation reaction with *E. coli* LysRS, its K_i value is only 1.4 μM (K_M for Lys is 1.6 μM), making LysRS unable to distinguish between these two substances. As a result, at least 12% of collagen (type I) expressed in embryonic chick tendon fibroblast cell culture contained thialysine, which altered the conformational stability of the collagen triple helix [210]. Selenalysine is also recognized by cell transport systems and LysRS, and subsequently incorporated into *E. coli* proteins, replacing in total about 14% of Lys residues in whole proteome [211]. Later it was demonstrated that mammalian cells can incorporate thialysine and selenolysine in their proteins in substitution of Lys, but proteins containing these analogs are found to be unstable and prone for rapid degradation [212].

There are numerous early studies on substrate specificity of LyRS by using various Lys analogs and surrogates (Fig. 5.18). It was shown that 5-hydroxylysine (Fig.

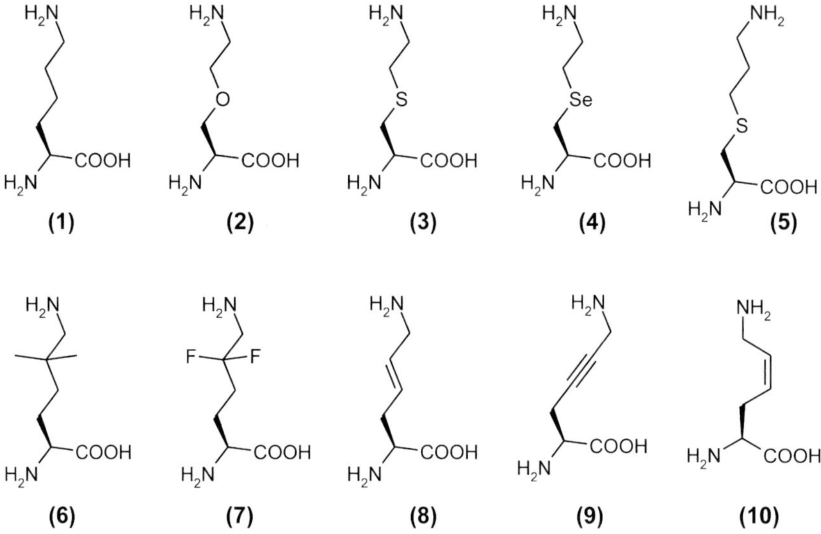

Fig. 5.18. Structures of some Lys (**1**) analogs and surrogates used to probe the substrate specificity of LysRS (extensively reviewed in [101]). Common names of the resented amino acids are 4-oxalysine (**2**), 4-thialysine (*S*-2-aminoethylcysteine, **3**), 4-selenalysine (**4**), 4-thiahomolysine (*S*-2-aminoethylhomocysteine, **5**), 5,5-dimethyllysine (**6**), 5,5-difluorolysine (**7**), *trans*-4-dehydrolysine (**8**), 2,6-diamino-4-hexynoic acid (**9**) and *cis*-4-dehydrolysine (**10**).

5.17, **8**) can be activated by LysRS, but was not transferred onto cognate $tRNA^{Lys}$, while 4-thiahomolysine (Fig. 5.18, **5**) and 4-selenahomolysine are not substrates in the activation reaction [101]. Interestingly, the activation of analogs presented in Fig. 5.18 requires a *trans*-like conformation of the amino acid side-chain since *trans*-4-dehydrolysine and 2,6-diamino-4-hexynoic acid are substrates in the ATP–PPi exchange reaction, whereas *cis*-4-dehydrolysine (Fig. 5.18, **10**) is not [213].

Since Lys residues in proteins are mainly distributed at surfaces, it is reasonable to expect that the LysRS-based orthogonal system that utilizes templates with optimized Lys codon distributions would most probably replace complicated suppressor-based approaches for chemical diversification of protein surfaces. Many biogenic amino acids listed in Fig. 3.12 on the surfaces of eukaryotic proteins arise from posttranslational modification of Lys residues. Because the enzymatic machinery for such Lys modification is not available in prokaryotes, no natural means exists to incorporate them into proteins synthesized in *E. coli*. On the other hand, by introduction of suitable analogs and surrogates it should be possible to achieve an old dream of chemical or enzymatic control of posttranslational modifications [189, 214] (see Section 7.8). In this context, LysRS should be an attractive target for engineering in order to relax or even exchange its substrate specificity in the desired direction, i.e. the accuracy of LysRS in selecting Lys against the other amino acids is less than average (D-factor ~ 130–1700) [100]. Such a remarkable pliancy in the *E. coli* lysine activation and charging system, as well as ribosomal tolerance for diverse amino acid types (see Section 7.2), might be used as a general vehicle for further chemical diversification of proteins based on Lys-derived analogs and surrogates.

In addition to the "intrinsically relaxed" substrate specificity, the existence of constitutive versus inducible as well as functionally equivalent, but structurally unrelated, LysRS forms might be also exploited for Lys-like amino acid repertoire expansion. Studies on thialysine-resistant *E. coli* strains enabled Zamecnik and coworkers to discover the existence of inducible class II LysRS forms [215]. Although cells generally posses a single AARS species for each amino acid, the existence of AARS duplicates among bacterial and other species is well documented [216]. One of these exceptions is LysRS from *E. coli* which posses a constitutive (a *lysS* gene locus in the *E. coli* genome) and an inducible (*lysU*) enzyme form. There is 88.5% sequence similarity between the two enzyme species, both have a very similar three-dimensional structures and in both the anticodon is a key element for efficient aminoacylation of $tRNA^{Lys}$. However, these two enzymes differ at the level of affinity for lysine and its toxic analog 4-thialysine – inducible LysRS has reduced capacity to recognize this analog, which explains the better resistance of *E. coli* strains carrying the *lysU* gene locus toward 4-thialysine under stress conditions (which facilitate *lysU* gene expression) [213]. This phenomena is similar to the species-specific differences toward azetidine-2-carboxylic acid aminoacylation found for ProRS (Section 5.2.1.5).

Continuous works on mapping of entire genomes revealed another big surprise in the field – genomes of Archaea, α-proteobacteria and spirochetes contain gene copies of class I LysRS enzyme; all other organisms contain exclusively structurally unrelated class II LysRS, with the exception of *Methanosarcinidae* whose genome contains copies of genes for both enzyme species [216]. This is a dramatic exception to the rule that the 20 AARS can be divided into two nonoverlapping classes, independent of their origin. Ibba and coworkers [217] also found that *Bacillus burgdorferii* class I LysRS (LysRS1) poorly recognizes 4-thialysine ($K_i = 1140 \pm 230$ μM) in comparison with *E. coli lysS*-encoded class II (LysRS2) ($K_i = 3.9 \pm 0.4$ μM). In general, a far wider range of analogs can be stably attached to $tRNA^{Lys}$ by LysRS2 than by LysRS1.

While any possible evolutionary rationale behind these processes remains to be explained, it is clear that in the case of the recognition of the canonical amino acid Lys, these observations teach us how nature performs the same cellular function using different structural patterns and molecular mechanisms. Conversely, the question remains whether dramatic differences in recognition of analogs either at the level of constitutive versus inducible and LysRS1 versus LysRS2 enzyme recognition might provide us with a more general key for generating novel substrate specificities.

5.3 *In Vitro* Chemical and Enzymatic tRNA Aminoacylation

Experimental intervention at all important levels of transmission of the genetic information might result in the engineering of the genetic code in terms of an expanded number of amino acids as the basic building blocks of proteins. At the level of the DNA/RNA sequence it is possible to manipulate the nature and length of the basic coding/decoding units (i.e. codons/anticodons). The aminoacylation reaction, being the interpretation level of the genetic code, is the most attractive point for experimental interventions. Thereby, ribozymes might become alternative for both enzymes and chemistry (i.e. AARS and chemical tRNA aminoacylations). Finally, the ribosome itself can be altered to affect the interpretation of the genetic code.

5.3.1
Chemical tRNA Acylation for Sense Codon Reassignment

Since codon–anticodon interactions are insensitive to the nature of the amino acid attached to the acceptor stem of the tRNA, the influence of the amino acid charging process is the most obvious way to introduce novel amino acids into the genetic code. This can be achieved either chemically, by the use of ribozymes or via the catalytic action of enzymes. Such tRNAs added in a suitably programmed *in vitro* translation system contribute directly in ribosome-mediated protein synthesis. In this way, an interpretational step in the flow of genetic information is efficiently avoided (i.e. *in vivo* selection, activation and aminoacylation of desired amino acid). Chemically altered tRNAs in the context of *in vitro* translation are still widely used to manipulate translation. In the early experiment of Lipmann and coworkers [218] it was possible to prepare Ala-tRNA^{Cys} by chemical desulfurization (Raney nickel) of Cys-tRNA^{Cys} (see Chapter 2; Fig. 2.1). The obtained misacylated tRNA could substitute Cys with Ala at all positions in hemoglobin, which was used as a model protein. The first report about an extension of this approach was published by Johnson and coworkers who used LysRS to synthesize Lys-tRNA^{Lys} in which the ε-amino group was then acetylated with *N*-acetoxysucccimide [219]. The resulting *N*-ε-acetyl-Lys-tRNA^{Lys} was successfully translated as a response to lysine codons into the hemoglobin sequence in the context of the rabbit reticulocyte cell-free system. This approach was later expanded by using lysine modification with fluorescent probes or with cross-linking reagents [220].

However, such direct chemical modifications of charged tRNAs are rather exceptional since this molecules contains numerous reactive groups and AARS are usually not suitable for enzymatic misacylations with such amino acids. To circumvent these problems Hecht developed more a general semisynthetic method for selective aminoacylation of tRNAs [221]. This strategy consists of enzymatic ligation of the aminoacyl-dinucleotide pCpA~AA (AA = amino acid) by T4RNA ligase to the 3′-end of a particular tRNA transcript which lacks the 3′-terminal pCpA sequence (Fig. 5.19) [222]. Alternatively the 3′-terminal pA of isolated and purified intact tRNA can be removed by *E. coli* RNase T. Chemically prepared pA~AA can be religated to such tRNA (lacking the 3′-terminal pA); tRNA aminoacylated in this way can be used for *in vitro* protein synthesis. Indeed, Hecht and coworkers demonstrated the utility of the misacylated tRNAs for effecting amino acid substitutions at predetermined sites in polypeptides. In this way, it was shown that Lys-tRNA^{Lys}

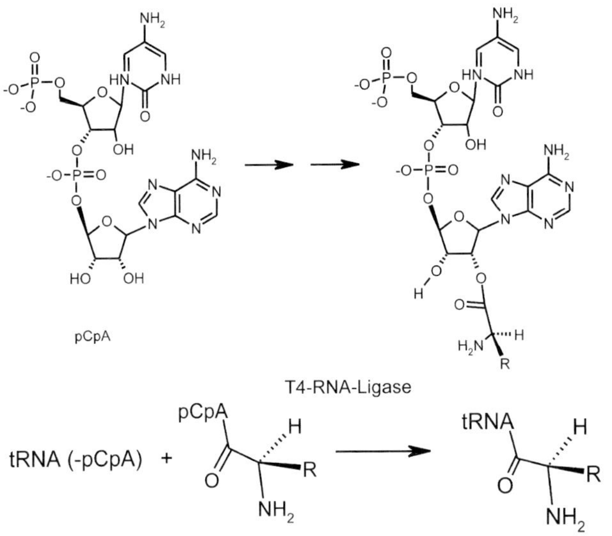

Fig. 5.19. Chemical tRNA aminoacylation. The construction of chemically aminoacylated tRNAs takes advantage of the fact that the constant feature of all tRNAs is a common acceptor sequence, –CCA. In the first step, controlled digestion allows the removal of two terminal nucleotides (–CA) in every type of tRNA. In the second step, the amino acid of interest can be chemically acylated to the dinucleotide p(d)CpA and then attached to the suitably prepared suppressor tRNA by T4 RNA ligase. Such tRNAs can be mainly used for *in vitro* protein synthesis and can only in special cases be used *in vivo*. pCpA = dinucleotide, R = amino acid side-chain.

acetylated at the N^{ε}-amino group can be incorporated into rabbit hemoglobin with nearly the same efficiency as the unmodified one [222]. The preparation of these misacylated (either suppressor or normal) tRNAs as a general method developed by Hecht and his group was further improved and modified by Brunner and coworkers [223]. They developed an approach which enables efficient full deprotection of pCpA and its subsequent ligation to the truncated tRNA in order to obtain aminoacyl-tRNA capable of serving as serve as a ribosomal A site donor.

Sisido and coworkers developed an interesting approach that was based on the use of rare codons that can be placed at any position in the target mRNA sequence [224]. Such codon optimization of the target gene sequence allows for position-specific or multiple-site mode incorporation of one or more different amino acid species. For example, the rarest arginine codon AGG (which contributes an average of less than 3% of the total $tRNA^{Arg}$ pool) is a suitable binding partner for chemically acylated $tRNA_{CCU}$. In this way, the incorporation of various photoactive amino acids [225] (*p*-aminophenylalanine, 2-anthrylalanine, 1-naphthylalanine, 2-naphthylalanine and *p*-biphenylalanine) was achieved. Takai and coworkers [226] inactivated $tRNA^{Asp}$ and $tRNA^{Phe}$ within crude S30 extracts from *E. coli* by antisense treatment or by digesting most of the tRNA without essential damage to the ribosomal activity. Using HIV-1 protease as a model protein in RNase-treated

S30 cell-free extracts with precharged amino acids, they succeeded in substituting aspartic acid and Phe residues with 2-napthylalanine and *p*-phenyl-azo-phenylalanine. However, they were not able to estimate the specific yield of fully substituted protein, probably because folding of the resulting protein was seriously impaired.

Kowal and Oliver [227] explored the use of nonassigned codons in *Micrococcus luteus* to circumvent problems associated with competition by release factors in suppression-based methodologies. In this microorganism, six codons are unassigned (or very rarely assigned), allowing one to evaluate the possibility to insert such codons into the DNA template for either position-specific or multiple-site incorporation of noncanonical amino acids. *In vitro* reaction in *M. luteus* lysate supplied with plasmid-encoded target genes and *E. coli* tRNAs for unassigned or rare codons led to their complete translation with Phe upon addition of preacylated tRNAs into the cell-free system devoid of deacylation problems.

Sense decoding methods recently gained importance for development of mRNA-based peptidomimetic display libraries (as an alternative to ribosome display), whereby noncanonical amino acids containing peptides are generated in a genetically encoded format [228]. In 2002, Roberts and coworkers [229] combined nonsense suppression and *in vitro* selection by creating a small mRNA-linked library of peptides and selecting for those containing noncanonical amino acids. Most recently, Forster and coworkers [230] combined a purified translation system free of AARS with chemo-enzymatically synthesized cognate tRNAs which enabled arbitrarily chosen coding triplets to be completely reassigned to noncanonical amino acids *in vitro*. There is great excitement about these novel methodological possibilities, since modified peptides are key pharmaceuticals for the treatment of a wide variety of diseases.

5.3.2
Ribozymes, Ribosomes and Missense Suppressions

The complicated chemistry associated with dinucleotide-mediated tRNA aminoacylation with noncanonical amino acids can be efficiently circumvented either by AARS engineering or by the design of suitable ribozymes. Such ribozymes should be able to transfer amino acids to the 3′-end of specific tRNA, i.e. useful in preparation tRNAs preacylated with noncanonical amino acids. Suga and coworkers recently provided proof-of-principle that ribozyme-mediated aminoacylation of $tRNA^{fMet}$ might be performed with Phe and Tyr analogs [231]. In this way, an additional possible tool for *in vitro* tRNAs aminoacylated with noncanonical amino acids was generated. Ribozymes capable of recognizing the acceptor stem and anticodon of tRNA should allow for the insertion of novel amino acids either in multiple-site or position-specific mode. Resin-immobilized ribozymes capable of aminoacylating a wide variety of tRNAs with different Phe analogs were also reported [232]. This methodology at the current level of development suffers from a series of shortcomings like inefficient amino acid transfer, complicated ribozyme construction (requiring, at least in part, chemical methods) and high concentrations of aminoacylated ribozymes.

The report of Dedkova and coworkers [233] opens a new avenue for manipulation of the ribosome structure as a possible way to expand the genetic code as well. Mutations in 23S rRNA in the region of the peptidyltransferase center of the ribosome and the helix 89 led to a conformational change, diminishing the normal mechanisms for discrimination between D- and L-amino acids in the ribosomal A site. Such modified ribosomes might exhibit altered translation properties, a feature that can be exploited for amino acid repertoire extension. The addition of suppressor $tRNA_{CUA}$ in a cell-free system containing such ribosomes allowed for the synthesis of luciferase with L- and D-Phe. These experiments open up the question of the role of ribosomal endogenous factors for the efficiency and fidelity of noncanonical amino acid incorporation. Indeed, little is known about the ribosome recognition elements that have to be altered to allow the introduction of versatile chemical functionalities into proteins. For example, Uhlenbeck and coworkers have shown that the affinities of the charged tRNAs for elongation factor Tu (EF-Tu) and ribosomal A site vary depending on the nature of the attached amino acid [234]. The phenomena of mischarged tRNA adaptability to the ribosomal A site was also extensively studied by Sisido and coworkers; they found that the ribosome is quite flexible in accepting a rather large repertoire of amino acids as substrates for translation [235], as shown in Fig. 5.21. This adaptability was confirmed by translation of various aromatic amino acid analogs, some carrying relatively large and bulky side groups. Thus, special noncanonical amino acids with linearly expanded aromatic groups (e.g. 2-naphtylalanine, *p*-biphenylalanine and *p*-phenylazo-phenylalanine) are more favorable than those with widely expanded or bent aromatic groups (e.g. 9-phenylanthrylalanine) (see also Section 7.2 and Fig. 7.1). Transfer of these components into the rabbit reticulocyte-based cell-free system resulted in essentially the same observations [236].

Finally, there is a recent report of a successful experiment regarding reassignment of coding triplets in the frame of codon families. Kwon and coworkers [237] have exploited the degeneracy of the genetic code to replace only a subset of Phe with the 2-naphthylalanine. The Phe is assigned to UUU and UUC codons, which are decoded by a single $tRNA^{Phe}$ isoacceptor containing the anticodon GAA. The UUU codon is recognized through a G·U wobble base interaction between the first nucleotide of the anticodon and the third nucleotide of the codon (see Section 3.8; Fig. 3.10). To enhance these interactions (through restoring standard Watson–Crick pairing) a mutant yeast $tRNA^{Phe}$ with the AAA anticodon was imported into *E. coli* cells to translate efficiently the UUU codon. This $tRNA^{Phe}$ was charged by 2-naphthylalanine through mutant yeast PheRS which has an extended substrate specificity. The translation of the analog in the target sequence is expected to be strongly biased towards the UUU codon. Such a reassignment of degenerate codons offers a promising alternative to strategies based on nonsense and frameshift suppression described below. However, this is essentially a missense suppression-based approach and therefore suffers from inherent problems of all these methodologies, i.e. reassignment is incomplete and canonical amino acids are often preferentially incorporated as a response to "reassigned" coding units during translation [62].

5.4 Novel Codon–Anticodon Base Pairs

5.4.1 *In Vitro* and *In Vivo* Frameshift Suppression of 4- and 5-base Codons

The mechanisms of reading frame maintenance are well documented as well as mutations in components of the protein synthesis apparatus that promote or inhibit switches into alternate reading frames. Early experiments from the 1960s and 1970s have demonstrated that such frameshift mutations can be externally suppressed. For example, mutant tRNA in *Salmonella* and yeast with extended anticodons can read nontriplet codons, although their read-through efficiency was comparatively low [238]. The work which shown that the efficiency of UAGA quadruplet decoding for $tRNA^{Leu}$ with the engineered $^{3'}AUCU^{5'}$ anticodon could be elevated to the range of 20–40% can be taken as an example of recent studies in this field [239]. These examples of naturally occurring +1 frameshift suppressions clearly demonstrate that the translational machinery could handle codon–anticodon pairs whose lengths were greater that standard triplets. As a consequence, frameshift suppressor tRNAs can efficiently incorporate various canonical amino acids *in vivo* [240].

Sisido was among first to exploit this feature as an efficient vehicle for noncanonical amino acid mutagenesis by chemical charging of a tRNA that contains a 4-base anticodon followed by *in vitro* synthesis in the presence of an mRNA that contains the corresponding 4-base codon [241]. A combination of 4-base codons AGGN and the corresponding anticodons NCCU was used to examine the scope and possibilities for *in vitro* frameshift suppression for incorporation of special canonical amino acids. Indeed, streptavidin was substituted with 2-anthranylalanine by using all of the combinations of quadruplets: AGGG, AGGA, AGGC and AGGU. The target gene sequence was optimized in a way that its mRNA contains extra downstream stop codons to terminate nonframeshifted products, i.e. to abort protein synthesis in the absence of frameshift suppression [242], as shown in Fig. 5.20A. This strategy takes advantage of the fact that following insertion of quadruplets at position of rare codons, frameshift suppressor tRNA decoding would work efficiently since there is less competition from endogenous tRNAs present in prokaryotic and eukaryotic cell-free translation systems [243].

Sisido and coworkers also pioneered the use of quadruplet codons for the site-specific insertion of one or two different amino acid analogs into a single protein *in vitro*. For example, streptavidin mRNA containing CGGG and AGGU quadruplets (Fig. 5.20B) was successfully translated in the presence of aminoacyl $tRNA_{CCCG}$ and aminoacyl-$tRNA_{ACCU}$ charged with ε-(7-nitrobenz-2-oxa-1,3-diazl-4-yl)lysine (CGGG signal in mRNA) and 2-naphthylalanine (AGGU signal in mRNA) [244]. The general aspects related to the limits of anticodon size that would help to identify efficient tRNA suppressors of 2-, 3-, 4-, 5- or even 6-base codons have been recently the subject of detailed studies [242]. For example, it was

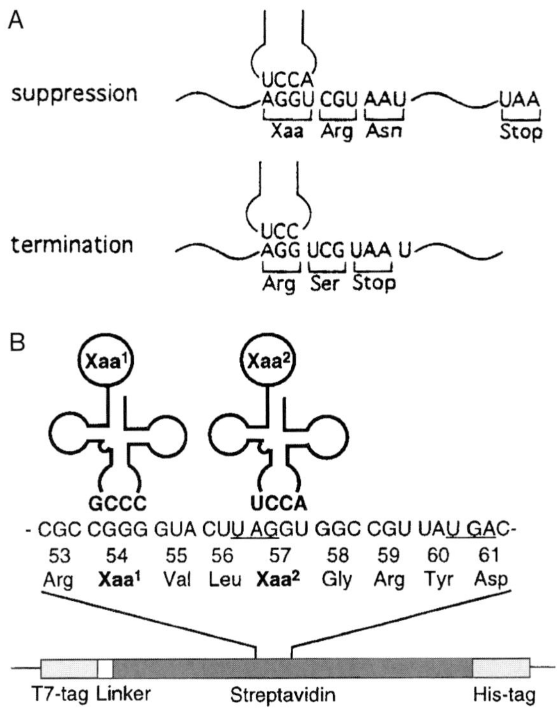

Fig. 5.20. The use of quadruplets in the context of *in vitro* protein biosynthesis. (A) Efficient suppression yield of intact protein labeled with the desired amino acid. In the absence of efficient suppression the downstream stop codons terminate the translation of the nonframeshifted product. (B) Simultaneous incorporation of two noncanonical amino acids in streptavidin by suppression of two different 4-base codons. This provided a general strategy for the introduction of special noncanonical amino acids at multiple sites into a single protein (reproduced from [241, 244] with permission of American Chemical Society. Copyright 1996 and 1999).

demonstrated that pentaplets (i.e. 5-base codons) are indeed suitable for incorporation of special canonical amino acids into proteins [245].

Extensive examinations using *in vivo* selection procedures for the determination of optimal lengths of extended codon–anticodon pairs have shown that the translation apparatus indeed permits decoding of codons consisting of 3–5 bases and that each codon type requires different tRNAs [246]. However, 6-base codons are incompatible with the translation machinery and pentaplets cannot be suppressed with the same efficiency as quadruplets. In other words, although able to tolerate various codon lengths, the *E. coli* translation machinery shows a preference for anticodon loops that interact with quadruplet codons. Such limits in codon size are most probably governed by the ribosome itself, where the tRNA acts a sort of "molecular ruler" that measures out codon size during translation [247].

A series of experiments performed in Sisido's laboratory convincingly showed that the ribosomal machinery could handle codon–anticodon pairs that are greater

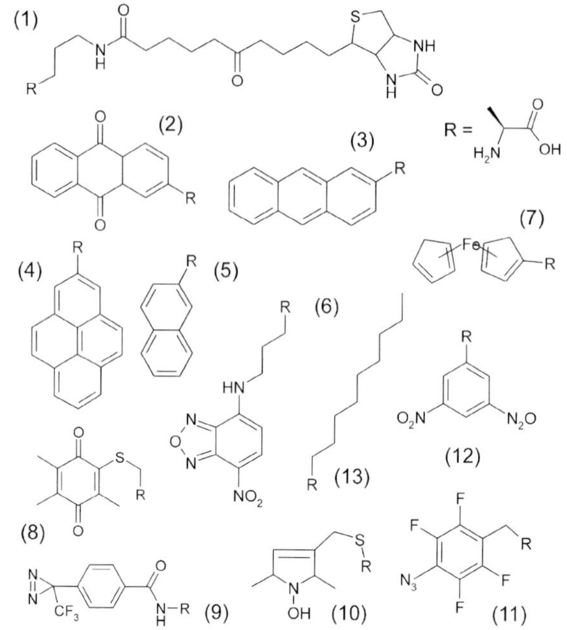

Fig. 5.21. Part of the chemical diversity of noncanonical amino acids that are chemically attached onto tRNA with sense, nonsense or extended codons. Amino acid names: biotinyl-alanine (**1**), antraquinonyl-alanine (**2**), 2-anthrylalanine (**3**), 2-pyrenylalanine (**4**), 2-naphtylalanine (**5**), Lys(NBD) (nitrobenz-oxo-diazolyl-lysine) (**6**), ferrocenyl-alanine (**7**), trimethyl-dioxo-(cyclohexa-1,4-dienylsulfanyl)-butyric acid (**8**) trifluoromethyl-diazirin-benzoylamino-propionic acid (**9**), 4-azido.tetrafluorophenylalanine (**11**), dinitrophenylalanine (**12**) and 2-aminododecanoic acid (**13**).

that 3 nucleotides in length. Although it was expected that the basic advantage of this strategy would be the circumvention of competition with release factors, the rules that govern the different suppression efficiencies for different quadruplets are unclear as well. For example, it was argued that some sites in mRNA are more "shiftier" than others [246]. Therefore, this strategy has been limited both by the overall read-through efficiency of the multiple suppressor tRNAs as well as by the occasional apparent misfolding of nascent protein containing modified amino acids. Finally, the yield of such reactions is always poor for at least two chief reasons: (i) a cyclic tRNA appears as a byproduct of chemical tRNA charging and inhibits the protein synthesis, and (ii) the tRNA charged with noncanonical amino acid has very short life-time and its recharging is not possible [248].

To avoid at least part of these and similar problems Schultz and coworkers transferred the quadruplet suppression strategy *in vivo* in order to enable intracellular protein labeling. An orthogonal AARS:tRNA pair was derived from archeal $tRNA^{Lys}$ sequences capable of frameshift suppression of the quadruplet codon AGGA *in vivo*. This pair proved to be compatible with the amber orthogonal system and enables simultaneous incorporation of homoglutamine and *O*-methyltyrosine at distinct positions within myoglobin [249].

5.4.2
Toward a Third, Noncanonical Base Pair in DNA

The genetic information of contemporary cells is stored in a linear sequence of four nucleotides in the biochemically more or less inert DNA molecules. They form stable double helices via weak interactions between only two base pairs (A:T and G:C) and exhibit resistance to hydrolytic cleavage (because it is missing a 2′-OH group in the *β*-D-2-deoxyribose). In almost all living organisms, DNA as the major information carrier replicate itself, whereas the interpretation of the information depends on the application of a code to translate the four-letter alphabet to the 20-letter alphabet of proteins. Therefore, the generation of truly artificial genetic systems or synthetic organisms would require additional and chemically different coding units generated by introduction of new base pairs which would in turn encode additional amino acids. The first steps in this direction would be to explore the possibilities to employ a new alphabet in order to change/modulate/reprogram the already existing replication, transcription and translation machineries of the living cells. The establishment of a translational system that would work with triplets containing noncanonical base pairs would avoid the difficulties of frameshifting with expanded codons as well as problems associated with suppression-based methods (Section 5.5.4) during translation.

Benner and coworkers [250] demonstrated that template-directed isoguanosine (*iso*G) and isocytidine (*iso*C) incorporation into DNA and RNA by appropriate polymerases is possible (Fig. 5.22). Chemically synthesized mRNA containing the modified nucleotides enabled direct site-specific insertion of 3-iodotyrosine into a peptide by using an *in vitro* translation system. This system was supplemented with a chemically synthesized mRNA containing the modified (*iso*C)AG codon at the site of interest and a suppressor tRNA containing the complementary CU(*iso*G) anticodon. Unfortunately, this newly generated *iso*C:*iso*G pair was found to be unstable in water solutions. As an alternative to orthogonal hydrogen bonding between nucleosides, nonhydrogen bonding nucleotide analogs have been examined [251]. These bases pair together via specific hydrophobic or van der Waals force interactions. In the laboratories of Romesberg [252] and Schultz [253], more than 30 hydrophobic bases have been designed, synthesized and characterized in order to create base pairs that do not rely on hydrogen bonding, but on hydrophobic interactions. Polymerase recognition, nucleotide strand incorporation and a certain degree of chain extension have been demonstrated in many of studies systems.

Hirao and coworkers [254] succeeded in greatly expanding the number of synthetic base pairs capable of transcription into mRNA. The next step was to demonstrate their utility for the incorporation of noncanonical amino acids into proteins. This achievement was recently reported by Kimoto and coworkers. They developed an *in vitro* transcription/translation system where DNA with unusual base pairs can be transcribed into RNA molecules and subsequently participate in protein

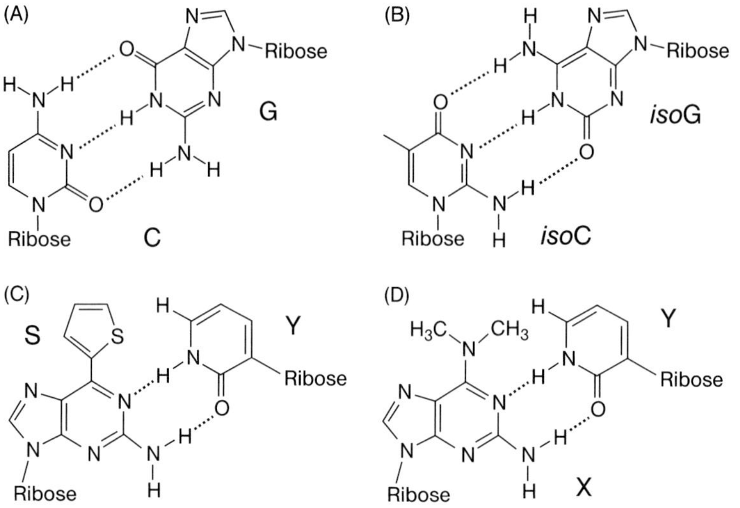

Fig. 5.22. Some canonical and noncanonical nucleotide base pairs. (A) Canonical G:C base pair. (B) Noncanonical *iso*C:*iso*G pair which exhibit a different (orthogonal) hydrogen bonding arrangement. Noncanonical base pairs Y:S (C) and Y:X (D) were designed on the basis of hydrogen-bonding pattern and shape complementarity. C = cytosine, G = guanosine, *iso*G = isoguanosine, *iso*C = isocytidine, S = 2-amino-6-(2-thienyl)purine, Y = pyridin-2-one, X = 2-amino-6-dimethylaminopurine.

synthesis on ribosome [255]. They combined the concepts of shuffled hydrogen bonds and van der Waals interactions to develop a base pair that could be polymerized into the RNA transcript. In particular, the transcription of the noncanonical base pair S:Y (Fig. 5.22) by T7 RNA polymerase yielded mRNA with the YAG coding triplet. In the transcription/translation *E. coli* cell-free system, the YAG codon from mRNA was recognized by the CUS anticodon of a yeast $tRNA^{Tyr}$ enzymatically charged with 3-chlorotyrosine, which resulted in its site-specific translation in the protein sequence. The noncanonical nucleotide Y can be also paired with another purine analog X (Fig. 5.22), transcribed into RNA and subsequently programmed to encode the desired noncanonical amino acid in the translation process. These efforts in the design and construction of noncanonical base pairs can be seen as a competition between many groups (e.g. Benner, Eaton, Kool, Hirao, Romesberg, Yokoyama and Schultz) which certainly will yield further rapid progress in this field (for a comprehensive review, see [256]).

5.5 Stop Codon Takeover

Codon reassignment in nature among different organisms occurs usually from one canonical amino acid to another. For example, vertebrate mitochondria read AUA as Met, while in the vertebrate cytosol this codon is assigned to code for isoleucine (see Section 4.7; Fig. 4.4). Although Osawa [257] postulated that "central to codon reassignment is the principle that a codon cannot have two assignments simultaneously, because this would be lethal to an organism", an exceptional and unique example is the UGA codon that can be decoded as the termination signal, Sec, Trp and Cys. Other examples have been identified as well, such as the universal CUG leucine codon which is translated as serine in some *Candida* species [24]. A coding unit capable of assigning a few distinct amino acids was termed a "polysemous codon [258]. Such diversity in the assignment of certain codons in the universal code suggests the possibility of experimental accommodation of additional amino acids into proteins.

The capacity for position-specific incorporation of desired noncanonical amino acid in traditional methods based on auxotrophism is dictated by sequence composition. Excellent models for position specific replacements are relatively rare for amino acids such as Trp, Cys and Met, especially when complemented with site-directed mutagenesis. However, for more frequent residues such as Lys or Leu, this approach might become time consuming and massive residue exchange through the structure is usually harmful for the target protein. In this case, the degeneracy of the three termination codons UAA, UAG and UGA (also termed stop or nonsense codons) can be used in order to gain a position-specific mode of incorporation, i.e. signal termination in the translation of the desired protein can be achieved by one stop codon while the remaining two ("blank") could be used uniquely to encode a specific noncanonical amino acid. Such an in-frame stop codon takeover approach has already been used for Sec and Pyl in the context of the previously discussed preprogrammed re-coding (see Section 3.10; Fig. 3.11). Although suppressions were not originally thought to be a tool for noncanonical amino acid incorporation, they are nowadays virtually the most "popular" (although not most efficient) methods for the insertion of desired amino acids into specific locations of proteins.

5.5.1
The Concept of Suppression in Protein Translation

Early experiments identified bacterial strains able to perform misincorporation of amino acid into proteins; these phenomena proved to be strain specific [259]. Cells carrying a nonsense (stop) mutation in the middle of a vital genetic message are viable since incomplete polypeptide released from the ribosome due to premature chain termination is prevented. Possible harmful effects are reversed by a second genetic change at different genes of mutant tRNAs and these are called suppressor mutations. These genetic changes might take place in tRNA anticodons or, more rarely, in some other tRNA identity elements. Suppressor tRNAs can insert different canonical amino acids in response to the *nonsense* (when one of the three terminator codons appears in the mRNA) or *missense* codons (i.e. alteration of one sense codon to another so that different amino acids are determined) or *frameshift* mutation in the parent gene. In this way, the normal functions of these termination or sense codons are suppressed by tRNAs with changed identity [238].

The first documented example of tRNA-mediated suppression was read-through of the UAG terminating codon in the coat protein of the RNA phage R17 by the Tyr suppressor $tRNA^{Tyr}$ with changes in the anticodon. As a response to the UAG terminal signal, Tyr residues were translated into the protein sequence of the coat protein of the RNA phage R17 [260]. The suppression of opal (UAA) and ochre (UGA) termination coding triplets is also mediated by mutant tRNAs. The existence of suppressor tRNAs raised the question of how they compete with release factors for termination codons, especially UAA and UGA that appear quite frequently at the end of genes in the genome. Nevertheless, some *E. coli* strains harboring high levels of natural amber suppressor tRNAs are viable and exhibit no significant differences in growth rates when compared with the native strains [36]. On the other hand, it was argued that most of the *E. coli* essential genes are not terminated by UAG [81]. However, it still remains mysterious why the UGA and UAG suppressor tRNAs do not generate an unacceptable number of abnormally long proteins. Obviously, there is no advantage for a normal cell to harbor suppressor mutations, even if only a small fraction of proteins would be nonfunctional. Therefore, suppressions, either *nonsense* or *missense*, should be regarded as exceptional natural phenomena, which are tolerated in some strains without adverse effects on the growth of the host organism. The response of a particular suppressor tRNA to a particular nonsense codon (read-through) can vary as much as 10-fold depending on its location in an mRNA – a phenomenon known as the "context effect". There are also suppressor tRNAs capable of masking the effects of certain frameshift mutations created by the insertion of nucleotides and subsequently able to restore a reading frame [259].

5.5.2
Chemical Aminoacylation of Amber Suppressor tRNA

Suppressor tRNA that contains the anticodon CUA can be prepared either by chemical modification of some commercially available tRNAs (e.g. yeast $tRNA^{Phe}$) or by runoff transcription [261] of a suppressor tRNA gene. The commonly used Hecht approach (Fig. 5.19) for chemical tRNA aminoacylation [221, 222] was further improved by Chamberlin [262], Schultz [263] and coworkers. First, the replacement of cytidine with deoxycytidine in pCpA simplified the synthesis and eliminated another reactive 2′-OH group without any significant effects on biological activity. Secondly, they protected the amino acid amine with suitable cleavable protecting groups [264]. In order to generate translationally competent aminoacyl-tRNA, deprotection of the attached amino acid could be done either chemically or photolytically. In the meantime, additional alternative approaches for chemical aminoacylation emerged [265], including the use of ribozymes [231].

The addition of chemically charged tRNA to a translation system programmed with a gene containing nonsense suppression sites results in the incorporation of the special noncanonical amino acid at the corresponding position in the protein. In this way, Schultz, Chamberlin, Dougherty [163, 266, 267] and others, in a series of *in vitro* experiments using termination suppressor tRNAs, extended and demon-

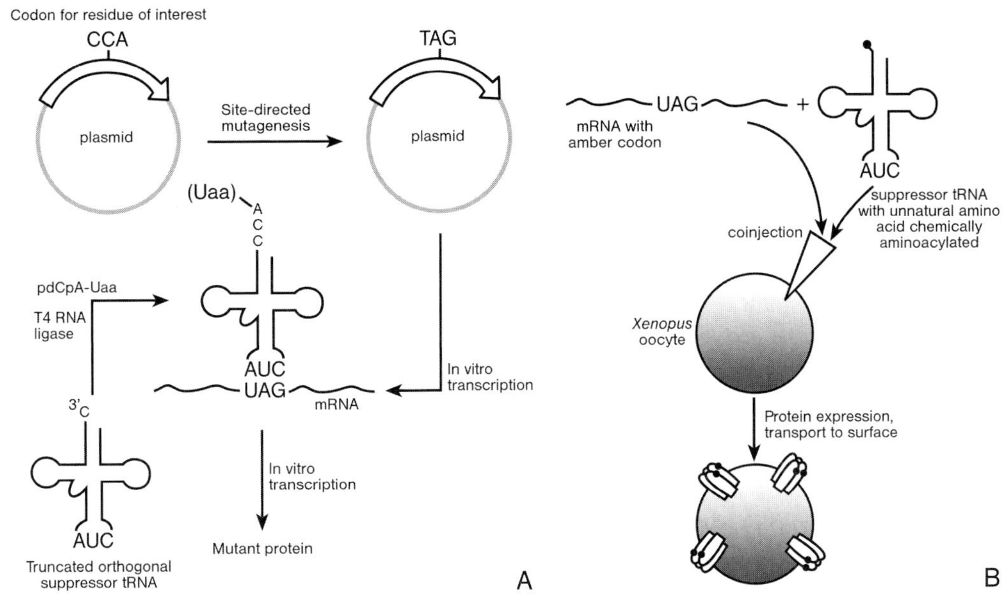

Fig. 5.23. Flow chart of the *in vitro* suppression method and its application in mammalian cells. (A) An *in vitro* transcription/translation system usually based on *E. coli* S30 lysate was required for transcription and translation of the target gene carrying an in-frame UGA termination codon [163]. Typically, the gene of interest is placed under a strong T7 promotor and chemically acylated suppressor tRNAs externally added. (B) The basic protocol for the import (by injection, transfection or electroporation) into mammalian cells of a chemically amino-acylated suppressor tRNAs and target mRNA gene with an in-frame UAG codon [271]. Minute amounts of the labeled protein generated in this way are sufficient to be indirectly detected, because the generated biological effects can be measured in situ using electrophysiological methods. (Reproduced from [86] with permission.)

strated an extraordinary capacity of the ribosome to allow for translation of more than 100 special noncanonical amino acids (part of this diversity is presented in Fig. 5.21). There are at least four essential requirements for the successful application of suppression methodology that needed to be fulfilled: (i) the generation of a suppressible amber (UAG) mutation in the gene of interest by conventional site-directed mutagenesis, (ii) the design of efficient suppressor tRNA, (iii) chemical acylation of suppressor tRNA, and (iv) a suitable *in vitro* protein synthesis system to enable efficient read-through of the termination UAG codon at optimal Mg^{2+} concentrations (Fig. 5.23A). The suppressor tRNA must be recognized by the translation apparatus sufficiently well to suppress the termination codon with high efficiency. On the other hand, such tRNA should not be acylated or deacylated by any of the *E. coli* AARS present in the cell-free lysate [81].

Site-directed incorporation of 3-iodotyrosine into the 16-residue peptide product translated *in vitro* with rabbit reticulocyte system was reported by Chamberlin and coworkers [262]. Similarly, by using yeast amber suppressor tRNA, Noren and coworkers [263] succeeded in incorporating several Phe analogs. This position-specific replacement strategy was first postulated by Kwok and Wong [268], who showed that *E. coli* PheRS aminoacylates yeast $tRNA^{Phe}$ with an efficiency of less than 1% when compared to that of *E. coli* $tRNA^{Phe}$. Thus, the amber suppressor tRNAs derived from yeast $tRNA^{Phe}$ were especially suitable for incorporation experiments in the *E. coli*-based *in vitro* coupled transcription/translation system. These tRNAs efficiently insert the desired amino acid into the β-lactamase taken as a target protein in response to the UAG coding signal on its mRNA. After the suppression reaction in the Zubay transcription/translation system, 2.8–7.5 $\mu g\ ml^{-1}$ of the replaced β-lactamase was produced, which is about 15–20% of the native protein yield (30–45 μg) [269]. Additional improvements of the system in order to increase protein yield have been reported. These include release factor inactivation by mild preheating of an S30 lysate derived from a temperature-sensitive *E. coli* strain or use of more efficient suppressor tRNAs [270]. Numerous noncanonical amino acids have been incorporated site-specifically into various proteins. They can be roughly divided into several groups such as conformationally restricted amino acids, photoaffinity and spin labels, caged amino acids, amino acids whose side-chains have changed acidity, nucleophilicity or hydrogen-bonding properties, and even α-hydroxy amino acids. A tiny portion of this diversity is presented in Fig. 5.21.

A particularly interesting approach to expand the scope of chemical tRNA aminoacylation is co-injection of amber suppressor tRNA and target mRNA containing an amber codon at a given position into *Xenopus* oocytes [271]. In this way, site-directed incorporation of desired noncanonical amino acids was achieved in eukaryotes by using suppressor tRNAs derived from yeast $tRNA^{Phe}$ and *Tetrahymena* $tRNA^{Gln}$ [272]. This approach proved to be useful for structure–function studies of integral membrane proteins, such a ligand-gated ion channels (Fig. 5.23B). Similarly, RajBhandary and coworkers imported amber and ochre suppressor tRNAs derived from *E. coli* initiator tRNAs into mammalian COS1 cells. Such tRNAs with two distinct suppressor activities are capable of introducing two different noncanonical amino acids into mammalian tissue simultaneously [273]. Other examples include suppressor tRNA import via microinjection of Chinese hamster ovary cells and via electroporation of several mammalian cell lines [274].

5.5.3
In Vitro versus *In Vivo* Translation for Extension of the Amino Acid Repertoire

Test tubes with cell-free extracts with chemically aminoacylated tRNAs demonstrated remarkable ribosome plasticity in terms of number, and structural and sterical properties, along with chemical diversity of amino acids that can be successfully translated into proteins. This chemical diversity can be in a great part lost by attempts to transfer into the *in vivo* methodology. A recent example of an "orthogonal pair" derived from the *M. jannaschii* $tRNA_{CUA}{}^{Tyr}$/TyrRS orthogonal system shows that this system in the context of an *in vivo* platform allows for entry in the translation mainly of amino acids that can be regarded as Tyr or Phe analogs, or extended aromatic systems (Section 5.1.5; Fig. 5.7). Obviously, to make fairly exotic and huge amino acids such as ε-(7-nitrobenz-2-oxa-1,3-diazl-4-yl)-L-lysine, or the biotinyl-alanine (Fig. 5.21), bioavailable intracellularly, experimental intervention for a "rational re-design" of many components in the living cell is necessary. The range of the intervention is such that it would require simultaneous remodeling and noninvasive changes for the amino acid transport into cells, and to find a way to avoid metabolic activation or modifications, as well as to pass translation editing check points of intact cellular physiology. For example, the introduction of new base pairs that would expand the genetic alphabet in living organisms would also require its recognition by a series of specific enzymes beyond the polymerases required to replicate them.

In vivo gene expression and subsequent protein synthesis is carried out by means of cell physiology, and is surrounded by cell walls and membranes. In contrast, cell-free *in vitro* protein synthesis is an attractive alternative to *in vivo* approaches for amino acid repertoire expansion in bacterial or yeast fermentation as well as mammalian or insect cell culture, since it provides an open system. Thus, *in vitro* coupled transcription/translation, which is devoid of the problems associated with the mechanisms of cellular physiological function, could be optimized as a platform for engineering the genetic code with a significantly wider spectrum than in living organisms. This, along with other reasons, is why there has been a significant interest in the development of high-yield cell-free translation systems. Various improvements, including optimized batch reactions, addition of chaperones, replenishment of consumable factors, continuous flow or optimization of the relative concentrations of various lysate components, etc., have been extensively described in current literature. For example, a recently reported *in vitro* translation system consisting of only ribosomes, and initiation and elongation factors was used to incorporate several amino acids into desired peptides simultaneously [230]. Thus, using sophisticated *in vitro* systems like PURE [275], where addition of all components can be precisely controlled, it should be possible to introduce the desired amino acid either in the position-specific or multiple-site mode. Therefore, novel generations of even more sophisticated *in vitro* systems with improved performance in terms of protein yield and control of translation conditions would certainly serve as attractive platforms for amino acid repertoire expansion.

5.5.4 General Limits of Suppression-based Approaches

The use of misacylated suppressor tRNAs has provided proteins containing various special noncanonical amino acids at single predetermined positions. Most of these studies involved a read-through of the UAG codon. The main drawback of all suppressor-based approaches is the limited capacity of chemically misacylated synthetic suppressor tRNAs charged with noncanonical amino acids to fully decode (suppress, read-through) nonsense (triplet or quadruplet) codons. This is not surprising since the suppressions are exceptional phenomena which are *context depen-*

dent, i.e. each desired position in a protein sequence cannot always be suppressed. It has been observed and documented that *release*, i.e. intervention of ribosomal *release factors* [276], is so effective at some sites that the natural function of the amber stop codon is not suppressed even when the suppressor tRNA is charged with the amino acid that normally appears at that point in the wild-type protein [227]. Problems of stability of mRNA containing terminating codons as well as protein stability with the incorporated amino acids are also unresolved issues.

Suppression efficiency depends largely on the nature of the amino acid to be incorporated into target proteins. For example, Nle can be incorporated into T4 lysozyme with 100% suppression efficiency, while α-methylleucine only shows 14% efficiency [277]. Suppression efficiencies cannot be reliably predicted, only empirical data are available (e.g. suppression with tRNAs with attached polar amino acids are generally weak). Bypassing these "side-effects" can at least be in part achieved by the generation of a completely unique quadruplet composed of canonical bases for particular AARS:tRNA pairs with orthogonal function [241] or by choosing nontoxic amino acids like substituted analogs of aromatic amino acids. More recently, Frankel and Roberts [278] described a selection approach to identify coding triplets especially susceptible to suppression. However, the most efficient way to completely circumvent these limits is to switch to the use of sense codons in the context of codon-optimized target gene sequences or to label at such positions where read-through is high (in best case about 70%) [279].

The generally impaired cellular viability upon *in vivo* introduction of orthogonal translation components is not the only "side-effect" of these experimental interventions. The efficiency of translation can also be substantially reduced, resulting in decreased yields of the desired protein mutants. For example, in rich media, *E. coli* with heterologous $tRNA_{CUA}{}^{Tyr}$/wtTyrRS expresses dihydrofolate reductase in yields of 67 mg L^{-1}, while by transfer in designed minimal medium the yield drops by more than 95% (2.6 mg L^{-1}) [82]. Nevertheless, this approach might be useful in those studies where minute amounts of labeled proteins are necessary in the context of the living cells [274]. It should be kept in mind that such experiments have always been performed with model proteins like didydrofolate reductase, streptavidin, luciferase, smaller coiled peptides, lactamase, lysozyme or GFP, while "difficult" proteins like single-chain antibodies are most probably not permissive of such incorporations at all. In addition, suppression-based approaches that incorporate amino acids such as ε-(7-nitrobenz-2-oxa-1,3-diazl-4-yl)-L-lysine or dansyllysine might be useful only for the chemical diversification of protein surfaces. On the other hand, there are a large number of widely used, efficient and much simpler protocols for chemical modifications of the protein surfaces (see Chapter 1). Therefore, *in vitro* suppression-based approaches will not be of much use in biological sciences until the currently available protocols for incorporations are considerably simplified.

The identities of amino acids and their nature show a certain correlation with their suppression efficiencies. For example, larger hydrophobic amino acids were incorporated with higher efficiency than smaller or charged ones [163]. Other factors that might dramatically influence suppression efficiency are polarity, stereochemisty and amino acid geometry. For example, D-amino acids and β-amino acids are not substrates for translation by nonsense suppression methodologies [269].

5.6 *In Vivo* Nonsense Suppression-based Methods

5.6.1 In Search for Orthogonal tRNA

tRNA as the carrier of adapter function in translation is the most attractive target for orthogonality engineering. The most important requirement for the successful generation of orthogonal tRNA is the strict absence of cross-reactivity with other components the host protein synthesis apparatus. However, this requires a detailed knowledge of the tRNA identity rules (see Section 2.6). Experimental attempts at a rational design of both positive and negative recognition elements of tRNAs are generally difficult, and often rely on serendipitous findings. Sometimes it is necessary to search for identity elements outside the anticodon as suitable markers for design of efficient suppressor tRNA. This clearly shows how intricate the "surgery" on sophisticated tRNA structures should be to make them suitable for incorporation experiments [280]. Nevertheless, the search for such orthogonal pairs in the context of *in vivo* and *in vitro* suppression methodologies can be summarized as follows. First, orthogonal amber suppressor $tRNA_{CUA}$ has to be generated that should be able to deliver the noncanonical amino acid in response to a UAG codon in the mRNA encoding the protein of interest. Secondly, orthogonal AARS has to be designed by using mutagenesis and screening approaches. Such an orthogonal AARS should recognize exclusively related orthogonal tRNA, but not any of the endogenous tRNAs. Finally, a library of mutants of both orthogonal tRNAs and AARS has to be screened in order to find orthogonal AARS:tRNA pairs capable of activating and transferring the desired noncanonical amino acid preferentially over the canonical ones [81].

An efficient method based on the clever design of mutant libraries and selection cycles for the generation of orthogonal tRNA with reduced cross-reactivity was developed in Schultz's laboratory [281]. At the heart of the method is the combination of positive and negative selection as follows. In negative selection, amber termination codons are introduced at permissive sites into the barnase gene. Any suppressor tRNA mutant from a library which is charged by endogenous AARS translates barnase and induces cell death. Cells containing orthogonal tRNA survive this cycle; survivors are then submitted to a positive selection cycle. This time β-lactamase with in-frame amber codons is used as a selection marker (i.e. ampicillin resistance). Clones which contain orthogonal tRNA that cannot be recognized by endogenous AARS are ampicillin sensitive. In summary, the selection yields only those suppressor tRNAs that are not substrates for endogenous AARS, but can be charged with newly generated (i.e. orthogonal) AARS (Section 5.1.5; Fig. 5.7). These orthogonal tRNA are also compatible with the host translation machin-

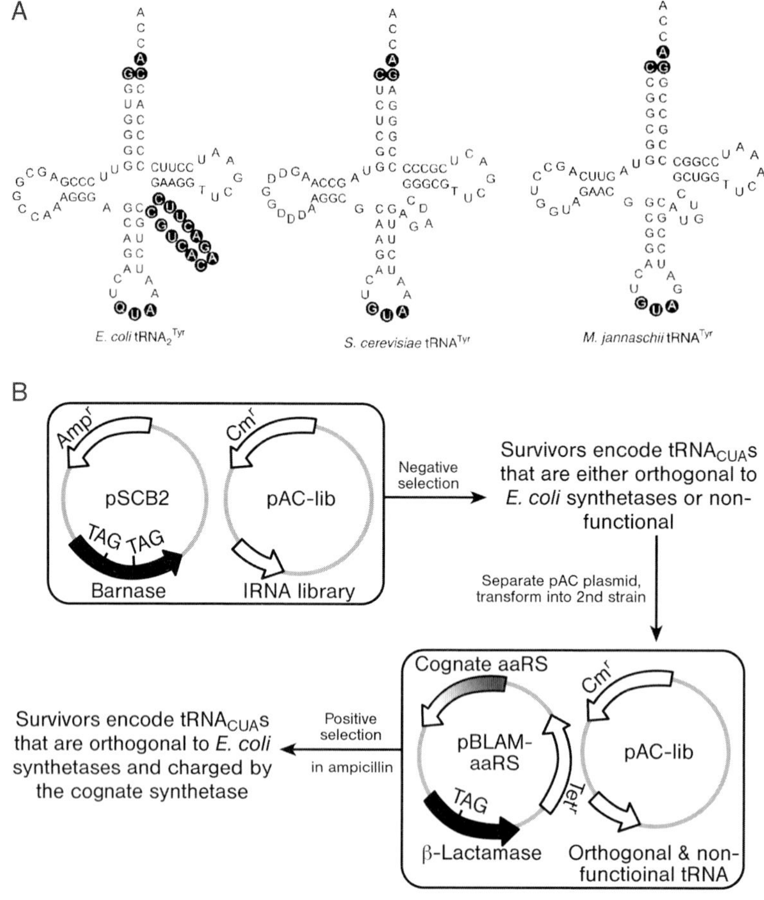

Fig. 5.24. Interspecies differences and the design of orthogonal tRNA (A). Comparison of $tRNA^{Tyr}$ from different species with major identity elements shaded in black. Note that prokaryotic $tRNA^{Tyr}$ is different from yeast and archeal ones: it has a long variable arm and a G1:G72 pair (C1:G72 in yeast and Archea). (B) The basic scheme for the negative and positive selection procedures used to generate orthogonal tRNAs. This approach consists of a combination of negative and positive selections with a mutant tRNA library. (Reproduced from [86] with permission.)

ery, i.e. they are functional in protein biosynthesis (Fig. 5.24) and have low background activity towards native enzyme.

5.6.2
Species-specific Aminoacylation Features and Orthogonal AARS:tRNA Pairs

At the level of the interpretation of the genetic code, i.e. aminoacylation, orthogonality is defined as a lack of cross-reactivity between heterologous AARS and tRNAs with the natural host endogenous synthetases, amino acids and tRNAs. The first systematic efforts to a generate truly orthogonal (i.e. 21st) AARS:tRNA pair were attempted with *E. coli* $tRNA^{Gln}$:GlnRS as a starting point [282]. The availability of three-dimensional structures of the *E. coli* GlnRS:$tRNA_2^{Gln}$ pair [283] allowed careful inspection of all interactions in the system. This enabled the design of several mutant suppressor tRNAs able to serve as adapters for *in vivo* protein synthesis. The tRNAs derived in this way were indeed not substrates for any endogenous AARS, including GlnRS that provides an important orthogonal component for one nontoxic *in vivo* aminoacylation system. Unfortunately, parallel attempts to design an orthogonal GlnRS did not yield a mutant enzyme capable of charging orthogonal tRNA better than the native enzyme [284]. Thus, efforts to create an *E. coli* GlnRS:$tRNA_2^{Gln}$ orthogonal pair even when heterologs of $tRNA_2^{Gln}$ from yeast were included, had only limited success, since no mutant yeast GlnRS could be evolved to efficiently charge orthogonal tRNA. There are two general difficulties in any attempt to create 21 AARS:tRNA pairs in the context of a living cell. First, such designed enzymes will selectively drift from living cells due to their redundancy; they can be maintained only under experimentally imposed selective pressure. Second, there is the problem specifying which component from the remaining tRNAs and enzymes from the cellular pool directly interfere with the engineered enzyme [19]. For example, RajBhandary and coworkers [285] initially found that import of yeast TyrRS into *E. coli* is lethal since this enzyme misayclates *E. coli* $tRNA^{Pro}$ with Tyr. Therefore, yeast TyrRS was modified so that it would no longer mischarge the *E. coli* $tRNA^{Pro}$. Only a mutant (generated by error-prone polymerase chain reaction) with very low rates of misacylation of *E. coli* $tRNA^{Pro}$ could serve as an orthogonal AARS.

The development of orthogonal AARS:tRNA pairs proved to be very successful when both the components, i.e. suppressor tRNA and AARS, were imported from another organism [281]. It is now clear that exploitation of the species specificity of aminoacylation is the most promising route to design orthogonal pairs. For example, after analysis of biochemical data available for $tRNA^{Tyr}$:TyrRS from a variety of organisms, Wang and coworkers [286] found that *M. jannaschii* $tRNA_{CUA}{}^{Tyr}$:TyrRS might serve as a valuable orthogonal pair for recombinantly expressed target proteins in *E. coli*. The basis for orthogonality in the Tyr systems can be derived from fundamentally different modes of recognition of $tRNA^{Tyr}$ by TyrRS in Archaea/eukaryote and eubacteria [287] (Fig. 5.24A). In this way, cross-aminoacylation in such a hybrid translation system is not possible. In addition, *M. jannaschii* TyrRS does not have any documented editing mechanisms and its cognate $tRNA^{Tyr}$ can be efficiently transformed into suppressor tRNA without a high decline in aminoacylation efficiency [82]. Finally, the *E. coli* host proved to be remarkably tolerant toward upon the import of such a pair into its cytoplasm. Obviously, species-specific features of aminoacylation (especially differences between eubacteria and eukaryotes versus Archaea) provide the best material for the design of orthogonal pairs. The AARS from mammalian cells are often poorly expressed in *E. coli*; in contrast, their counterparts for Archaea are usually efficiently expressed in active form in this bacterium [86].

5.6.3
Orthogonal AARS:tRNA Pairs in *E. coli*

Chemical aminoacylations are difficult and complex experiments, and *in vitro* expression platforms still suffer from low yields in spite of the use of improved cell-free systems for preparative-scale synthesis of proteins carrying amino acid analogs. In addition, the suppressor tRNAs selected should be purified readily and in sufficiently large enough quantities, while *in vitro* transcripts lack some base modifications which are required for maximum activity and specificity of tRNAs. All these certainly sped up efforts for the *in vivo* transfer of suppression-based methodologies. Although efforts to create an *E. coli*-based efficient GlnRS:$tRNA_2^{Gln}$ orthogonal pair failed, these experiments yielded a methodological breakthrough in the development of an efficient system for positive/negative selection of mutant AARS enzymes and tRNAs [19] (Figs. 5.7 and 5.24). They also brought into focus other important aspects and issues that should be taken into consideration in host strain engineering. For example, amino acid uptake proved to be a critical issue (Section 5.1.2). Most of the translationally active noncanonical amino acids actually are on average the same size as canonical amino acids. Thus, it is not surprising that "exotic" amino acids such as *p*-phenylazophenylalanine or amino acids with strong side-chain reactivity are hardly acceptable as substrates of cellular amino acid uptake systems. Therefore, the systematic generation of libraries in order to search for cell-permeable amino acids represents an important future research avenue [288].

Extrachromosomally controlled heterologous expression systems in microorganisms are especially practical as platforms for amino acid repertoire expansion due to easy maintenance, propagation and handling. These expression systems can be programmed either for the position-specific or multiple-site mode of noncanonical amino acid translation by supplementation with additional translation components. These goals can be achieved in the context of living cells only if four basic requirements are fulfilled: (i) the availability of coding units programming the noncanonical amino acid translation, (ii) tRNAs capable of decoding these units, (iii) an enzyme capable of charging the tRNA, and (iv) both the AARS and its corresponding tRNA should be free from cross-aminoacylation and compatible with the host translation machinery and physiology [81].

Furter [289] developed the first general approach towards successful implementation of these criteria for position-specific *in vivo* incorporation of noncanonical amino acids. The Phe analog *p*-fluorophenylalanine was incorporated at position 5 as a response to the UAG coding triplet in the mRNA sequence of recombinantly expressed dihydrofolate reductase. This was achieved by importing yeast PheRS/amber suppressor $tRNA^{Phe}$ into the *E. coli* expression host. There is almost no cross-reactivity between the yeast and *E. coli* PheRS since identity elements in their tRNAs evolved in a completely different manner [268]. The *E. coli* strain used was a Phe auxotrophic one and displays resistance to *p*-fluorophenylalanine, i.e. it harbors a PheRS in the chromosome that excludes *p*-fluorophenylalanine from charging onto the cognate $tRNA^{Phe}$. For yeast PheRS, *p*-fluorophenylalanine is almost as good a substrate as Phe, i.e. both of them compete as substrates for yeast PheRS. This would lead to a "leakage" of the system, since the UAG coding triplet at position 5 of dihydrofolate reductase would be read as both Phe and *p*-fluorophenylalanine. This problem was circumvented by a careful experimental design of the fermentation, resulting in replacement of 64–75% at the amber position in dehydrofolate reductase. This was indeed the first demonstration of the successful introduction of a new redundant aminoacylation pathway and a new design of the hybrid translation system. In the context of carefully calibrated fermentation, such an expression host allows the preferential incorporation of a noncanonical amino acid over the canonical one (Section 5.1.4.2).

A step further in this direction was made by use of the *M. jannaschii* $tRNA_{CUA}{}^{Tyr}$/TyrRS orthogonal pair in *E. coli* as an expression host for position-specific incorporation of *O*-methyltyrosine into dihydrofolate reductase with yields of about 2 mg L^{-1} culture [82]. This was demonstrated in the case of GFP as well, where a variety of Tyr analogs were incorporated into the chromophore of this protein [89]. The *M. jannaschii* $tRNA_{CUA}{}^{Tyr}$/TyrRS pair imported into *E. coli* represents a technically much more advanced hybrid translation system than the one used by Furter as it is allows more stringent control of analog incorporation. In other words, orthogonality is almost fully achieved, since amber positions on the mRNA are translated with great fidelity into target proteins with minimal suppression-associated toxic effects. Finally, use of substituted analogs of aromatic amino acids as well as extended aromatic systems proved to be a great advantage for this system [86].

With such *in vivo* expression systems at hand, it remains to be seen how efficiently they will be extended for use in molecular biology and biochemistry. Limited success in experiments with GlnRS/$tRNA^{Gln}$ poses a dilemma whether all synthetase/tRNA systems are generally suited for such design. In addition, the orthogonal suppressor tRNA/synthetase pair for each specific noncanonical amino acid has to be generated through a series of complicated mutagenesis and selection cycles. It remains also to be checked whether the *in vivo* transfer resolves the general problems associated with context-dependent suppression phenomena (i.e. the problem of equal suppressability of all positions in a desired protein sequence). In comparison with the natural 21st AARS PylRS (see Section 3.10), which inserts Pyl with more that 75% efficiency [290], artificial AARS with designed substrate specificity translate UAG codons with less than 20% efficiency [291]. The *in vivo* approach also introduces additional difficulties associated with metabolic toxicity and bioavailability of the desired amino acids, i.e. their transfer into the cytoplasm.

A "consensus suppressor strategy" was developed recently to generate other efficient orthogonal pairs for *E. coli* host cells [291]. This strategy is necessary since the *M. jannaschii*-based pair is unable to decode UGA (opal) or the quadruplet, AGGA. Based on this approach, two orthogonal pairs with suppression efficiencies comparable to those from the system derived from *M. jannaschii* were developed. First, the LeuRS from *Methanobacterium thermoautotrophicum* and *Halobacterium* $tRNA^{Leu}$ were designed to serve as an orthogonal pair in *E. coli.* Second, GlnRS from achaebacterium *Pyrococcus horikoshii*, which was paired with orthogonal tRNAs derived from archeal sequences, was used to generate an amber suppression orthogonal pair for the *E. coli* host cells [292]. More recently, *E. coli* LeuRS:tRNA^{Leu} was used to generate an orthogonal pair capable of selectively incorporating *O*-methyltyrosine and α-aminoacpryllic acid in response to the amber nonsense codon in *E. coli* [193]. Other reported amber suppression orthogonal pairs for use in *E. coli* are *Saccharomyces cerevisiae* AspRS:tRNA^{Asp} [287] and mutant yeast TyrRS with tRNA^{fMet} from *E. coli* [285].

5.6.4
Orthogonal Pairs in Yeast and Mammalian Cells

Due to the large differences in both sequence identity and identity elements of tRNAs between prokaryotes and eukaryotes, most orthogonal AARS:tRNA pairs that have been reported for eukaryotes are derived from bacterial enzymes and tRNAs. This fact was also used in order to develop yeast orthogonal pairs specific for the desired noncanonical amino acids. This clever and complicated procedure enabled the identification of suitable orthogonal AARS:tRNA pairs, and evolution of the enzyme substrate specificity was reported by Schultz and coworkers (a detailed description is available in [293]). A special strain of *S. cerevisiae* that contains the transcriptional activator protein GAL4 with amber nonsense codons, together with markers for histidine and uracil auxotrophy (HIS3 and URA3), was used for this purpose. In addition, the *lacZ* reporter gene (β-galactosidase) serves in this system as an additional marker to colormetrically identify active AARS from inactive ones (similar to GFP in the bacterial system).

Positive selection provides for clones expressing active AARS:tRNA pairs as follows. Suppression of amber codons leads to the production of full-length GAL4, which in turn drives transcription of genomic GAL4 responsive HIS3, URA3 and *lacZ* reporter genes (expression of HIS3 and URA3 complements them in this strain). On the other hand, addition of 5-fluoroorotic acid, which is converted into a toxic product by URA3, results in the death of cells expressing active AARS:tRNA pairs. In the absence of the unnatural amino acid, this serves as a negative selection to remove synthetases specific for endogenous amino acids. The selected *E. coli* tRNA^{Tyr}/TyrRS amber suppressor pair is orthogonal in yeast, and enabled specific and selective introduction of various amino acids into target proteins [294].

However, suppressor *E. coli* tRNA^{Tyr} was not orthogonal in mammalian cells (due to differences in the transcription of tRNA genes in mammalians and bacteria) and thus was not suitable for transfer of the optimized AARS:tRNA pairs directly to mammalian cells [295]. To circumvent this problem, Yokoyama and coworkers [96] screened a small collection of specifically designed active site variants of *E. coli* TyrRS in a wheat germ translation system and discovered a mutant AARS that charges onto suppressor tRNA derived from *B. stearothermophilus* tRNA^{Tyr}. This orthogonal pair was used to incorporate 3-iodotyrosine into proteins in mammalian cells. Zhang and coworkers used a *B. subtilis* TrpRS:tRNA^{Trp} pair which is naturally orthogonal in mammalian cells to generate an amber suppressor pair capable of preferential incorporation of 5-hydroxytrytophan [133] in mammalian proteins (Section 5.2.1.1).

Generation of a complete set of suppression-based orthogonal AARS:tRNA pairs for decoding amber, ochre and opal nonsense codons was reported by RajBhandary and coworkers [285]. First, they designed an orthogonal pair for use in mammalian cells based on *E. coli* GlnRS and amber suppressor tRNAs derived from *E. coli* $tRNA^{Gln}$ and the mammalian initiator $tRNA^{Met}$. Later they extended this strategy to show that amber, ochre and opal suppressor tRNAs, derived from *E. coli* glutamine tRNA, can suppress related termination codons in a reporter mRNA in mammalian cells [296]. Since the activity of each suppressor tRNA was dependent upon the expression of *E. coli* GlnRS, none of the suppressor tRNAs was mischarged by any of the 20 AARS in the mammalian cytoplasm [297]. The most active suppressor tRNAs can be used in combination to simultaneously suppress two or three termination codons in target mRNA, opening in perspectives for *in vivo* position-specific/multisite incorporation [298]. However, it should always be kept in mind that that nonsense mutations in mammalian genes are often the cause of a variety of diseases [299] (e.g. muscular dystrophy, xeroderma pigmentosum, etc.) and constitutive expression of suppressor tRNAs is expected to generate adverse effects in mammalian cells. For that reason the expression of suppressor tRNA function in these cells should be tightly regulated.

5.7 Outlook and Visions

5.7.1 Coupling Reprogrammed Translation with Metabolic Engineering

5.7.1.1 Catalytic Promiscuity and Synthetic Capacity Extension of Metabolic Pathways

In terms of code engineering as a research field, the following aspects of metabolic engineering are especially attractive: (i) improvement of tailor-made protein expression and yields, (ii) extension of the product/substrate range of existing metabolic pathways, and (iii) *de novo* design of pathways capable for the efficient intracellular generation and production of noncanonical amino acids. In routine incorporation experiments the noncanonical amino acid of interest is added into the growth media and subsequently transferred into the cytoplasm by cellular uptake machinery. As an alternative to this procedure, a pathway for the biosynthesis of the desired noncanonical amino acid can be engineered, imported and integrated into the cellular metabolism. In this way host cells would be able to generate the desired amino acid from simple carbon sources. For many years, metabolic engineering has allowed to generate or design a novel endogenous supply route of substances of interest (vitamins, hormones, metabolic intermediates) using gene expression tools [300]. The traditional biotechnological engineering of Trp biosynthetic pathways mainly addressed enhanced production of its useful derivatives [301] or similar amino acids. For example, the Trp-producing mutant of *Corynebacterium glutamicum* was engineered to produce tyrosine or phenylalanine in abundance by genetic manipulation of the common aromatic amino acid biosynthetic pathway [302]. Until recently there were no reported attempts to produce elevated amounts of Trp-based noncanonical amino acid analogs and surrogates for expanded scope of protein biosynthesis (i.e. their incorporation into recombinant proteins).

The last steps in Trp biosynthesis include indole generation and its condensation with serine into L-Trp [303]. This final step is performed by Trp synthase by a well-characterized reaction mechanism that includes direct transfer of the indole intermediate between α and β subunits through a "tunnel" in the enzyme complex ("channeling effect") [304]. In fact, pyridoxal phosphate-dependent Trp synthase which catalyzes the β-substitution reaction on indole is one of the best-studied enzymatic systems in biochemistry [305]. One remarkable property of Trp synthase, essential for amino acid derivatization, is its rather broad substrate specificity – even amino acids structurally and chemically not related to Trp can be synthesized by using this enzyme [306]. Indeed, most of the indole analogs/surrogates in the "noncoding" level of the Trp genetic code presented in Fig. 5.8 (Section 5.2.1.1) are efficient substrates for this enzyme (but related amino acids are not substrates for endogenous TrpRS). Thus, in a fermentation medium provided with a variety of (natural or synthetic) indole analogs/surrogates and a controlled expression system (in the context of the SPI method), related amino acids analogs can be synthesized intracellularly and subsequently incorporated into target proteins, as was demonstrated recently [307].

A similar semisynthetic approach for the production of noncanonical amino acids by metabolic engineering of the Cys biosynthetic pathway was reported by Maier [308]. The final step in the Cys biosynthetic pathway is catalyzed by *O*-acetylserine sulfhydrylase, a pyridoxal phosphate-dependent enzyme which catalyzes the β-substitution reaction on *O*-acetylserine. The catalytic promiscuity of this enzyme is reflected in its broad substrate specificity in a similar manner to Trp synthase. This feature combined with intracellular deregulation of the Cys biosynthetic pathway enables the biosynthesis of amino acid derivatives with diverse side-chains with interesting chemical functionality. For example, fermentation media supplied with toxic substances such as azide, cyanide or triazole allowed for their biotransformation into amino acids such as azidoalanine, cyanoalanine and triazole-1-yl-alanine (Fig. 5.25, **11**, **12** and **8**). High production yields of noncanonical amino acids using this approach are reported, which opens a fairly good perspective to make a further step and couple such reprogrammed metabolic processes with reprogrammed protein translation apparatus for tailor-made protein production on an industrial scale.

5.7.1.2 Importing Natural Product Metabolic Pathways and Possible *De Novo* Design

Nature is a much richer source for such amino acid substrates than synthetic chemistry itself. Plants are the most excellent chemists among living entities due to their remarkable capacity to produce a wide variety of secondary metabolites such as alkaloids, terpenes and tannins to protect themselves from predators, para-

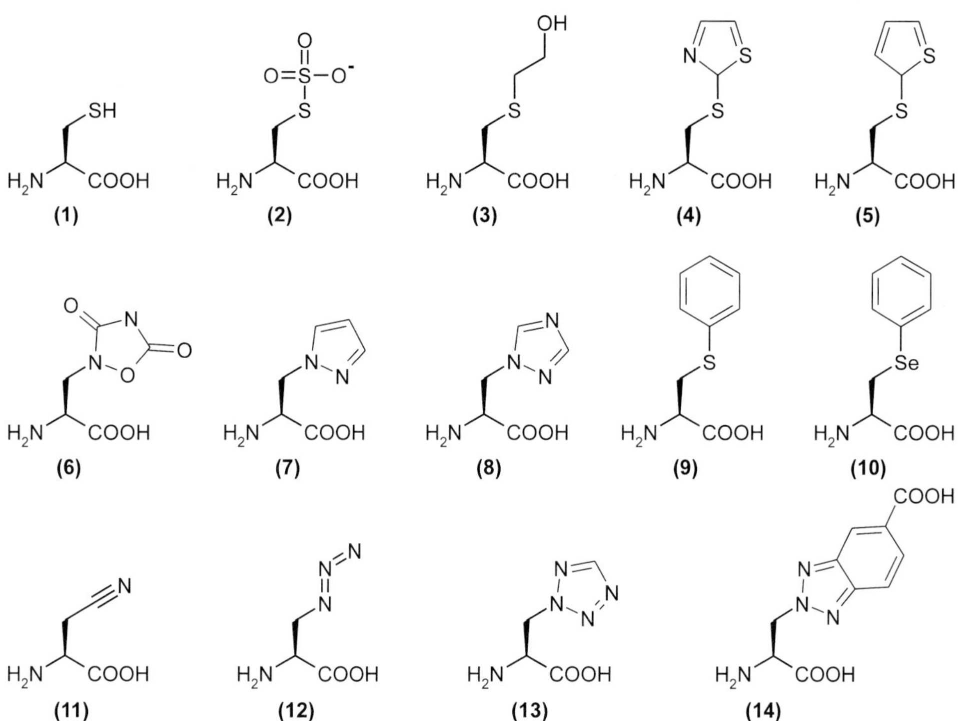

Fig. 5.25. Amino acids generated by metabolic engineering coupled with the catalytic promiscuity of *O*-acetylserine sulfur hydrase as a last enzyme in the Cys biosynthetic pathway. Canonical amino acid Cys (**1**) is natural product of the intracellular catalytic activity of this enzyme. Amino acids such as azidoalanine (**12**) (Section 5.2.1.5; Fig. 5.15) are useful for protein surface diversification by Staudinger ligation or for "click" chemistry of proteins (see Section 7.8.2), cyanoalanine (**11**) might serve as biophysical marker (IR probe), whereas *S*-phenylcysteine (**9**) is an interesting building block in inhibitor design for AIDS therapy. Common names of other amino acids are: *S*-sulfocysteine (**2**), *S*-hydroxyethylcysteine (**3**), *S*-thiazole-2-yl-Cys (**4**), *S*-thien-2-yl-Cys (**5**), 1,2,4-oxadiazolidinedionyl-alanine (**6**), pyrazole-1-yl-alanine (**7**), triazole-1-yl-alanine (**8**), tetrazole-2-yl-alanine (**13**), *S*-phenyl-Cys (**9**), phenyl-selenocysteine (**10**) and 5-carboxybenzotria-zole-2-yl-alanine (**14**). (Data taken from [308].)

sites and infection by viruses. Therefore, it is not surprising that among more than 700 known biogenic amino acids, at least 300 originate from plants [205, 309]. These represent a large natural reservoir of potential substrates for protein biosynthesis, i.e. for entry into the genetic code [28, 310]. It should be kept in mind that all these pathways are genetically encoded and metabolically regulated. Therefore, by means of molecular genetics it should be possible to supply entire cells with such heterologous metabolic pathways, making them capable of generating the desired product. For example, erythromycin and carotenoid pathways have been introduced into *E. coli* [311, 312, 313], which enable production of these substances from simple carbon sources.

A step further would be to combine these possibilities with the reprogrammed translation apparatus. The combination of such metabolic engineering and *in vivo* incorporation of noncanonical amino acids was indeed demonstrated by Mehl and coworkers [314]. Gene loci (*papA*, *papB* and *papC*) encoding production of *p*-aminophenylalanine (pAF) as a metabolic intermediate identified in *Streptomyces venezuele* were "borrowed" and put into *E. coli* that harbors the imported and engineered *M. jannashii* $tRNA_{CUA}$Tyr/TyrRS 21st pair specific for pAF. Genes for these enzymes were imported by a low-copy plasmid into *E. coli*, making this microbe capable of the intracellular production of pAF from chorismate. The intracellularly generated pAF that does not interfere with cell metabolism and physiology was incorporated into proteins by the use of a corresponding orthogonal pair. It is reasonable to expect that future engineering experiments will yield cells capable of biosynthesizing and subsequently translating other interesting amino acids containing methylated, acetylated, alkylated, glycosylated, etc., side-chains.

It is conceivable that heterologous target gene expression systems which utilize expression hosts supplied with orthogonal components will be considerably improved by importing already existing or designed metabolic pathways. In addition to bacteria, yeast and insects, even mammalian cells might offer another option as potential hosts for the heterologous production of amino acids and its coupling with reprogrammed translation (Fig. 5.26). The already discussed natural product amino acids that have been identified so far certainly represent only a small fraction of all chemically possible amino acids. This diversity can be further expanded to generate a pool of amino acids of anthropogenic origin whose chemical and structural diversity transcends those found in nature. For example, by combination of rational pathway assembly with directed evolution one might construct novel amino acid biosynthetic pathways either in prokaryotic or eukaryotic cells [315, 316]. There should be no doubt that the combination of all these possibilities with reprogrammed translation in one unique system is, in fact, the future avenue toward efficient industrial production of tailor-made proteins.

5.7.2
Shuttle Orthogonal Pair and Hybrid Translation Systems with Codon Capture

Life as we know it, with its conserved code, radiated from the last common ancestor in many directions and code invasion with novel amino acids could in the best case be achieved at species-specific level. At this stage of our experimental expertise, it is difficult to imagine how such invasion could spread out through all life kingdoms. Therefore, the most relevant questions for code engineering are as follows. Is it possible to insert a new amino acid into the existing repertoire despite a complex genome encoding thousands of highly evolved proteins? Are there general mechanisms (other than preprogrammed re-coding) by which such "amino acid–codon takeover" can be efficiently made? Which organisms are best suitable as research models for efficient artificial invasion of the existing code with novel building blocks? For example, the resemblance between *Mycoplasma* (with the smallest the genome among all free-living organisms) and mitochondria is well docu-

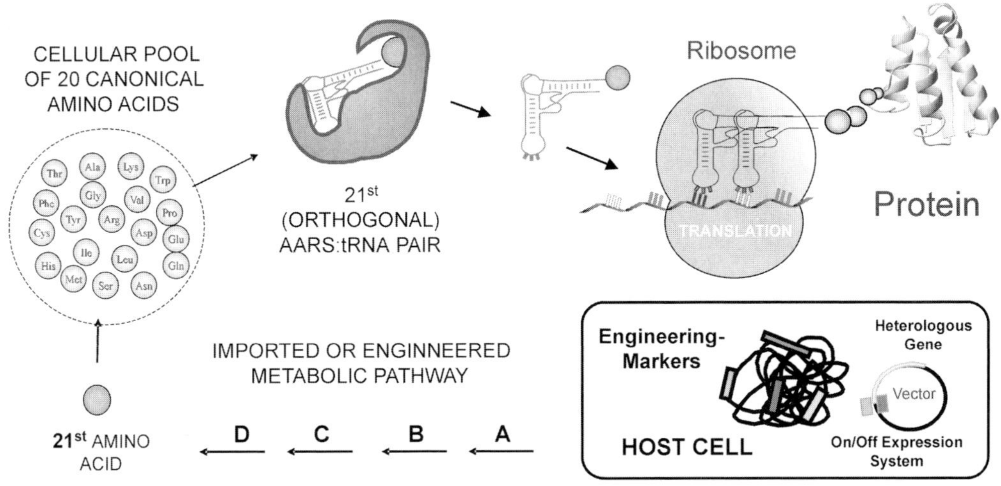

Fig. 5.26. Coupling metabolic engineering with reprogrammed protein translation. Combined together, natural product chemistry and code engineering could lead to larger-scale screening for suitable amino acids as "candidates" for entry into the genetic code from the intermediary metabolism of various species. The natural biosynthetic pathways can be imported and integrated into the metabolism of host cells in order to generate intracellular production of a desired noncanonical amino acid (which should be an exclusive substrate for the 21st or orthogonal AARS:tRNA pair). These pathways can be attached to the existing ones in host cells and further modified, reduced, extended or optimized for balanced synthesis of the desired substrate. However, the ultimate goal is *in vivo* evolution of novel synthetic pathways capable of generating substrate diversity to an extent far the beyond natural pathways.

mented (see Section 4.8.1). Could it be used as a favorable feature for code expansion or even for the design altered codes with overwhelmingly new codon–amino acids associations? Are there organisms with fully unassigned codons? What about artificial (e.g. virus-based) replicative systems? What would be the best starting point when attempting to design a novel "anthropogenic" code based on an entirely new set of chemicals?

On the other hand, it is reasonable to expect that practical applications either in research or industry will dictate to a great extent the current and future development and trends in genetic code engineering. The main focus will be on the development of efficient *in vitro* or *in vivo* hybrid translation systems, capable of producing elevated levels of tailor-made proteins. The most conceivable way to reach high levels of production of tailor-made therapeutic proteins or novel materials would be the experimental design of specialized prokaryotic or eukaryotic cells capable of synthesizing protein variants with a high fidelity. Metabolism, bioenergetics, synthesis and supply routes, as well as amino acid pools of such cells, should be controllable, tightly regulated and properly balanced. Obviously, coupling metabolic engineering with code engineering in the context of systems biology has a great

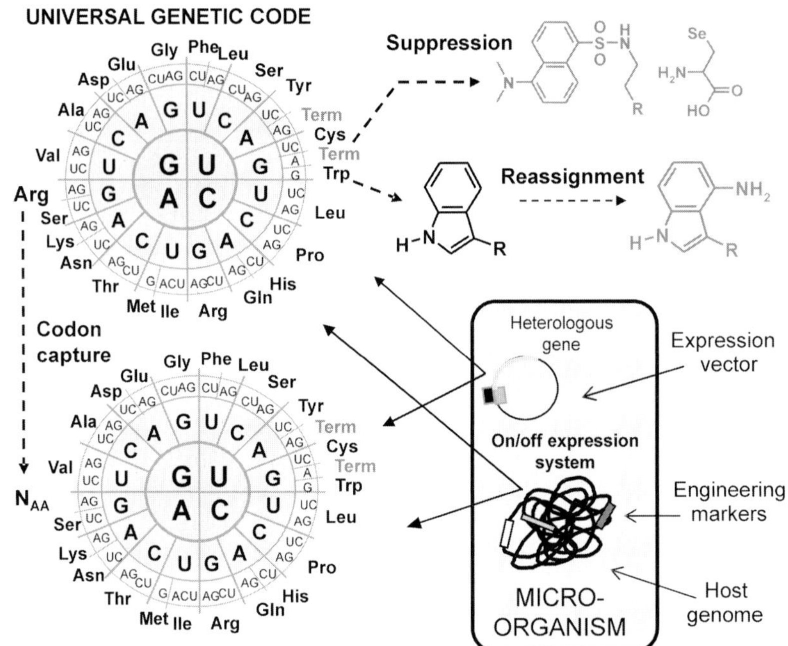

Fig. 5.27. Codon capture as a vehicle to expand the coding capacities of the universal genetic code. Microorganisms used for genetic code engineering to date are, in fact, forced to make read-through of coding units (sense, nonsense and nontriplet), named reassignments or suppressions, mainly at the level of a single protein and rarely at the proteome-wide level. To replace this ambiguous re-coding, the existing genetic code should be tailored (or target engineered) with codon capture, which is defined as permanent reassignments or novel assignment of particular coding units either at the level of the expression system or at the level of a whole living cell (i.e. proteome-wide). The Arg coding triplets AGA and AGG were chosen as examples for permanent reassignment (codon capture) to a hypothetical novel amino acid (NAA). However, any other combination is also conceivable as well. (Reproduced from [3] with permission.)

future. However, *in vivo* expression of individual tailor-made proteins as tags in mammalian cells would be also highly desirable. In this case, the amount of labeled proteins does not necessarily need to be elevated, since minimal amounts of the tailored protein can easily be detected, e.g. by highly sensitive fluorescence assays. Finally, the capability to shuttle such systems between cells of different life kingdoms (archeal, eubacterial and eukaryotic) should be of a great advantage. Indeed, a recent report about the use of the *M. jannaschii* $tRNA_{CUA}^{Tyr}$/TyrRS orthogonal system in yeast, *B. subtilis* $tRNA_{CUA}^{Trp}$/TrpRS or the introduction of the *E. coli* GlnRS/$tRNA^{Gln}$ orthogonal pair into mammalian cells is only the first step in this direction [86, 279].

Engineering of the genetic code at these initial phases of field development clearly shows that it is possible to insert new amino acids into the existing repertoire despite a complex genome encoding thousands of highly evolved proteins. In most cases, this was readily possible since the goal was not to produce "unnatural organisms" [317], but rather to generate a reliable and efficient system which works with an expanded amino acid repertoire [3, 62]. This system is compatible with host translation machinery which serves only as a platform for codon-reassignment experiments. However, currently available methods for amino acid repertoire expansion are characterized by ambiguous read-throughs, suppressions, codon reassignments, etc., which are achieved by changes in the genotype and physiology of the living cells, and by controlled experimental conditions. This is, in fact, possible due to the malleability of the genetic code [318]. However, the goal is permanent reassignments or novel assignments of the coding units, i.e. codon capture [319], either at the level of the expression system or at the level of the living cell (Fig. 5.27, see also Section 6.4.1).

This codon capture should be (i) nonlethal and (ii) an intrinsic property of the system [3]. Expression host cells with such nonlethal codon capture (e.g. an organism without seriously impaired viability upon insertion of a novel amino acid in its repertoire) would be the most advanced system to expand the coding capacities of the universal genetic code *in vivo*. Codon capture can be achieved by permanent reassignment of coding units in the frame of the existing structure of the universal genetic code. Another, much more challenging, venture is to introduce novel coding units by the generation of novel nucleic acid base pairs, as discussed earlier. In both cases, protein translation would be enriched with truly novel canonical amino acids. Such codon capture in an organism with a tailor-made (or target-engineered) genetic code and with an intrinsically expanded amino acid repertoire is one of the most challenging aspects in future experimental works in the field of the code engineering.

References

1 Ayala, F. J. (1978). Mechanisms of evolution. *Scientific American* **239**, 56–69.
2 Wang, P., Fichera, A., Kumar, K. and Tirrell, D. A. (2004). Alternative translations of a single RNA message: an identity switch of (2*S*,3*R*)-4,4,4-trifluorovaline between valine and isoleucine codons. *Angewandte Chemie International Edition* **43**, 3664–3666.
3 Budisa, N. (2004). Prolegomena to future efforts on genetic code engineering by expanding its amino acid repertoire. *Angewandte Chemie International Edition* **43**, 3387–3428.
4 Copley, S. D. (2003). Enzymes with extra talents: moonlighting functions and catalytic promiscuity. *Current Opinion in Chemical Biology* **7**, 265–272.
5 Richmond, M. H. (1962). Effect of amino acid analogues on growth and protein synthesis in microorganisms. *Bacteriological Reviews* **26**, 398–420.
6 Vaughan, M. and Steinberg, D. (1959). The specificity of protein synthesis. *Advances in Protein Chemistry* **XIV**, 116–173.
7 Fruton, J. S. (1963). Chemical aspects of protein synthesis. In *The Proteins: Composition, Structure, and Function* (Neurath, H., ed.), vol. I, pp. 189–310. Academic Press, New York.
8 Trudinger, P. A. and Cohen, G. N.

(1956). Effect of 4-methyltryptophan on growth and enzyme systems of *Escherichia coli*. *Biochemical Journal* **62**, 488–491.

9 Rabinovitz, M., Olson, M. E. and Greenberg, D. M. (1954). Independent antagonism of amino acid incorporation into protein. *Journal of Biological Chemistry* **210**, 837–849.

10 Fowden, L., Lewis, D. and Tristam, H. (1967). Toxic amino acids – their action as antimetabolites. *Advances in Enzymology and Related Areas of Molecular Biology* **29**, 89–163.

11 Wheatley, D. N., Inglis, M. S. and Malone, P. C. (1986). The concept of the intracellular amino acid pool and its relevance in the regulation of protein metabolism, with particular reference to mammalian cells. *Current Topics in Cellular Regulation* **28**, 107–182.

12 Kirk, K. L. (1991). *Biochemistry of Halogenated Organic Compounds*. Plenum, New York.

13 Anfinsen, C. B. and Corley, L. G. (1969). An active variant of staphylococcal nuclease containing norleucine in place of methionine. *Journal of Biological Chemistry* **244**, 5149–5152.

14 Gilles, A. M., Marliere, P., Rose, T., Sarfati, R., Longin, R., Meier, A., Fermandjian, S., Monnot, M., Cohen, G. N. and Barzu, O. (1988). Conservative replacement of methionine by norleucine in *Escherichia coli* adenylate kinase. *Journal of Biological Chemistry* **263**, 8204–8209.

15 Bogosian, G., Violand, B. N., Dorwardking, E. J., Workman, W. E., Jung, P. E. and Kane, J. F. (1989). Biosynthesis and incorporation into protein of norleucine by *Escherichia coli*. *Journal of Biological Chemistry* **264**, 531–539.

16 Budisa, N., Moroder, L. and Huber, R. (1999). Structure and evolution of the genetic code viewed from the perspective of the experimentally expanded amino acid repertoire *in vivo*. *Cellular and Molecular Life Sciences* **55**, 1626–1635.

17 Janecek, J. and Rickenberg, H. V. (1964). Incorporation of beta-2-thienylalanine into beta-galactosidase of *Escherichia coli*. *Biochimica et Biophysica Acta* **81**, 108–112.

18 Cowie, D. B., Cohen, G. N., Bolton, E. T. and DeRobichon-Szulmajster, R. H. (1959). Amino acid analog incorporation into bacterial proteins. *Biochimica et Biophysica Acta* **34**, 39–46.

19 Liu, D. R. and Schultz, P. G. (1999). Progress toward the evolution of an organism with an expanded genetic code. *Proceedings of the National Academy of Sciences of the USA* **96**, 4780–4785.

20 Buechter, D. D., Paolella, D. N., Leslie, B. S., Brown, M. S., Mehos, K. A. and Gruskin, E. A. (2003). Co-translational incorporation of *trans*-4-hydroxyproline into recombinant proteins in bacteria. *Journal of Biological Chemistry* **278**, 645–650.

21 Kim, W. Y., George, A., Evans, M. and Conticello, V. P. (2004). Cotranslational incorporation of a structurally diverse series of proline analogues in an *Escherichia coli* expression system. *ChemBioChem* **5**, 928–936.

22 Wood, J. M. (1988). Proline porters effect the utilization of proline as nutrient or osmoprotectant for bacteria. *Journal of Membrane Biology* **106**, 183–202.

23 Gowrishankar, J. (1985). Identification of osmoresponsive genes in *Escherichia coli*: evidence for participation of potassium and proline transport systems in osmoregulation. *Journal of Bacteriology* **164**, 434–445.

24 Tuite, M. F. and Santos, M. A. S. (1996). Codon reassignment in *Candida* species: an evolutionary conundrum. *Biochimie* **78**, 993–999.

25 Norbeck, J. and Blomberg, A. (1997). Metabolic and regulatory changes associated with growth of *Saccharomyces cerevisiae* in 1.4 M NaCl. *Journal of Biological Chemistry* **272**, 5544–5554.

26 Weissman, A. and Koe, B. K. (1967). M-fluorotyrosine convulsions and mortality – relationship to catechola-

mine and citrate metabolism. *Journal of Pharmacology and Experimental Therapeutics* **155**, 135–144.

27 O'Hagan, D., Schaffrath, C., Cobb, S. L., Hamilton, J. T. G. and Murphy, C. D. (2002). Biosynthesis of an organofluorine molecule – a fluorinase enzyme has been discovered that catalyses carbon–fluorine bond formation. *Nature* **416**, 279–279.

28 Rosenthal, G. (1982). *Plant Non-protein Amino and Imino Acids. Biological, Biochemical and Toxicological Properties.* Academic Press, New York.

29 Richmond, M. H. (1959). Incorporation of canavanine by *Staphylococcus aureus. Biochemical Journal* **73**, 261–264.

30 DeMarco, C., Busiello, V., Digirolamo, M. and Cavallini, D. (1977). Selenaproline and protein-synthesis. *Biochimica et Biophysica Acta* **478**, 156–166.

31 Ibba, M. and Söll, D. (1999). Quality control mechanisms during translation. *Science* **286**, 1893–1897.

32 Koide, H., Yokoyama, S., Katayama, Y., Muto, Y., Kigawa, T., Kohno, T., Takusari, H., Oishi, M., Takahashi, S., Tsukumo, K., Sasaki, T., Miyake, T., Fuwa, T., Kawai, G. and Miyazawa, T. (1994). Receptor-binding affinities of human epidermal growth factor variants having unnatural amino acid residues in position-23. *Biochemistry* **33**, 7470–7476.

33 Bae, J. H., Rubini, M., Jung, G., Wiegand, G., Seifert, M. H. J., Azim, M. K., Kim, J. S., Zumbusch, A., Holak, T. A., Moroder, L., Huber, R. and Budisa, N. (2003). Expansion of the genetic code enables design of a novel "gold" class of green fluorescent proteins. *Journal of Molecular Biology* **328**, 1071–1081.

34 Budisa, N., Minks, C., Alefelder, S., Wenger, W., Dong, F. M., Moroder, L. and Huber, R. (1999). Toward the experimental codon reassignment *in vivo*: protein building with an expanded amino acid repertoire. *FASEB Journal* **13**, 41–51.

35 Minks, C. (1999). In vivo *Einbau nicht-natürlicher Aminosäuren in rekombinante Proteine.* PhD Thesis. Technische Universität, München.

36 Neidhardt, F. C. (1987). Escherichia coli *and* Salmonella typhimurium*: Cellular and Molecular Biology.* American Society of Microbiology, Washington, DC.

37 Minks, C., Alefelder, S., Moroder, L., Huber, R. and Budisa, N. (2000). Towards new protein engineering: *in vivo* building and folding of protein shuttles for drug delivery and targeting by the selective pressure incorporation (SPI) method. *Tetrahedron* **56**, 9431–9442.

38 Cowie, D. B. and Cohen, G. N. (1957). Biosynthesis by *Escherichia coli* of active altered proteins containing selenium instead of sulfur. *Biochimica et Biophysica Acta* **26**, 252–261.

39 Rennert, O. M. and Anker, H. S. (1963). On incorporation of 5′,5′,5′-trifluoroleucine into proteins of *Escherichia coli. Biochemistry* **2**, 471–476.

40 Fenster, E. D. and Anker, H. S. (1969). Incorporation into polypeptide and charging on transfer ribonucleic acid of amino acid analog 5′,5′,5′-trifluoroleucine by leucine auxotrophs of *Escherichia coli. Biochemistry* **8**, 269–274.

41 Rennert, O. M. and Anker, H. S. (1964). Effect of 5′,5′,5′-trifluoroleucine on number of mouse leukaemias. *Nature* **203**, 1256–1257.

42 Budisa, N., Pipitone, O., Siwanowicz, I., Rubini, M., Pal, P. P., Holak, T. A. and Gelmi, M. L. (2004). Efforts toward the design of "Teflon" proteins: *in vivo* translation with trifluorinated leucine and methionine analogues. *Chemistry and Biodiversity* **1**, 1465–1475.

43 Hendrickson, W. A. and Ogata, C. M. (1997). Phase determination from multiwavelength anomalous diffraction measurements. *Macromolecular Crystallography A* **276**, 494–523.

44 Wong, J. T. F. (1983). Membership mutation of the genetic code – loss of fitness by tryptophan. *Proceedings of the National Academy of Sciences of the USA Biological Sciences* **80**, 6303–6306.

81 Wang, L. and Schultz, P. G. (2002). Expanding the genetic code. *Chemical Communications*, 1–11.

82 Wang, L., Brock, A., Herberich, B. and Schultz, P. G. (2001). Expanding the genetic code of *Escherichia coli*. *Science* **292**, 498–500.

83 Paddon, C. J. and Hartley, R. W. (1987). Expression of *Bacillus amyloliquefaciens* extracellular ribonuclease (barnase) in *Escherichia coli* following an inactivating mutation. *Gene* **53**, 11–19.

84 Axe, D. D., Foster, N. W. and Fersht, A. R. (1996). Active barnase variants with completely random hydrophobic cores. *Proceedings of the National Academy of Sciences of the USA* **93**, 5590–5594.

85 Magliery, T. J., Pastrnak, M., Anderson, J. C., Santoro, S. W., Herberich, B., Meggers, E., Wang, L. and Schultz, P. G. (2003). *In vitro* tools and *in vivo* engineering: incorporation of unnatural amino acids into proteins. In *Translation Mechanisms* (Lapointe, J. and Brakier-Gingras, L., eds), pp. 95–114. Landes Biosciences, George-town, TX.

86 Wang, L. and Schultz, P. G. (2005). Expanding the genetic code. *Angewandte Chemie International Edition* **44**, 34–66.

87 Santoro, S. W., Wang, L., Herberich, B., King, D. S. and Schultz, P. G. (2002). An efficient system for the evolution of aminoacyl-tRNA synthetase specificity. *Nature Biotechnology* **20**, 1044–1048.

88 Pastrnak, M. and Schultz, P. G. (2001). Phage selection for site-specific incorporation of unnatural amino acids into protein *in vivo*. *Bioorganic and Medicinal Chemistry* **9**, 2373–2379.

89 Wang, L., Xie, J. M., Deniz, A. A. and Schultz, P. G. (2003). Unnatural amino acid mutagenesis of green fluorescent protein. *Journal of Organic Chemistry* **68**, 174–176.

90 Zhang, Z. W., Wang, L., Brock, A. and Schultz, P. G. (2002). The selective incorporation of alkenes into proteins in *Escherichia coli*. *Angewandte Chemie International Edition* **41**, 2840–2842.

91 Wang, L., Zhang, Z. W., Brock, A. and Schultz, P. G. (2003). Addition of the keto functional group to the genetic code of *Escherichia coli*. *Proceedings of the National Academy of Sciences of the USA* **100**, 56–61.

92 Chin, J. W., Santoro, S. W., Martin, A. B., King, D. S., Wang, L. and Schultz, P. G. (2002). Addition of *p*-azido-L-phenylalanine to the genetic code of *Escherichia coli*. *Journal of the American Chemical Society* **124**, 9026–9027.

93 Chin, J. W., Martin, A. B., King, D. S., Wang, L. and Schultz, P. G. (2002). Addition of a photocross-linking amino acid to the genetic code of *Escherichia coli*. *Proceedings of the National Academy of Sciences of the USA* **99**, 11020–11024.

94 Wang, L., Brock, A. and Schultz, P. G. (2002). Adding L-3-(2-naphthyl)alanine to the genetic code of *E. coli*. *Journal of the American Chemical Society* **124**, 1836–1837.

95 Hamano-Takaku, F., Iwama, T., Saito-Yano, S., Takaku, K., Monden, Y., Kitabatake, M., Söll, D. and Nishimura, S. (2000). A mutant *Escherichia coli* tyrosyl-tRNA synthetase utilizes the unnatural amino acid azatyrosine more efficiently than tyrosine. *Journal of Biological Chemistry* **275**, 40324–40328.

96 Sakamoto, K., Hayashi, A., Sakamoto, A., Kiga, D., Nakayama, H., Soma, A., Kobayashi, T., Kitabatake, M., Takio, K., Saito, K., Shirouzu, M., Hirao, I. and Yokoyama, S. (2002). Site-specific incorporation of an unnatural amino acid into proteins in mammalian cells. *Nucleic Acids Research* **30**, 4692–4699.

97 Tang, Y. and Tirrell, D. A. (1999). Stabilization of leucine zipper coiled coils by introduction of trifluoroleucine. *Abstracts of Papers of the American Chemical Society* **218**, 416.

98 Calendar, R. and Berg, P. (1966). Catalytic properties of tyrosyl ribonucleic acid synthetases from

Escherichia coli and *Bacillus subtilis*. *Biochemistry* **5**, 1690– and.

99 Kobayashi, T., Nureki, O., Ishitani, R., Yaremchuk, A., Tukalo, M., Cusack, S., Sakamoto, K. and Yokoyama, S. (2003). Structural basis for orthogonal tRNA specificities of tyrosyl-tRNA synthetases for genetic code expansion. *Nature Structural Biology* **10**, 425–432.

100 Jakubowski, H. and Goldman, E. (1992). Editing of errors in selection of amino acids for protein synthesis. *Microbiological Reviews* **56**, 412–429.

101 Freist, W. and Gauss, D. H. (1995). Lysyl-tRNA synthetase. *Biological Chemistry* **376**, 451–472.

102 Dayhoff, M. O. (1972). *Atlas of Protein Sequence and Structure 5*. National Biomedical Research Foundation, Washington, DC.

103 Lakowitz, J. R. (1999). *Protein Fluorescence*, 2nd edn. Kluwer, New York.

104 Ross, J. B. A., Szabo, A. G. and Hogue, C. W. V. (1997). Enhancement of protein spectra with tryptophan analogs: fluorescence spectroscopy of protein–protein and protein–nucleic acid interactions. *Fluorescence Spectroscopy* **278**, 151–190.

105 Pal, P. P. and Budisa, N. (2004). Designing novel spectral classes of proteins with tryptophan-expanded genetic code. *Biological Chemistry* **385**, 893–904.

106 Senear, D. F., Mendelson, R. A., Stone, D. B., Luck, L. A., Rusinova, E. and Ross, J. B. A. (2002). Quantitative analysis of tryptophan analogue incorporation in recombinant proteins. *Analytical Biochemistry* **300**, 77–86.

107 Ross, J. B. A., Rusinova, E., Luck, L. A. and Rousslang, K. W. (2000). Spectral enhancement of proteins by *in vivo* incorporation of tryptophan analogues. In *Trends in Fluorescence Spectroscopy* (Lakowitz, J. R., ed.), vol. 6, pp. 17–42. Plenum, New York.

108 Wong, C. Y. and Eftink, M. R. (1998). Incorporation of tryptophan analogues into staphylococcal nuclease, its V66W mutant, and delta 137–149 fragment: spectroscopic studies. *Biochemistry* **37**, 8938–8946.

109 Wolfenden, R. and Radzicka, A. (1986). How hydrophilic is tryptophan. *Trends in Biochemical Sciences* **11**, 69–70.

110 Cornette, J., Cease, K. B., Margalit, H., Spouge, J. L., Berzofsky, J. A. and DeLisi, C. (1987). Hydrophobicity scales and computational techniques for detecting amphipathic structures in proteins. *Journal of Molecular Biology* **195**, 659–685.

111 Trinquier, G. and Sanejouand, Y. H. (1998). Which effective property of amino acids is best preserved by the genetic code? *Protein Engineering* **11**, 153–169.

112 Dougherty, D. A. (1996). Cation interactions in chemistry and biology: a new view of benzene, Phe, Tyr, and Trp. *Science* **271**, 163–168.

113 Huber, R., Romisch, J. and Paques, E. (1990). The crystal and molecular structure of human annexin V, an anticoagulant protein that binds to calcium and membranes. *EMBO Journal* **9**, 3867–3874.

114 Golbik, R., Fischer, G. and Fersht, A. R. (1999). Folding of barstar C40A/C82A/P27A and catalysis of the peptidyl-prolyl *cis*/*trans* isomerization by human cytosolic cyclophilin (Cyp18). *Protein Science* **8**, 1505–1514.

115 Henikoff, S. and Henikoff, J. G. (1992). Amino acid substitution matrices from protein blocks. *Proceedings of the National Academy of Sciences of the USA* **89**, 10915–10919.

116 Crick, F. H. C. (1968). Origin of genetic code. *Journal of Molecular Biology* **38**, 367–379.

117 Doublie, S., Bricogne, G., Gilmore, C. and Carter, C. W. (1995). Tryptophanyl-transfer-RNA synthetase crystal-structure reveals an unexpected homology to tyrosyl-transfer-RNA synthetase. *Structure* **3**, 17–31.

118 Yu, Y. D., Liu, Y. Q., Shen, N., Xu, X., Xu, F., Jia, J., Jin, Y. X., Arnold, E. and Ding, J. P. (2004). Crystal structure of human tryptophanyl-tRNA synthetase catalytic fragment – insights into substrate recognition,

tRNA binding, and angiogenesis activity. *Journal of Biological Chemistry* **279**, 8378–8388.

119 Carter, C. W. (2004). Tryptophanyl-tRNA synthetase. In *Aminoacyl-tRNA Synthetases* (Ibba, M., Francklyn, C. and Cusack, S., eds), pp. 99–110. Landes Bioscience, Georgetown, TX.

120 Sever, S., Rogers, K., Rogers, M. J., Carter, C. and Söll, D. (1996). *Escherichia coli* tryptophanyl-tRNA synthetase mutants selected for tryptophan auxotrophy implicate the dimer interface in optimizing amino acid binding. *Biochemistry* **35**, 32–40.

121 Budisa, N., Alefelder, S., Bae, J. H., Golbik, R., Minks, C., Huber, R. and Moroder, L. (2001). Proteins with beta-(thienopyrrolyl)alanines as alternative chromophores and pharmaceutically active amino acids. *Protein Science* **10**, 1281–1292.

122 Loidl, G., Musiol, H. J., Budisa, N., Huber, R., Poirot, S., Fourmy, D. and Moroder, L. (2000). Synthesis of beta-(1-azulenyl)-L-alanine as a potential blue-colored fluorescent tryptophan analog and its use in peptide synthesis. *Journal of Peptide Science* **6**, 139–144.

123 Bae, J. H., Alefelder, S., Kaiser, J. T., Friedrich, R., Moroder, L., Huber, R. and Budisa, N. (2001). Incorporation of beta-selenolo[3,2-*b*]pyrrolyl-alanine into proteins for phase determination in protein X-ray crystallography. *Journal of Molecular Biology* **309**, 925–936.

124 Schlesinger, S. and Schlesinger, M. J. (1967). Effect of amino acid analogues on alkaline phosphatase formation in *Escherichia coli K-12*: substitution of triazolealanine for histidine. *Journal of Biological Chemistry* **242**, 3369–3378.

125 Xu, Z. J., Love, M. L., Ma, L. Y. Y., Blum, M., Bronskill, P. M., Bernstein, J., Grey, A. A., Hofmann, T., Camerman, N. and Wong, J. T. F. (1989). Tryptophanyl-tRNA synthetase from *Bacillus subtilis*: characterization and role of hydrophobicity in substrate recognition. *Journal of Biological Chemistry* **264**, 4304–4311.

126 Budisa, N., Pal, P. P., Alefelder, S., Birle, P., Krywcun, T., Rubini, M., Wenger, W., Bae, J. H. and Steiner, T. (2004). Probing the role of tryptophans in *Aequorea victoria* Green Fluorescent Proteins with an expanded genetic code. *Biological Chemistry* **385**, 191–202.

127 Kidder, G. W. and Dewey, V. C. (1955). Inhibition of tetrahymena by a new tryptophan analog. *Biochimica et Biophysica Acta* **17**, 288–298.

128 Pardee, A. B., Shore, V. G. and Prestidge, L. S. (1956). Incorporation of azatryptophan into proteins of bacteria and bacteriophage. *Biochimica et Biophysica Acta* **21**, 406–407.

129 Brawerman, G. and Ycas, M. (1957). Incorporation of the amino acid analog tryptazan into the protein of *Escherichia coli*. *Archives of Biochemistry and Biophysics* **68**, 112–117.

130 Pratt, E. A. and Ho, C. (1974). Incorporation of fluorotryptophans into protein in *Escherichia coli*, and their effect on induction of beta-galactosidase and lactose permease. *Federation Proceedings* **33**, 1463–1463.

131 Budisa, N., Rubini, M., Bae, J. H., Weyher, E., Wenger, W., Golbik, R., Huber, R. and Moroder, L. (2002). Global replacement of tryptophan with aminotryptophans generates non-invasive protein-based optical pH sensors. *Angewandte Chemie International Edition* **41**, 4066–4069.

132 Rubini, M. (2004). Noncanonical amino acid in proteins. *PhD Thesis*, TU Munich.

133 Zhang, Z. W., Alfonta, L., Tian, F., Bursulaya, B., Uryu, S., King, D. S. and Schultz, P. G. (2004). Selective incorporation of 5-hydroxytryptophan into proteins in mammalian cells. *Proceedings of the National Academy of Sciences of the USA* **101**, 8882–8887.

134 Betts, M. J. and Russell, R. B. (2003). Amino acid properties and consequences of substitutions. In *Bioinformatics for Geneticists* (Barnes, M. R. and Gray, I. C., eds), pp. 289–316. Wiley-VCH, Weinheim.

135 Munier, R. L. and Sarrazin, G. (1963). Substitution totale de la 3-

fluorotyrosine a la Tyrosine dans les Proteines d'*Escherichia coli*. *Comptes Rendus Hebdomadaires des Seances de L'Academie des Sciences* **256**, 3376–3378.

136 Lu, P., Jarema, M. and Mosser, K. (1976). *Lac*-repressor with 3-fluorotyrosine substitution for NMR studies. *Federation Proceedings* **35**, 1456–1456.

137 Ring, M., Armitage, I. M. and Huber, R. E. (1985). Meta-fluorotyrosine substitution in beta-galactosidase – evidence for the existence of a catalytically active tyrosine. *Biochemical and Biophysical Research Communications* **131**, 675–680.

138 Ring, M. and Huber, R. E. (1993). The properties of beta-galactosidases (*Escherichia coli*) with halogenated tyrosines. *Biochemistry and Cell Biology* **71**, 127–132.

139 Brooks, B., Phillips, R. S. and Benisek, W. F. (1998). High-efficiency incorporation *in vivo* of tyrosine analogues with altered hydroxyl acidity in place of the catalytic tyrosine-14 of delta(5)-3-ketosteroid isomerase of *Comamonas (Pseudomonas) testosteroni*: effects of the modifications on isomerase kinetics. *Biochemistry* **37**, 9738–9742.

140 Pal, P. P., Bae, J. H., Azim, M. K., Hess, P., Friedrich, R., Huber, R., Moroder, L. and Budisa, N. (2005). Structural and spectral response of *Aequorea victoria* green fluorescent proteins to chromophore fluorination. *Biochemistry* **44**, 3663–3672.

141 Xiao, G. Y., Parsons, J. F., Tesh, K., Armstrong, R. N. and Gilliland, G. L. (1998). Conformational changes in the crystal structure of rat glutathione transferase M1-1 with global substitution of 3-fluorotyrosine for tyrosine. *Journal of Molecular Biology* **281**, 323–339.

142 Bae, J. H., Pal, P. P., Moroder, L., Huber, R. and Budisa, N. (2004). Crystallographic evidence for isomeric chromophores in 3-fluorotyrosyl-green fluorescent protein. *ChemBioChem* **5**, 720–722.

143 de Prat Gay, G., Duckworth, H. W. and Fersht, A. R. (1993). Modification of the amino acid specificity of tyrosyl-tRNA synthetase by protein engineering. *FEBS Letters* **317**, 167–171.

144 Praetorius-Ibba, M., Stange-Thomann, N., Kitabatake, M., Ali, K., Söll, I., Carter, C. W., Ibba, M. and Söll, D. (2000). Ancient adaptation of the active site of tryptophanyl-tRNA synthetase for tryptophan binding. *Biochemistry* **39**, 13136–13143.

145 Goldstein, J. A., Cheung, Y. F., Marletta, M. A. and Walsh, C. (1978). Fluorinated substrate analogs as stereochemical probes of enzymatic-reaction mechanisms. *Biochemistry* **17**, 5567–5575.

146 Armstrong, M. D. and Lewis, J. D. (1951). The toxicity of *ortho*-fluorophenyl-DL-alanine and *para*-fluorophenyl-DL-alanine for the rat. *Journal of Biological Chemistry* **188**, 91–95.

147 Yoshida, A. (1960). Studies on the mechanism of protein synthesis – incorporation of *para*-fluorophenylalanine into alpha-amylase of *Bacillus subtilis*. *Biochimica et Biophysica Acta* **41**, 98–103.

148 Yoshikawa, E., Fournier, M. J., Mason, T. L. and Tirrell, D. A. (1994). Genetically engineered fluoropolymers – synthesis of repetitive polypeptides containing *p*-fluorophenylalalanine residues. *Macromolecules* **27**, 5471–5475.

149 Minks, C., Huber, R., Moroder, L. and Budisa, N. (2000). Noninvasive tracing of recombinant proteins with "fluorophenylalanine-fingers". *Analytical Biochemistry* **284**, 29–34.

150 Kothakota, S., Mason, T. L., Tirrell, D. A. and Fournier, M. J. (1995). Biosynthesis of a periodic protein containing 3-thienylalanine – a step toward genetically engineered conducting polymers. *Journal of the American Chemical Society* **117**, 536–537.

151 Regan, L. (1995). Protein design: novel metal-binding sites. *Trends in Biochemical Sciences* **20**, 280–285.

152 Schmitt, J., Hess, H. and Stunnenberg, H. G. (1993). Affinity purification of histidine-tagged proteins. *Molecular Biology Reports* **18**, 223–230.
153 Francklyn, C. and Arnez, J. (2004). Histidyl-tRNA Synthetases. In *Aminoacyl-tRNA Synthetases* (Ibba, M., Francklyn, C. and Cusack, S., eds), pp. 135–149. Landes Biosciences, Georgetown, TX.
154 Beiboer, S. H. W., vandenBerg, B., Dekker, N., Cox, R. C. and Verheij, H. M. (1996). Incorporation of an unnatural amino acid in the active site of porcine pancreatic phospholipase A_2. Substitution of histidine by 1,2,4-triazole-3-alanine yields an enzyme with high activity at acidic pH. *Protein Engineering* **9**, 345–352.
155 Klein, D. C., Weller, J. L., Kirk, K. L. and Hartley, R. W. (1977). Incorporation of 2-fluoro-L-histidine into cellular protein. *Molecular Pharmacology* **13**, 1105–1110.
156 Jackson, D. Y., Burnier, J., Quan, C., Stanley, M., Tom, J. and Wells, J. A. (1994). A designed peptide ligase for total synthesis of ribonuclease-A with unnatural catalytic residues. *Science* **266**, 243–247.
157 Ikeda, Y., Kawahara, S., Taki, M., Kuno, A., Hasegawa, T. and Taira, K. (2003). Synthesis of a novel histidine analogue and its efficient incorporation into a protein *in vivo*. *Protein Engineering* **16**, 699–706.
158 Reimer, U., Scherer, G., Drewello, M., Kruber, S., Schutkowski, M. and Fischer, G. (1998). Side-chain effects on peptidyl-prolyl *cis/trans* isomerization. *Journal of Molecular Biology* **279**, 449–460.
159 Guichou, J. F., Patiny, L. and Mutter, M. (2002). Pseudo-prolines (Psi Pro): direct insertion of Psi Pro systems into cysteine containing peptides. *Tetrahedron Letters* **43**, 4389–4390.
160 Renner, C., Alefelder, S., Bae, J. H., Budisa, N., Huber, R. and Moroder, L. (2001). Fluoroprolines as tools for protein design and engineering. *Angewandte Chemie International Edition* **40**, 923–925.
161 Budisa, N., Minks, C., Medrano, F. J., Lutz, J., Huber, R. and Moroder, L. (1998). Residue-specific bioincorporation of non-natural, biologically active amino acids into proteins as possible drug carriers: structure and stability of the perthiaproline mutant of annexin V. *Proceedings of the National Academy of Sciences of the USA* **95**, 455–459.
162 Deming, T. J., Fournier, M. J., Mason, T. L. and Tirrell, D. A. (1997). Biosynthetic incorporation and chemical modification of alkene functionality in genetically engineered polymers. *Journal of Macromolecular Science-Pure and Applied Chemistry* A34, 2143–2150.
163 Mendel, D., Cornish, V. W. and Schultz, P. G. (1995). Site-directed mutagenesis with an expanded genetic code. *Annual Review of Biophysics and Biomolecular Structure* **24**, 435–462.
164 Busiello, V., Digirolamo, M., Cini, C. and DeMarco, C. (1980). Beta-selenaproline as competitive inhibitor of proline activation. *Biochimica et Biophysica Acta* **606**, 347–352.
165 Katz, E. and Goss, W. A. (1959). Controlled biosynthesis of actinomycin with sarcosine. *Biochemical Journal* **73**, 458–465.
166 Rose, G., Geselowitz, A., Lesser, G., Lee, R. and Zehfus, M. (1985). Hydrophobicity of amino acid residues in globular proteins. *Science* **229**, 834–838.
167 Vogt, W. (1995). Oxidation of methionyl residues in proteins: tools, targets, and reversal. *Free Radicals in Biology and Medicine* **18**, 93–105.
168 Shacter, E. (2000). Quantification and significance of protein oxidation in biological samples. *Drug Metabolism Reviews* **32**, 307–326.
169 Brot, N. and Weissbach, H. (1983). Biochemistry and physiological role of methionine sulfoxide residues in proteins. *Archives of Biochemistry and Biophysics* **223**, 271–281.
170 Freitas, J. O., Termignoni, C., Borges, D. R., Sampaio, C. A. M., Prado, J. L. and Guimaraes, J. A. (1981). Methionine aminopeptidase

associated with liver-mitochondria and microsomes. *International Journal of Biochemistry* **13**, 991–997.

171 Bradshaw, R. A. (2004). Methionine aminopeptidase 2 inhibition: anti-angiogenesis and tumour therapy. *Expert Opinion on Therapeutic Patents* **14**, 1–6.

172 Alix, J. H. (1982). Molecular aspects of the *in vivo* and *in vitro* effects of ethionine, an analog of methionine. *Microbiological Reviews* **46**, 281–295.

173 Dyer, H. M. (1938). Evidence of the physiological specificity of methionine in regard to the methylthiol group: the synthesis of S-ethylhomocysteine (ethionine) and a study of its availability for growth. *Journal of Biological Chemistry* **124**, 519–524.

174 Yoshida, A. (1958). Studies on the mechanism of protein synthesis – bacterial alpha-amylase containing ethionine. *Biochimica et Biophysica Acta* **29**, 213–214.

175 Black, A. L. and Kleiber, M. (1955). The recovery of norleucine from casein after administering norleucine-3-C14 to intact cows. *Journal of the American Chemical Society* **77**, 6082–6083.

176 Crepin, T., Schmitt, E., Mechulam, Y., Sampson, P. B., Vaughan, M. D., Honek, J. F. and Blanquet, S. (2003). Use of analogues of methionine and methionyl adenylate to sample conformational changes during catalysis in *Escherichia coli* methionyl-tRNA synthetase. *Journal of Molecular Biology* **332**, 59–72.

177 Deniziak, M. A. and Barciszewski, J. (2001). Methionyl-tRNA synthetase. *Acta Biochimica Polonica* **48**, 337–350.

178 Blanquet, S., Crepin, T., Mechulam, Y. and Schmitt, E. (2003). Methionyl-tRNA synthetases. In *Aminoacyl-tRNA Synthetases* (Ibba, M., Francklyn, C. and Cusack, S., eds), pp. 47–59. Landes Biosciences, Georgetown, TX.

179 Fersht, A. R. and Dingwall, C. (1979). Editing mechanism for the methionyl-tRNA synthetase in the selection of amino acids in protein synthesis. *Biochemistry* **18**, 1250–1256.

180 Jakubowski, H. (2000). Translational incorporation of S-nitrosohomocysteine into protein. *Journal of Biological Chemistry* **275**, 21813–21816.

181 Duewel, H., Daub, E., Robinson, V. and Honek, J. F. (1997). Incorporation of trifluoromethionine into a phage lysozyme: implications and a new marker for use in protein F-19 NMR. *Biochemistry* **36**, 3404–3416.

182 Vaughan, M. D., Cleve, P., Robinson, V., Duewel, H. S. and Honek, J. F. (1999). Difluoromethionine as a novel F-19 NMR structural probe for internal amino acid packing in proteins. *Journal of the American Chemical Society* **121**, 8475–8478.

183 Kiick, K. L., Saxon, E., Tirrell, D. A. and Bertozzi, C. R. (2002). Incorporation of azides into recombinant proteins for chemoselective modification by the Staudinger ligation. *Proceedings of the National Academy of Sciences of the USA* **99**, 19–24.

184 Link, A. J., Vink, M. K. S. and Tirrell, D. A. (2004). Presentation and detection of azide functionality in bacterial cell surface proteins. *Journal of the American Chemical Society* **126**, 10598–10602.

185 Budisa, N., Huber, R., Golbik, R., Minks, C., Weyher, E. and Moroder, L. (1998). Atomic mutations in annexin V – thermodynamic studies of isomorphous protein variants. *European Journal of Biochemistry* **253**, 1–9.

186 Besse, D., Budisa, N., Karnbrock, W., Minks, C., Musiol, H. J., Pegoraro, S., Siedler, F., Weyher, E. and Moroder, L. (1997). Chalcogen-analogs of amino acids their use in X-ray crystallographic and folding studies of peptides and proteins. *Biological Chemistry* **378**, 211–218.

187 Müller, S. (1997). Design neuer Selenoproteine. *PhD Thesis*, Ludwig-Maximilian-University.

188 Feeney, J., McCormick, J. E., Bauer, C. J., Birdsall, B., Moody, C. M., Starkmann, B. A., Young, D. W., Francis, P., Havlin, R. H., Arnold, W. D. and Oldfield, E. (1996). F-19

nuclear magnetic resonance chemical shifts of fluorine containing aliphatic amino acids in proteins: studies on *Lactobacillus casei* dihydrofolate reductase containing (2*S*,4*S*)-5-fluoroleucine. *Journal of the American Chemical Society* **118**, 8700–8706.

189 Hortin, G. and Boime, I. (1983). Applications of amino acid analogs for studying co-translational and posttranslational modifications of proteins. *Methods in Enzymology* **96**, 777–784.

190 Apostol, I., Levine, J., Lippincott, J., Leach, J., Hess, E., Glascock, C. B., Weickert, M. J. and Blackmore, R. (1997). Incorporation of norvaline at leucine positions in recombinant human hemoglobin expressed in *Escherichia coli*. *Journal of Biological Chemistry* **272**, 28980–28988.

191 Englisch-Peters, S., von der Haar, F. and Cramer, F. (1990). Fidelity in the aminoacylation of tRNA(Val) with hydroxy analogues of valine, leucine, and isoleucine by valyl-tRNA synthetases from *Saccharomyces cerevisiae* and *Escherichia coli*. *Biochemistry* **29**, 7953–7958.

192 Englisch, S., Englisch, U., von der Haar, F. and Cramer, F. (1986). The proofreading of hydroxy analogues of leucine and isoleucine by leucyl-tRNA synthetases from *E. coli* and yeast. *Nucleic Acids Research* **14**, 7529–7539.

193 Wu, N., Deiters, A., Cropp, T. A., King, D. S. and Schultz, P. G. (2004). A genetically encoded photocaged amino acid. *Journal of the American Chemical Society* **126**, 14306–14307.

194 Freist, W. and Gauss, D. H. (1995). Threonyl-tRNA synthetase. *Biological Chemistry* **376**, 213–224.

195 Fersht, A. R. (1981). Enzymic editing mechanisms and the genetic code. *Proceedings of the Royal Society of London Series B Biological Sciences* **212**, 351–379.

196 Fukai, S., Nureki, O., Sekine, S., Shimada, A., Tao, J. S., Vassylyev, D. G. and Yokoyama, S. (2002). Structural basis for double-sieve discrimination of L-valine from L-isoleucine and L-threonine by the complex of tRNA(Val) and valyl-tRNA synthetase. *Cell* **103**, 353–362.

197 Porter, T. H., Smith, S. C. and Shive, W. (1977). Inhibition of valine utilization by cyclo-butaneglycine. *Archives of Biochemistry and Biophysics* **179**, 266–271.

198 Kohno, T., Kohda, D., Haruki, M., Yokoyama, S. and Miyazawa, T. (1990). Nonprotein amino acid furanomycin, unlike isoleucine in chemical structure, is charged to isoleucine tRNA by isoleucyl-tRNA synthetase and incorporated into protein. *Journal of Biological Chemistry* **265**, 6931–6935.

199 Wang, P., Tang, Y. and Tirrell, D. A. (2003). Incorporation of trifluoroisoleucine into proteins *in vivo*. *Journal of the American Chemical Society* **125**, 6900–6906.

200 Tanaka, K., Tamaki, M. and Watanabe, S. (1969). Effect of furanomycin on the synthesis of isoleucyl-tRNA. *Biochimica et Biophysica Acta* **195**, 244–255.

201 Schimmel, P., Taob, J. and Hill, J. (1998). Aminoacyl tRNA synthetases as targets for new anti-infectives. *FASEB Journal* **12**, 1599–1609.

202 Rosenthal, G. A. (1977). The biological effects and mode of action of L-canavanine, a structural analogue of L-arginine. *Quarterly Review of Biology* **52**, 155–178.

203 Thomas, D. A., Rosenthal, G. A., Gold, D. V. and Dickey, K. (1986). Growth inhibition of a rat colon tumor by L-canavanine. *Cancer Research* **46**, 2898–2903.

204 Eriani, G. and Cavarelli, J. (2003). Arginyl-tRNA synthetase. In *Aminoacyl-tRNA Synthetases* (Ibba, M., Francklyn, C. and Cusack, S., eds), pp. 3–12. Landes Biosciences, Georgetown, TX.

205 Hunt, S. (1991). Non-protein amino acids. In *Amino acids, proteins and nucleic acids* (Dey, P., Harborne, J. and Rogers, L., eds), vol. 5, pp. 55–137. Academic Press, New York.

206 Rosenthal, G. A., Reichhart, J. M. and Hoffmann, J. A. (1989). L-canavanine incorporation into vitellogenin and macromolecular conformation. *Journal of Biological Chemistry* **264**, 13693–13696.

207 Moncada, S., Palmer, R. M. and Higgs, E. A. (1991). Nitric oxide: physiology, pathophysiology, and pharmacology. *Pharmacology Review* **43**, 109–142.

208 Liaudet, L., Feihl, F., Rosselet, A., Markert, M., Hurni, J. M. and Perret, C. (1996). Beneficial effects of L-canavanine, a selective inhibitor of inducible nitric oxide synthase, during rodent endotoxaemia. *Clinical Science* **90**, 369–77.

209 Schmidt, H. H., Baeblich, S. E., Zernikow, B. C., Klein, M. M. and Bohme, E. (1999). L-Arginine and arginine analogues: effects on isolated blood vessels and cultured endothelial cells. *British Journal of Pharmacology* **101**, 145–151.

210 Christner, P., Yankowski, R. L., Benditt, M. and Jimenez, S. A. (1996). Alteration in the conformational stability of collagen caused by the incorporation of the lysine analogue *S*-2-aminoethyl-cysteine. *Biochimica et Biophysica Acta* **1294**, 37–47.

211 Cini, C., Busiello, V., Di Girolamo, M. and De Marco, C. (1981). *In vivo* incorporation of selenalysine in *Escherichia coli* proteins and its effects on cell growth. *Biochimica et Biophysica Acta* **678**, 165–171.

212 Di Girolamo, M., Busiello, V., Di Girolamo, A., De Marco, C. and Cini, C. (1987). Degradation of thialysine- or selenalysine-containing abnormal proteins in CHO cells. *Biochemistry International* **15**, 971–980.

213 Blanquet, S., Plateau, P. and Onesti, S. (2004). Class II lysyl-tRNA synthetases. In *Aminoacyl-tRNA Synthetases* (Ibba, M., Francklyn, C. and Cusack, S., eds), pp. 227–241. Landes Biosciences, Georgetown, TX.

214 Hortin, G. and Boime, I. (1983). Markers for processing sites in eukaryotic proteins: characterization with amino acid analogs. *Trends in Biochemical Sciences* **8**, 320–323.

215 Hirshfield, I. N., Tomford, J. W. and Zamecnik, P. C. (1972). Thiosine-resistant mutants of *Escherichia coli* K-12 with growth-medium-dependent lysyl-tRNA synthetase activity. II. Evidence for an altered lysyl-tRNA synthetase. *Biochimca et Biophysica Acta* **259**, 344–356.

216 Jester, B. C., Levengood, J. D., Roy, H., Ibba, M. and Devine, K. M. (2003). Nonorthologous replacement of lysyl-tRNA synthetase prevents addition of lysine analogues to the genetic code. *Proceedings of the National Academy of Sciences of the USA* **100**, 14351–14356.

217 Levengood, J., Ataide, S. F., Roy, H. and Ibba, M. (2004). Divergence in noncognate amino acid recognition between class I and class II lysyl-tRNA synthetases. *Journal of Biological Chemistry* **279**, 17707–17714.

218 Chapeville, F., Ehrenstein, G. V., Benzer, S., Weisblum, B., Ray, W. J. and Lipmann, F. (1962). On role of soluble ribonucleic acid in coding for amino acids. *Proceedings of the National Academy of Sciences of the USA* **48**, 1086–1092.

219 Johnson, A. E., Woodward, W. R., Herbert, E. and Menninger, J. R. (1976). Nepsilon-acetyllysine transfer ribonucleic acid: a biologically active analogue of aminoacyl transfer ribonucleic acids. *Biochemistry* **15**, 569–575.

220 Johnson, A. E. and Cantor, C. R. (1977). Affinity labeling of multi-component systems. *Methods in Enzymology* **46**, 180–194.

221 Hecht, S. M., Alford, B. L., Kuroda, Y. and Kitano, S. (1978). "Chemical aminoacylation" of tRNA's. *Journal of Biological Chemistry* **253**, 4517–4520.

222 Heckler, T. G., Chang, L. H., Zama, Y., Naka, T., Chorghade, M. S. and Hecht, S. M. (1984). T4 RNA-ligase mediated preparation of novel chemically misacylated phenylalanine transfer-RNA. *Biochemistry* **23**, 1468–1473.

223 Baldini, G., Martoglio, B., Schachenmann, A., Zugliani, C. and Brunner, J. (1988). Mischarging

Escherichia coli tRNAPhe with L-4′-[3-(trifluoromethyl)-3H-diazirin-3-yl]phenylalanine, a photoactivatable analogue of phenylalanine. *Biochemistry* **27**, 7951–7959.

224 Hohsaka, T., Sato, K., Sisido, M., Takai, K. and Yokoyama, S. (1993). Adaptability of nonnatural aromatic amino acids to the active center of the *Escherichia coli* ribosomal A-site. *FEBS Letters* **335**, 47–50.

225 Hohsaka, T., Sato, K., Sisido, M., Takai, K. and Yokoyama, S. (1994). Site-specific incorporation of photofunctional nonnatural amino-acids into a polypeptide through *in vitro* protein biosynthesis. *FEBS Letters* **344**, 171–174.

226 Kanda, T., Takai, K., Hohsaka, T., Sisido, M. and Takaku, H. (2000). Sense codon-dependent introduction of unnatural amino acids into multiple sites of a protein. *Biochemical and Biophysical Research Communications* **270**, 1136–1139.

227 Kowal, A. K. and Oliver, J. S. (1997). Exploiting unassigned codons in *Micrococcus luteus* for tRNA-based amino acid mutagenesis. *Nucleic Acids Research* **25**, 4685–4689.

228 Roberts, R. W. and Szostak, J. W. (1997). RNA–peptide fusions for the *in vitro* selection of peptides and proteins. *Proceedings of the National Academy of Sciences of the USA* **94**, 12297–12302.

229 Li, S., Millward, S. and Roberts, R. (1992). *In vitro* selection of mRNA display libraries containing an unnatural amino acid. *Journal of the American Chemical Society* **124**, 9972–9973.

230 Forster, A. C., Tan, Z. P., Nalam, M. N. L., Lin, H. N., Qu, H., Cornish, V. W. and Blacklow, S. C. (2003). Programming peptidomimetic syntheses by translating genetic codes designed *de novo*. *Proceedings of the National Academy of Sciences of the USA* **100**, 6353–6357.

231 Bessho, Y., Hodgson, D. R. W. and Suga, H. (2002). A tRNA aminoacylation system for non-natural amino acids based on a programmable ribozyme. *Nature Biotechnology* **20**, 723–728.

232 Murakami, H., Bonzagni, N. J. and Suga, H. (2002). Aminoacyl-tRNA synthesis by a resin-immobilized ribozyme. *Journal of the American Chemical Society* **124**, 6834–6835.

233 Dedkova, L. M., Fahmi, N. E., Golovine, S. Y. and Hecht, S. M. (2003). Enhanced D-amino acid incorporation into protein by modified ribosomes. *Journal of the American Chemical Society* **125**, 6616–6617.

234 LaRiviere, F. J., Wolfson, A. D. and Uhlenbeck, O. C. (2001). Uniform binding of aminoacyl-tRNAs to elongation factor Tu by thermodynamic compensation. **294**, 165–168.

235 Hohsaka, T. and Sisido, M. (2002). Incorporation of non-natural amino acids into proteins. *Current Opinion in Chemical Biology* **6**, 809–815.

236 Hohsaka, T., Kajihara, D., Ashizuka, Y., Murakami, H. and Sisido, M. (1999). Efficient incorporation of nonnatural amino acids with large aromatic groups into streptavidin in *in vitro* protein synthesizing systems. *Journal of the American Chemical Society* **121**, 34–40.

237 Kwon, I., Kirshenbaum, K. and Tirrell, D. A. (2003). Breaking the degeneracy of the genetic code. *Journal of the American Chemical Society* **125**, 7512–7513.

238 Watson, J. D., Hopkins, N. H., Roberts, J. W., Steitz, J. A. and Weiner, A. M. (1988). *Molecular Biology of the Gene*. Benjamin Cummings, Amsterdam.

239 Moore, B., Persson, B. C., Nelson, C. C., Gesteland, R. F. and Atkins, J. F. (2000). Quadruplet codons: implications for code expansion and the specification of translation step size. *Journal of Molecular Biology* **298**, 195–209.

240 Atkins, J. F., Weiss, R. B., Thompson, S. and Gesteland, R. F. (1991). Towards a genetic dissection of the basis of triplet decoding, and its natural subversion: programmed reading frame shifts and hops. *Annual Review of Genetics* **25**, 201–228.

241 Hohsaka, T., Ashizuka, Y., Murakami, H. and Sisido, M. (1996). Incorporation of nonnatural amino acids into streptavidin through *in vitro* frame-shift suppression. *Journal of the American Chemical Society* **118**, 9778–9779.

242 Hohsaka, T. (2004). Incorporation of nonnatural amino acids into proteins through extension of the genetic code. *Bulletin of the Chemical Society of Japan* **77**, 1041–1049.

243 Hohsaka, T., Ashizuka, Y., Taira, H., Murakami, H. and Sisido, M. (2001). Incorporation of nonnatural amino acids into proteins by using various four-base codons in an *Escherichia coli in vitro* translation system. *Biochemistry* **40**, 11060–11064.

244 Hohsaka, T., Ashizuka, Y., Sasaki, H., Murakami, H. and Sisido, M. (1999). Incorporation of two different nonnatural amino acids independently into a single protein through extension of the genetic code. *Journal of the American Chemical Society* **121**, 12194–12195.

245 Hohsaka, T., Ashizuka, Y., Murakami, H. and Sisido, M. (2001). Five-base codons for incorporation of nonnatural amino acids into proteins. *Nucleic Acids Research* **29**, 3646–3651.

246 Magliery, T. J., Anderson, J. C. and Schultz, P. G. (2001). Expanding the genetic code: selection of efficient suppressors of four-base codons and identification of "shifty" four-base codons with a library approach in *Escherichia coli*. *Journal of Molecular Biology* **307**, 755–769.

247 Anderson, J. C., Magliery, T. J. and Schultz, P. G. (2002). Exploring the limits of codon and anticodon size. *Chemistry and Biology* **9**, 237–244.

248 Yamanaka, K., Nakata, H., Hohsaka, T. and Sisido, M. (2004). Efficient synthesis of nonnatural mutants in *Escherichia coli* S30 *in vitro* protein synthesizing system. *Journal of Bioscience and Bioengineering* **97**, 395–399.

249 Anderson, J. C., Wu, N., Santoro, S. W., Lakshman, V., King, D. S. and Schultz, P. G. (2004). An expanded genetic code with a functional quadruplet codon. *Proceedings of the National Academy of Sciences of the USA* **101**, 7566–7571.

250 Bain, J. D., Switzer, C., Chamberlin, A. R. and Benner, S. A. (1992). Ribosome-mediated incorporation of a nonstandard amino acid into a peptide through expansion of the genetic code. *Nature* **356**, 537–539.

251 Endo, M., Mitsui, T., Okuni, T., Kimoto, A., Hirao, I. and Yokoyama, S. (2004). Unnatural base pairs mediate the site-specific incorporation of an unnatural hydrophobic component into RNA transcripts. *Bioorganic and Medicinal Chemistry Letters* **14**, 2593–2596.

252 Henry, A. A., Olsen, A. G., Matsuda, S., Yu, C. Z., Geierstanger, B. H. and Romesberg, F. E. (2004). Efforts to expand the genetic alphabet: identification of a replicable unnatural DNA self-pair. *Journal of the American Chemical Society* **126**, 6923–6931.

253 Yu, C., Henry, A. A., Romesberg, F. E. and Schultz, P. G. (2002). Polymerase recognition of unnatural base pairs. *Angewandte Chemie International Edition* **41**, 3841–3844.

254 Hirao, I., Ohtsuki, T., Fujiwara, T., Mitsui, T., Yokogawa, T., Okuni, T., Nakayama, H., Takio, K., Yabuki, T., Kigawa, T., Kodama, K., Nishikawa, K. and Yokoyama, S. (2002). An unnatural base pair for incorporating amino acid analogs into proteins. *Nature Biotechnology* **20**, 177–182.

255 Kimoto, M., Endo, M., Mitsui, T., Okuni, T., Hirao, I. and Yokoyama, S. (2004). Site-specific incorporation of a photo-crosslinking component into RNA by T7 transcription mediated by unnatural base pairs. *Chemistry and Biology* **11**, 47–55.

256 Bergstrom, D. E. (2004). Orthogonal base pairs continue to evolve. *Chemistry and Biology* **11**, 18–20.

257 Osawa, S., Jukes, T. H., Watanabe, K. and Muto, A. (1992). Recent evidence for evolution of the genetic code. *Microbiological Reviews* **56**, 229–264.

258 Suzuki, T., Ueda, T. and Watanabe, K. (1997). The "polysemous" codon –

a codon with multiple amino acid assignment caused by dual specificity of tRNA identity. *EMBO Journal* **16**, 1122–1134.

259 Murgola, E. J. (1985). tRNA, suppression, and the code. *Annual Review of Genetics* **19**, 57–80.

260 Garen, A. (1968). Sense and nonsense in the genetic code. *Science* **160**, 149–159.

261 Milligan, J. F. and Uhlenbeck, O. C. (1989). Synthesis of small RNAs using T7 RNA polymerase. *Methods in Enzymology* **180**, 51–62.

262 Bain, J. D., Glabe, C. G., Dix, T. A., Chamberlin, A. R. and Diala, E. S. (1989). Biosynthetic site-specific incorporation of a non-natural amino acid into a polypeptide. *Journal of the American Chemical Society* **111**, 8013–8014.

263 Noren, C. J., Anthonycahill, S. J., Griffith, M. C. and Schultz, P. G. (1989). A general method for site-specific incorporation of unnatural amino acids into proteins. *Science* **244**, 182–188.

264 Noren, C. J., Anthony-Cahill, S. J., Suich, D. J., Noren, K. A., Griffith, M. C. and Schultz, P. G. (1990). *In vitro* suppression of an amber mutation by a chemically aminoacylated transfer RNA prepared by runoff transcription. *Nucleic Acids Research* **18**, 83–88.

265 Ninomiya, K., Kurita, T., Hohsaka, T. and Sisido, M. (2003). Facile aminoacylation of pdCpA dinucleotide with a nonnatural amino acid in cationic micelle. *Chemical Communications* **17**, 2242–2243.

266 Steward, L. E. and Chamberlin, A. R. (1998). Protein engineering with nonstandard amino acids. In *Protein Synthesis – Methods and Protocols* (Martin, R., ed.), vol. 77, pp. 325–354. Humana Press, Totowa, NJ.

267 Petersson, E. J., Brandt, G. S., Zacharias, N. M., Dougherty, D. A. and Lester, H. A. (2003). Caging proteins through unnatural amino acid mutagenesis. *Methods in Enzymology* **360**, 258–273.

268 Kwok, Y. and Wong, J. T. F. (1980). Evolutionary relationship between *Halobacterium cutirubrum* and eukaryotes determined by use of aminoacyl-tRNA synthetases as phylogenetic probes. *Canadian Journal of Biochemistry* **58**, 213–218.

269 Cornish, V. W., Mendel, D. and Schultz, P. G. (1995). Probing protein structure and function with an expanded genetic code. *Angewandte Chemie International Edition* **34**, 621–633.

270 Short, G. F., Golovine, S. Y. and Hecht, S. M. (1999). Effects of release factor 1 on *in vitro* protein translation and the elaboration of proteins containing unnatural amino acids. *Biochemistry* **38**, 8808–8819.

271 Nowak, M. W., Gallivan, J. P., Silverman, S. K., Labarca, C. G., Dougherty, D. A. and Lester, H. A. (1998). *In vivo* incorporation of unnatural amino acids into ion channels in *Xenopus* oocyte expression system. *Methods in Enzymology* **293**, 504–529.

272 Dougherty, D. A. (2002). Unnatural amino acids as probes of protein structure and function. *Current Opinion in Chemical Biology* **4**, 645–452.

273 Kohrer, C., Xie, L., Kellerer, S., Varshney, U. and Rajbhandary, U. L. (2001). Import of amber and ochre suppressor tRNAs into mammalian cells: a general approach to site-specific insertion of amino acid analogues into proteins. *Proceedings of the National Academy of Sciences of the USA* **98**, 14310–14315.

274 Monahan, S. L., Lester, H. A. and Dougherty, D. A. (2003). Site-specific incorporation of unnatural amino acids into receptors expressed in mammalian cells. *Chemistry and Biology* **10**, 573–580.

275 Shimizu, Y., Inoue, A., Tomari, Y., Suzuki, T., Yokogawa, T., Nishikawa, K. and Ueda, T. (2001). Cell-free translation reconstituted with purified components. *Nature Biotechnology* **19**, 751–755.

276 Buckingham, R. H., Grentzmann, G. and Kisselev, L. (1997).

Polypeptide chain release factors. *Molecular Microbiology* **24**, 449–456.

277 Ellman, J., Mendel, D., Anthonycahill, S., Noren, C. J. and Schultz, P. G. (1991). Biosynthetic method for introducing unnatural amino acids site specifically into proteins. *Methods in Enzymology* **202**, 301–336.

278 Frankel, A. and Roberts, R. W. (2003). *In vitro* selection for sense codon suppression. *RNA* **9**, 780–786.

279 Hendrickson, T. L. and Schimmel, P. (2003). Transfer RNA-dependent amino acid discrimination by aminoacyl-tRNA synthetase. In *Translation Mechanisms* (Lapointe, J. and Brakier-Gingras, L., eds), pp. 35–69. Landes Bioscience, Georgetown, TX.

280 Schimmel, P. and Söll, D. (1997). When protein engineering confronts the tRNA world. *Proceedings of the National Academy of Sciences of the USA* **94**, 10007–10009.

281 Wang, L. and Schultz, P. G. (2001). A general approach for the generation of orthogonal tRNAs. *Chemistry and Biology* **8**, 883–890.

282 Liu, D. R., Magliery, T. J. and Schultz, P. G. (1997). Characterization of an "orthogonal" suppressor tRNA derived from *E. coli* $tRNA_2^{Gln}$. *Chemistry and Biology* **4**, 685–691.

283 Rould, M. A., Perona, J. J., Söll, D. and Steitz, T. A. (1989). Structure of *E. coli* glutaminyl-tRNA synthetase complexed with $tRNA^{Gln}$ and ATP at 2.8 Å resolution. *Science* **246**, 1135–1142.

284 Liu, D. R., Magliery, T. J., Pasternak, M. and Schultz, P. G. (1997). Engineering a tRNA and aminoacyl-tRNA synthetase for the site-specific incorporation of unnatural amino acids into proteins *in vivo*. *Proceedings of the National Academy of Sciences of the USA* **94**, 10092–10097.

285 Kowal, A. K., Kohrer, C. and RajBhandary, U. L. (2001). Twenty-first aminoacyl-tRNA synthetase-suppressor tRNA pairs for possible use in site-specific incorporation of amino acid analogues into proteins in eukaryotes and in eubacteria. *Proceedings of the National Academy of Sciences of the USA* **98**, 2268–2273.

286 Wang, L., Magliery, T. J., Liu, D. R. and Schultz, P. G. (2000). A new functional suppressor tRNA/aminoacyl-tRNA synthetase pair for the *in vivo* incorporation of unnatural amino acids into proteins. *Journal of the American Chemical Society* **122**, 5010–5011.

287 Pastrnak, M., Magliery, T. J. and Schultz, P. G. (2000). A new orthogonal suppressor tRNA/aminoacyl-tRNA synthetase pair for evolving an organism with an expanded genetic code. *Helvetica Chimica Acta* **83**, 2277–2286.

288 Koide, K., Finkelstein, J. M., Ball, Z. and Verdine, G. L. (2001). A synthetic library of cell-permeable molecules. *Journal of the American Chemical Society* **123**, 398–408.

289 Furter, R. (1998). Expansion of the genetic code: site-directed *p*-fluorophenylalanine incorporation in *Escherichia coli*. *Protein Science* **7**, 419–426.

290 Schimmel, P. and Beebe, K. (2004). Genetic code sizes pyrrolysine. *Nature* **431**, 257–258.

291 Anderson, J. C. and Schultz, P. G. (2003). Adaptation of an orthogonal archaeal leucyl-tRNA and synthetase pair for four-base, amber, and opal suppression. *Biochemistry* **42**, 9598–9608.

292 Santoro, S. W., Anderson, J. C., Lakshman, V. and Schultz, P. G. (2003). An archaebacteria-derived glutamyl-tRNA synthetase and tRNA pair for unnatural amino acid mutagenesis of proteins in *Escherichia coli*. *Nucleic Acids Research* **31**, 6700–6709.

293 Chin, J. W., Cropp, T. A., Chu, S., Meggers, E. and Schultz, P. G. (2003). Progress toward an expanded eukaryotic genetic code. *Chemistry and Biology* **10**, 511–519.

294 Chin, J. W., Cropp, T. A., Anderson, J. C., Mukherji, M., Zhang, Z. and Schultz, P. G. (2003). An expanded eukaryotic genetic code. *Science* **301**, 964–967.

295 Cropp, A. and Schultz, P. G. (2004). An expanding genetic code. *Trends in Genetics* **20**, 625–630.

296 Köhrer, C., Yoo, J., Bennett, M., Schaack, J. and RajBhandary, U. L. (2003). A possible approach to site-specific insertion of two different unnatural amino acids into proteins in mammalian cells via nonsense suppression. *Chemistry and Biology* **10**, 1095–1102.

297 Kohrer, C., Sullivan, E. L. and Rajbhandary, U. L. (2004). Complete set of orthogonal 21st aminoacyl-tRNA synthetase-amber, ochre and opal suppressor tRNA pairs: concomitant suppression of three different termination codons in an mRNA in mammalian cells. *Nucleic Acids Research* **32**, 6200–6211.

298 Köhrer, C. and RajBhandary, U. L. (2004). Proteins carrying one or more unnatural amino acids. In *Aminoacyl-tRNA Synthetases* (Ibba, M., Francklyn, C. and Cusack, S., eds), pp. 353–364. Landes Bioscience, Georgetown, TX.

299 Atkinson, J. and Martin, R. (1994). Mutations to nonsense codons in human genetic disease: implications for gene therapy by nonsense suppressor tRNAs. *Nucleic Acids Research* **22**, 1327–1334.

300 Keasling, J. D. (1999). Gene-expression tools for the metabolic engineering of bacteria. *Trends in Biotechnology* **17**, 452–460.

301 Chavadej, S., Brisson, N., McNeil, J. N. and DeLuca, V. (1994). Redirection of tryptophan leads to production of low indole glucosinolate canola. *Proceedings of the National Academy of Sciences of the USA* **91**, 2166–2170.

302 Ikeda, M. and Katsumata, R. (1992). Metabolic engineering to produce tyrosine or phenylalanine in a tryptophan-producing *Corynebacterium glutamicum* strain. *Applied and Environmental Microbiology* **58**, 781–785.

303 Michal, G. (1999). *Biochemical Pathways: An Atlas of Biochemistry and Molecular Biology*. Wiley, New York.

304 Miles, E. W. (2001). Tryptophan synthase: a multienzyme complex with an intramolecular tunnel. *Chemical Record* **1**, 140–151.

305 Yanofsky, C. (1987). Tryptophan synthetase: its charmed history. *BioEssays* **6**, 133–137.

306 Phillips, R. S., Miles, E. W. and Cohen, L. A. (1984). Interactions of tryptophan synthase, tryptophanase, and pyridoxal phosphate with oxindolyl-L-alanine and 2,3-dihydro-L-tryptophan: support for an indolenine intermediate in tryptophan metabolism. *Biochemistry* **23**, 6228–6234.

307 Broos, J., Gabellieri, E., Biemans-Oldehinkel, E. and Strambini, G. B. (2003). Efficient biosynthetic incorporation of tryptophan and indole analogs in an integral membrane protein. *Protein Science* **12**, 1991–2000.

308 Maier, T. H. P. (2003). Semisynthetic production of unnatural L-alpha-amino acids by metabolic engineering of the cysteine-biosynthetic pathway. *Nature Biotechnology* **21**, 422–427.

309 Rosenthal, G. (1982). *Plant Non-protein Amino and Imino Acids. Biological, Biochemical and Toxicological Properties.* Academic Press, New York.

310 Evans, C. S. and Bell, E. A. (1980). Neuroactive plant amino acids and amines. *Trends in Neurosciences* **3**, 70–72.

311 Pfeifer, B. A., Admiraal, S. J., Gramajo, H., Cane, D. E. and Khostla, C. (2001). Biosynthesis of complex polyketides in a metabolically engineered strain of *E. coli*. *Science* **291**, 1790–1792.

312 Schmidt-Dannert, C., Umeno, D. and Arnold, F. H. (2000). Molecular breeding of carotenoid biosynthetic pathways. *Nature Biotechnology* **18**, 750–753.

313 Albrecht, M., Takaichi, S., Steiger, S., Wang, Z. Y. and Sandermann, G. (2000). Novel hydroxycarotenoids with improved antioxidative properties produced by gene combination in *E. coli*. *Nature Biotechnology* **18**, 843–846.

314 Mehl, R. A., Anderson, J. C., Santoro, S. W., Wang, L., Martin,

A. B., King, D. S., Horn, D. M. and Schultz, P. G. (2003). Generation of a bacterium with a 21 amino acid genetic code. *Journal of the American Chemical Society* **125**, 935–939.

315 Nielsen, J. (2001). Metabolic engineering. *Applied Microbiology and Biotechnology* **55**, 263–283.

316 Arnold, F. H. (2001). Combinatorial and computational challenges for biocatalyst design. *Nature* **409**, 253–257.

317 Service, R. F. (2003). Metabolic engineering: researchers create first autonomous synthetic life form. *Science* **299**, 640–640.

318 Schultz, D. W. and Yarus, M. (1996). On malleability in the genetic code. *Journal of Molecular Evolution* **42**, 597–601.

319 Jukes, T. H. and Osawa, S. (1989). Codon reassignment (codon capture) in evolution. *Journal of Molecular Evolution* **28**, 271–278.

6
Implications and Insights: From Reprogrammed Translation and Code Evolution to Artificial Life

Everything existing in the Universe is the fruit of chance and necessity.

[Democritus – Greek philosopher]

6.1
Code Engineering and Synthetic Biology

Engineering of the genetic code either by expansion of its existing repertoire or by the introduction of novel coding units is one of the many opportunities to design/redesign biological systems situated at the dynamic interface of physics, chemistry and biology. The chemistry of life is based on a defined number of generic monomeric building blocks (α-amino acids, nucleotides, metabolic intermediates, etc.). They are condensed into transient polymeric assemblies like polypeptides, polynucleotides, polysaccharides and natural products (such as polyketides) which represent entities of great variety, exhibiting biocatalytic flexibility and sophisticated functions [1]. In this context, there are two main challenges for code engineering. (i) The intentional diversification and even generation of novel generic classes of biological monomers by using the tools of synthetic chemistry, biocomputing and target-engineered metabolic circuits. (ii) The entry of novel α-amino acids in the genetic code in the context of reprogrammed protein translation. The emergence of designer molecules and cells carrying out novel properties optimized for user-defined environments marks the birth of synthetic biology. This will create tremendous opportunities for the biotechnology of the 21st century. It is not difficult to conceive that the main driving forces for development in the field would be the increase in demand for biological materials with tailored functions (such as pharmacological activity) or genetically encoded devices generated *in vivo* from programmed templates. In addition, the generation of molecules with novel biological, chemical and physical properties should allow us to influence, modulate and even create novel complex interactions and interaction networks in the extant or newly designed (synthetic) cells. From an academic viewpoint, such developments will bring an opportunity to achieve a better understanding of biological systems and processes at all levels of complexity.

Engineering the Genetic Code. Nediljko Budisa

ISBN: 3-527-31243-9

6.2
Novel Features of Protein Translation that have Emerged from Research in Code Engineering

If coding triplets of the genetic code represent specific code words, the question arises of how to redefine the message of such a code? The answer is to reprogram the translation apparatus by the manipulation of existing, or additional novel, translation components (see Chapter 5). This in turn again led to the question about the possible novel features of such reprogrammed protein translation machinery. In these early stages of the development of code engineering there is still a lack of precise terminology. This should not be a surprise because often not enough is known to permit accurate definitions for newly observed phenomena. The pragmatic strategy requires us to accept provisional, rough terminological characterizations which can provide leverage to the field's first developmental stages with taxonomic refinements emerging as the related facts become clear [1]. The novel features that have emerged from reprogramming of translation might be therefore roughly divided in four categories. (i) There is the novel possibility to manipulate the "identity" of the amino acids (i.e. the possibility to shift canonical/noncanonical amino acids between different coding units) by manipulation of the native translational components, especially by exploiting the phenomenon of enzyme catalytic promiscuity (Section 6.1.1). (ii) The delineation of the relationships between physicochemical properties of amino acids and their translation activity might yield a sort of "entrance index" in the genetic code for various noncanonical amino acids (Section 6.1.2). (iii) The possibility to establish novel links between translation, transcription and metabolism become evident (Section 6.1.4). (iv) There is an intriguing question about protein translation with amino acid analogs and surrogates in relation to the structural type of target proteins that needs to be answered (Section 6.1.3).

6.2.1
Code Malleability, Catalytic Promiscuity and the Amino Acid "Identity" Problem

Since the tRNA molecule is the central player in the protein translation process, the meaning of each particular codon can easily be altered by anticodon mutation of the tRNA. In fact, the ribosome matches codons with anticodons, but never checks directly whether tRNAs are charged with cognate, noncognate, canonical or noncanonical amino acids. The responsibility for linking specific amino acids with specific codons (via anticodon-bearing tRNAs) lies within the aminoacyl-tRNA synthetases (AARS), which can employ sophisticated proofreading mechanisms, but do not always recognize the anticodon specifically. In this context, the universal genetic code becomes malleable. Such malleability might, in addition, be achieved by simple manipulation of the enzyme activity of AARS by exploiting the promiscuity in their substrate recognition (see Section 5.1.4). Indeed, the lack of absolute substrate specificity, i.e. catalytic promiscuity, of AARS has been well documented in many systems for a long time and was very early recognized as a route for the expansion of the amino acid repertoire *in vivo*.

The term catalytic promiscuity describes the capacity of the enzyme to catalyze an adventitious secondary activity at the active site responsible for the primary activity [2]. The persistence of catalytic promiscuity in modern enzymes can be explained by the fact that such secondary activities usually do not affect the fitness of the organism and therefore there is no selective pressure to eliminate them. Most organisms did not evolve mechanisms such as editing against synthetic amino acids, because these substances are either not strong inhibitors or are not rate-limiting in aminoacylation. In the presence of strong selective pressure cells can indeed develop alternative enzyme isoforms such as constitutive and inducible LysRS gene copies found in *Escherichia coli* (see Section 5.2.1.10) or efficient editing mechanisms. These also explain why pairs of canonical amino acids with isosteric shapes (Val–Thr and Cys–Ser), the same molecular volume (Ile–Val and Pro–Cys) or very similar chemistry (Ser–Thr) are differentiated (i.e. edited) in the aminoacylation reaction. Indeed, almost perfect discrimination between closely related structures is the hallmark of biochemical systems. For example, homocysteine, a metabolic precursor of Met, is efficiently edited on the basis of size and chemistry (see Section 5.2.1.6), whereas synthetic Met analogs, such as trifluoromethionine or azidohomolalnine (see Fig. 5.15), are recognized in aminoacylation reactions and subsequently loaded onto $tRNA^{Met}$, i.e. their translation in the protein sequences is fully possible.

Catalytic promiscuity is often used by plants in their defense against different predators [3]. They produce multiple product mixtures ("cocktails") which are able to broaden the spectrum of their defense capacities. The ambiguous substrate recognition or the existence of enzyme secondary activities might become dominant at some point in evolutionary development, and the enzyme might be finally recruited to provide this novel activity exclusively. Such catalytic promiscuity should play a central role in the evolutionary model based on "statistical proteins" recently proposed by Pezo and coworkers [4, 5] (Section 6.4.1). It might provide a starting point for the evolution the AARS capable of gaining novel substrate specificity, i.e. to change the interpretation of the universal genetic code. In the simplest case such statistical proteins are generated by the catalysis of an additional aminoacylation reaction using substrate analogs. A further step is kinetic control of the reactants in the aminoaclyaton reaction as described in Section 5.1.4.3. Finally, the active site as well as the enzyme surface might further evolve to catalyze distinct aminoacylations, in this gaining way more exclusive substrate specificity.

The enzymatic specificity of AARS affected in this way leads to cognate tRNA charging with noncanonical amino acids. This leads to a change in the "identity" of the coding triplet during translation (i.e. codon reassignment). An intriguing question in this context relates to the possibility of one amino acid being a substrate for two AARS. This would allow for isoacceptors from two different tRNA species to be loaded with the same amino acid and to have the same "identity". Namely, it is well known that the relative concentrations of a tRNA and its cognate AARS are normally well balanced and crucial for maintenance of accurate aminoacylation [6–9]. This opens the perspective to use the kinetic knowledge about relative rates of competing aminoacylation reactions in order to expand the amino acid repertoire. Tirrel and coworkers [10] clearly demonstrated that norleucine and norvaline are not only substrates for MetRS, but also for LeuRS, with attenuated editing activity. Similarly, chlorinated and brominated Phe analogs were equally good substrates for engineered PheRS from *E. coli* as well as TyrRS from *Methanococcus jannashi* (see Fig. 5.11). In their more recent study, Tirrel and coworkers [6] provided an impressive example of how variations of cytosolic concentrations of AARS in bacterial expression hosts can allow a single RNA message to be read in different ways. In particular, the identity of the noncanonical amino acid (2*S*,3*R*)-4,4,4-trifluorovaline can be changed in protein translation by its assignment either to isoleucine or to valine codons, depending on the intracellular levels of the isoleucyl- or the valyl-tRNA synthetase in the bacterial expression host (Fig. 6.1). It is reasonable to expect that future experimental efforts to expand the amino acid repertoire will unravel additional or similar cases.

6.2.2
A Barrier between Allowable and Nonpermissive Amino Acids – An Index for Entry in the Genetic Code

The dramatic development of genetic code engineering in recent decades has led to the common acceptance of the idea that codon reassignment in protein translation can be bypassed under experimental conditions. In this context, the most intriguing question was whether it would be possible to disclose the rules or determinants that allow for an experimental extension of the coding properties of particular triplets. Assuming an adaptive nature of the code, the incorporation of various amino acid analogs using traditional selective pressure-based auxotrophic approaches is easy to conceptualize. Such simple codon reassignments occur by obeying the well-known chemistry rule "similar replaces similar". For example, the Met codon AUG could be reassigned to Ile or Leu, but never to Arg or Lys, as such reassignments would be detrimental. In other words, using "similar replaces similar" substitutions (i.e. modest alterations in the amino acid side-chain) minimizes the disruption of the resulting protein structure. Indeed, in the evolution of proteins, amino acid substitutions are found to occur more frequently between similar amino acids than between dissimilar ones [11]. This is also in agreement with the general "apolar in–polar out" principle in protein building and folding as elaborated in Section 4.5.

The reassignment or substitution among canonical and noncanonical amino acids should follow these principles as well. The role of a polar requirement in delineating these rules and principles is demonstrated by the chromatographic mobility test shown in Fig. 6.2. The polar requirement of the series of isosteric canonical and noncanonical amino acids starting with Met and its counterparts Nle (**2**), SeMet (**3**) and metoxinine (**4**) as well as Leu (**5**) and it analogs 5′,5′,5′-trifluroleucine (TFL, **6**) and azaleucine (**7**), revealed full agreement between amino acid polarities and their capacities for incorporation into proteins. Namely, neutral

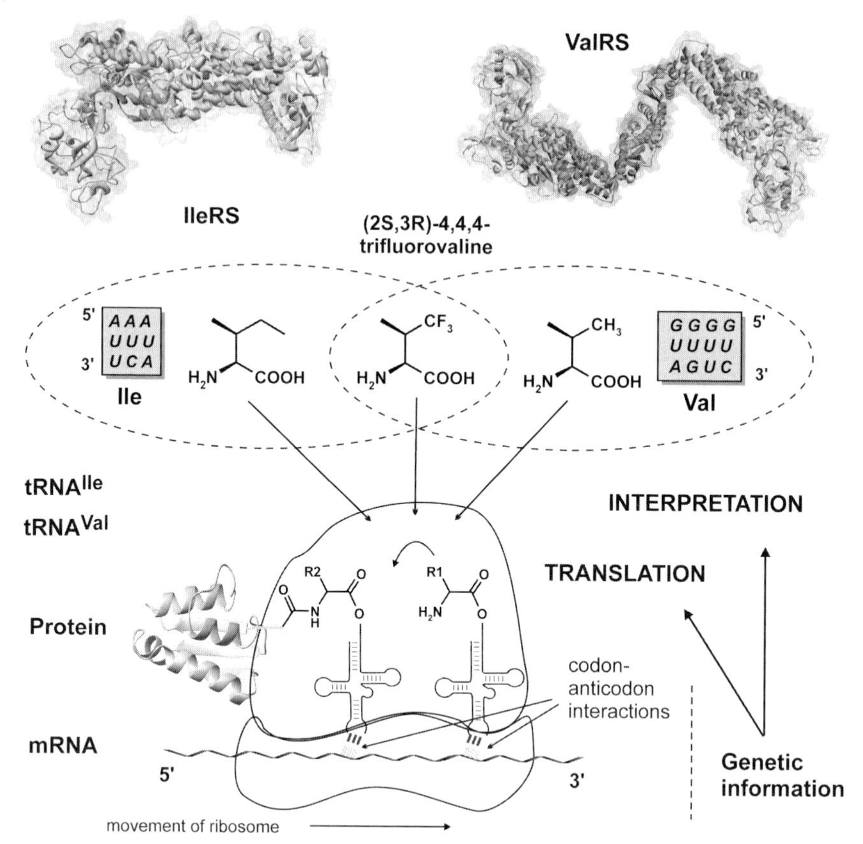

Figure 6.1. Tirrell–Kumar experiment [6]: kinetic control, translational fidelity and amino acid identity during protein synthesis. Both IleRS and ValRS are capable of activating (2*S*,3*R*)-4,4,4-trifluorovaline and transfer it onto cognate tRNAs. Competitive attachment of this amino acid to tRNAs in the cytosol depends on the intracellular concentrations of IleRS or ValRS. Elevated amounts of ValRS lead to reassignment of four valine codons, whereas three isoleucine codons are translated as trifluorovaline in cells outfitted with IleRS. In other words, the identity of the noncanonical amino acid (2*S*,3*R*)-4,4,4-trifluorovaline can be changed in protein translation by its assignment either to isoleucine or to valine codons. This represents a general strategy that allows a single RNA message to be translated in different ways by manipulation of cytosolic concentrations of AARS in expression hosts.

and hydrophobic (apolar) noncanonical amino acids like Nle, SeMet and TFL indeed serve as substitution substrates for the canonical amino acids Met and Leu. However, their isosteric, but hydrophilic (polar), counterparts metoxinine or azaleucine (**4** and **7**; Fig. 6.2) are hardly imaginable as members of the hydrophobic group of amino acids coded by XUX coding triplets. Not surprisingly, Marilere and coworkers [12] used azaleucine as a functional analog for Arg in order to suppress a lethal phenotype.

The main practical lesson from the current knowledge about the code as an adaptive structure is that novel reassignments of coding triplets are strongly corre-

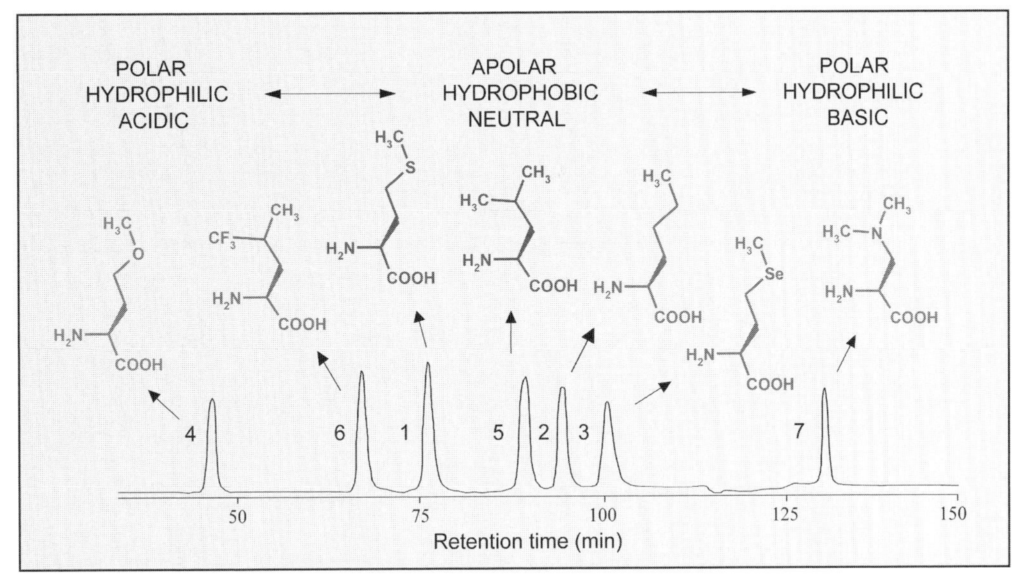

Figure 6.2. Chromatographic mobility of canonical amino acids as a function of their molecular size, polarity, hydrophobicity and p*I* values (experimental conditions are available in [1]). The amino acid property termed "polar requirement" (a measure which indicates whether an amino acid is hydrophobic or hydrophilic) is best preserved by the genetic code [13]. In this particular experiment, the polar requirement was determined for canonical and noncanonical amino acids of Met (**1**) and its isosteric analogs Nle (**2**), SeMet (**3**) and metoxinine (**4**) as well as leucine (**5**) and its isosteric analogs 5′,5′,5′-trifluroleucine (**6**) and azaleucine (**7**). Since they have almost identical molecular shapes, their separation is based mainly on differences in polarity and p*I* values. Hydrophobic amino acids Leu and Met are chromatographically positioned in the "neutral apolar area" close to the signals for Val, Ile and Phe (XUX codon family). Conversely, their isosteric analogs metoxinine (**4**) and azaleucine (**7**) are in the acidic and basic "polar areas". They can be exceptionally incorporated into particular proteins on single predefined sites; however, proteins cannot be globally substituted with hydrophilic (polar) azaleucine or metoxinine since it would have an adverse effect on protein folding. Therefore, the positions of chromatographic peaks for these noncanonical analogs revealed the rule for codon redefinition among similar amino acids. These simple chromatographic relationships are in full agreement with the intrinsic structure of the genetic code. They might be used for the prediction of which noncanonical amino acids are allowed the entry into the genetic code.

lated with the biophysical relationships among amino acids. Thus, the existence of a firm boundary between "allowable" and nonpermissive amino acids in the genetic code is obvious. This boundary can be delineated by simply measuring differences in their physicochemical properties (Fig. 6.2) or from substitution matrices calculated from observed replacement patterns in proteins [14]. It was already elaborated how the "polar requirement" index captures remarkably well an order in the code table. As shown in Fig. 4.2, the genetic code assigns amino acids to related triplets on the basis polarity/hydrophobicity. In this context, it would be interesting to see if there is a quantitative value (i.e. index) for the possibility of a desired synthetic amino acid to enter the genetic code. Future work on top-down approaches in code engineering should decide to which extent criteria other than the polar requirement (hydropathy, molecular volume, isoelectric point, position in the genetic code table and putative protein structure, etc.) might be more (or less) important in developing of such indices. Such indices would also be a valuable tool for the bottom-up design approaches as well (Sections 6.4.2 and 6.4.3). A novel "intrinsic" codon reassignment designed experimentally should be chemically plausible. Every amino acid "privileged" with entry in such a tailored genetic code must be optimally designed to fulfill the selective requirement for polarity, hydropathy, molecular volume, isoelectric point, position in the protein structure with codon optimized template, etc. The first practical experiments have been underway for more than a decade; carefully designed experimental selective pressures exploit the fact that there is room for an increase in the diversity in the code vocabulary. In this way, adaptive features of the code, transcending those found in nature, would continue to evolve in the laboratory.

It is imaginable that position-specific incorporation by suppression-based methods might exceptionally allow the introduction of such nonpermissive amino acids at predefined positions. Indeed, these methods mimic natural preprogrammed context-dependent re-coding (see Section 3.10) and will further evolve to allow position-specific incorporation, mostly at single amino acids regardless of their stereochemical properties (and plasticity of protein structures usually allows for tolerance in such interventions). On the other hand, the above codon reassignment rules and related indices that will be developed in future research will be fully valid in "real" code engineering, i.e. proteome-wide substitutions/insertions that would yield synthetic life with novel chemical possibilities.

6.2.3
Protein Structural Types and Amino Acid Substitution Capacity

Under the influence of various selective pressures, living cells evolved molecules of the most efficient size and shape for their function. Basic monomeric protein building blocks having an average size of an amino acid can be divided into 20 unambiguous types apparently fulfilling these criteria. One can speculate that with more building blocks used in protein synthesis, mistakes could be so frequent that they would seriously outweigh the advantage of having more types. In the analog/surrogate incorporation experiments the aminoacylation capacity itself is no guarantee for successful translation of a particular analog. For example, thiaproline and selenoproline (see Section 5.2.1.5) are both of recognized, selected and loaded onto cognate tRNAs in the aminoacylation reaction by ProRS, but only thiaproline can be fully translated into target proteins. In this particular case, selenoproline insertion into the target protein sequence initiates premature termination of target mRNA translation, via an unknown mechanism [15].

However, it would be even more interesting to find a case where one amino acid is a substrate for one protein substitution, but not for another. In this case, a novel feature of translation might be revealed – the highly speculative possibility

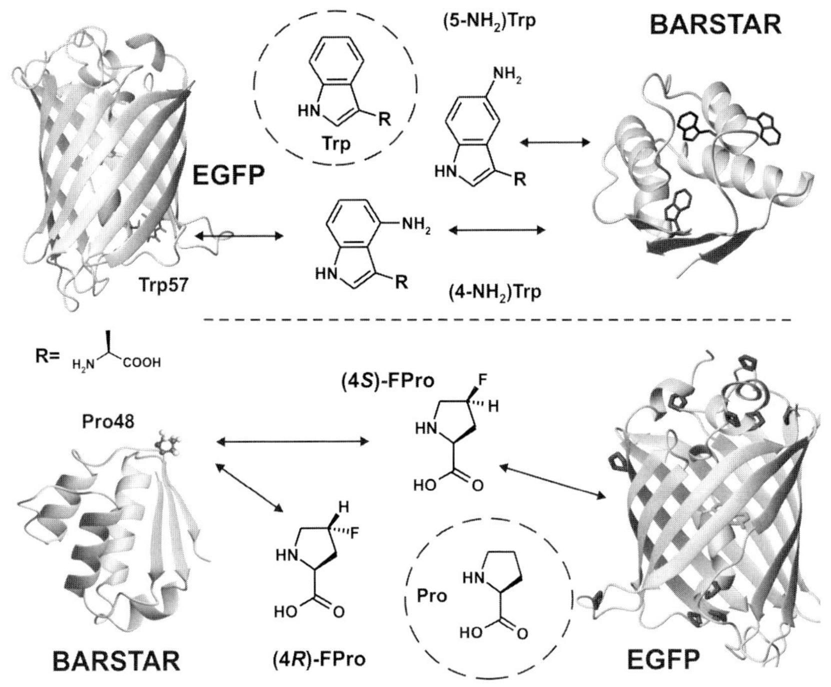

Figure 6.3. Protein structural types and substitution capacity during protein expression in bacterial host cells. The Trp analogs 4-aminotryptophan [(4-NH_2)Trp] and 5-aminotryptophan [(5-NH_2)Trp] can substitute all Trp residues (Trp53, Trp44 and Trp38) in barstar (α-helix/β-sheet protein), whereas a single Trp residue (Trp57) in EGFP (exclusively β-sheet protein) can be substituted only by (4-NH_2)Trp. Similarly, it is possible to express barstar variants that contain a single Pro48 replaced by *cis*-fluoroproline [(4*S*)-FPro] and *trans*-fluoroproline [(4*R*)-FPro], whereas Pro side-chains in EGFP can be replaced only by (4*S*)-FPro. This is not surprising since there are 10 Pro residues in the native structure of EGFP in *cis* and only one (Pro56) in *trans*. Obviously, the structural context is decisive criteria for such replacements. Interestingly, human recombinant annexin-V with predominant α-helices (containing five Pro residues in the *trans* conformation and a single Trp) is permissive for replacement with all the above-mentioned analogs.

that the structure, folding or even folding pathway of the target protein determines the translation capacity. Indeed, such examples were recently reported. Using the selective pressure incorporation (SPI) approach in heterologous expression in *E. coli* Trp auxotrophic host strains it was demonstrated that it is possible to incorporate 4-aminotryptophan, 5-aminotryptophan, 4-hydroxytryptophan and 5-hydroxytryptophan (Fig. 6.3) into barstar, a small protein inhibitor of RNase barnase [16]. However, under the same experimental conditions the expression of *Aequorea victoria* enhanced green fluorescent protein (EGFP) was possible only with 4-aminotryptophan [17]. The expression of related barstar variants undoubtedly indicates that it is principally possible to "pass" all these amino acids through the ribosome. Why then does their translational activity depend of the protein type used for expression experiments?

Another impressive example includes attempts to replace 10 Pro residues in EGFP with *cis*- and *trans*-fluoroprolines (Fig. 6.3). Since in the native structure of EGFP nine Pro side-chains occupy the *cis* conformation, while only one (Pro56) is present in the *trans* conformation, the expression was possible only with the *cis* analog, while all attempts to express protein variants with the *trans* analog failed [18]. However, under identical fermentation conditions using the same Pro auxotrophic *E. coli* strain, the single Pro48 (present in the *cis* conformation) of barstar can be fully replaced by both analogs [19]. Does it mean that there must be a stereochemical bias for ribosomal translation with Pro analogs and surrogates induced by the target protein structure? The reasons for such behavior are still unclear. It may be that certain, still unknown, editing mechanisms during the ribosomal synthetic cycles or cotranslational folding are at work. Studies of cotranslational or assisted (i.e. chaperone-mediated) protein folding as well as ribosome fidelity with various noncanonical amino acids are needed. Only these types of studies will provide novel insights into these still somewhat mysterious (but reproducible) observations and answer the questions to which extent they represent novel and general features of reprogrammed translation.

6.2.4 Building a Direct Link between Metabolism and Reprogrammed Translation

Organisms of all life kingdoms produce a variety of α-amino acids that are end-products of secondary metabolism, arise as intermediates of metabolic pathways or originate from the metabolism and detoxification of foreign substances. Their introduction into the host cells by metabolic engineering experiments as described in Section 5.7.1 might influence both aspects of metabolism, i.e. catabolism and anabolism. Other interesting questions are related to the possibilities of their regulation and integration into existing metabolic circuits. Finally, the effects of the activities of imported metabolic enzymes on cellular metabolism and regulated gene expression are still largely unknown.

Interestingly, the co-evolution theory (Section 6.3.3) postulates that the possibility for genetic code amino acid repertoire expansion took place via precursor amino acid recruitment from developing biosynthetic pathways. Indeed, the biosynthetic machinery for the production of amino acid precursors to proteins is remarkably diverse and the capacity to produce all canonical amino acids is not a feature of all living cells. For example, the absence of GlnRS, AsnRS and even CysRS in some Archaea is compensated for by the existence of indirect biosynthetic pathways that generate related aminoacyl-tRNAs. With respect to their origin, Francklyn [20] divided "proteingenic" amino acids into three major groups: "early" amino acids (Leu, Ile, Phe, Val, Gly, Arg, Ser, Glu, Pro, Thr, Trp, Asp, Cys, Lys, Met and His) are synthesized only by direct metabolic pathways, "middle" amino acids (Gln and Asn) are generated by a combination of direct and indirect pathways, and "late" amino acids such as selenocysteine (Sec) and pyrrolysine (Pyl) are synthesized exclusively by indirect pathways.

The most exciting practical questions in this context relate to the possibilities of using these mechanisms for practical code engineering. This means that an expansion of the coding capacities of the existing genetic code could be achieved if host microorganism cells are additionally provided with metabolic pathways that further modify amino acids attached to tRNA prior to its ribosome translation. Such an aminoacyl-tRNA should be acceptable as a substrate for elongation factor EF-Tu which usually prevents mischarged tRNAs from entering ribosomal translation cycles. If such an experiment were to be performed, it would generate an important novel link between aminoacyl-tRNA synthesis and amino acid biosynthesis. It would also be an attractive alternative to the already described approaches (see Section 5.7.1) for entry of novel amino acids into the genetic code directly from the intermediary metabolism.

6.3 The Amino Acid Repertoire and its Evolution

6.3.1 "Copernican Turn" and the Last Sacrosanct in Biochemistry

It is generally accepted that the genetic code evolved from an earlier form with fewer amino acids. Among the researchers in the field, however, there is no agreement about how this occurred. Some of them share the opinion that the genetic code is still evolving [21], whereas the majority regard it as a basic and invariant structure in all extant life forms established at the time of the universal common ancestor. From the experimentalist's point of view, departures in codon reassignment, although described as "recent evidence for the evolution of the genetic code" [22], are in fact functional adaptations. For example, the well-documented (CUG)Leu → Ser reassignment in some genes of *Candida cylindrica* is a useful adaptation to the stress response [23]. In addition, such codon reassignments are rather species specific and most of them are not disruptive at the level of the protein structure. Therefore, code engineers should consider such "recent evidence" as "shifting" of the canonical amino acid between different codons, since no known genetic code encodes an amino acid other than the canonical 20. This way of reasoning was named by Budisa and coworkers [24] as the "Copernican turn"; it postulates that the amino acid repertoire of the genetic code is conserved and of prime importance in reasoning about its natural and experimental evolution (see Section 4.7). Indeed, it remains true that no studied organism encodes amino acids other than the canonical 20 – a boundary that according to Ellington and Bacher [25] "remained relatively sacrosanct in organismal chemistry". Experimental attempts to develop heterologous expression systems with the possibility for a "standard" → "alternative" amino acid repertoire change at the level of single target proteins were successful in the last decade. Conversely, there are only a few attempts to evolve entirely new organism functions via systematic proteome-wide introductions of noncanonical amino acids. Protein sequences optimized on the basis of 20 amino acids will expectedly show poor performance upon substitution with noncanonical amino acids. Wong [21] speculated about the possibility "to mutate those sequences to induce an induced fit for the analog and in favorable instances even a misfit for the natural amino acids". The combination of traditional genetic methods and experimental evolution has already demonstrated that viruses and microorganisms are indeed effective models for the directed experimental evolution of features such as metabolic pathways [26], resistance [27], growth rate [28], etc. In fact, such a combination of methods was at the heart of the experiment for proteome-wide replacement of valine by α-aminobutyrate [29] (see Section 5.1.4.1).

Interestingly, α-aminobutyrate along with norleucine and norvaline represent a group of noncanonical amino acids which are abundantly generated in prebiotic syntheses, and are found in meteorites [30] and even in interstellar medium [31] (Fig. 6.3). Since their incorporation in proteins induces no adverse structural effects, it is difficult to explain their absence from the repertoire of the universal code. Even an experimental example for possible evolutionary advantage of the presence of such amino acids in cellular proteome is provided. Namely, Pezo and coworkers [5] reported the design of a nonreverting strain of *E. coli* with impaired editing activity of endogenous IleRS which under Ile starvation conditions achieved higher culture growth rates with analogs such as norvaline than the wild-type *E. coli* strain. Not surprisingly, the catalytic site of IleRS deprived of the editing function (see Section 5.1.4.5) is not capable to distinguish between Val, Cys, norvaline, *O*-methylthreonine or *O*-methylserine. Does it mean these primordial amino acids were first present in the early code and later eliminated from its repertoire? Although such and similar questions will most probably remain unanswered, current experiments have already shown that their "return" to the code structure is indeed plausible, possible and might even be highly desirable.

6.3.2
Spontaneous Terrestrial and Extraterrestrial Generation of Amino Acids

Geochemistry provided solid evidence that conditions for the appearance of life already existed on Earth around 4 billion years ago. The cells with the present-day code organization emerged relatively quickly after that event, either through convergent or divergent evolution in the context of the postulated (and widely accepted) prebiotic and RNA world [32]. Such fast establishment of the code with its rather sophisticated structure was also explained by the less widely accepted "Cosmic Panspermia" theory [33]. This theory was first proposed by the Swedish chemist Arrhenius [34], who postulated that the information necessary for life to start evolving was transmitted to the Earth. Regardless of the origin of life components, the concept of the spontaneous evolution of life is nowadays widely accepted. This process is based on assembling polymeric molecules from building blocks by means of predefined combination rules [35]. The number of monomeric building blocks is defined, while their transient assemblies (i.e. polymers) become entities of great variety, some of which are capable of function. The organized polymeric structures in living beings originate through self-organization under nonequilibrium conditions facilitated via self-enhancement or cooperativity [36]. These structures are templated and the choice of their basic building blocks is dictated by the need to implement biological functions carried out by them. In this process of assembling, evolution led to the accumulation of structures/functions on the basis of their selective advantages.

Almost all theories about the origin of life postulate simple organic molecules in aqueous solution to build a milieu called the "primordial soup". This concept originated in the first part of last century when Oparin and Haldane [37] proposed that the primordial atmosphere of the Earth, like that of the outer planets, was a reducing atmosphere. It contained almost no oxygen, and was rich in hydrogen and hydrogen donors such as methane and ammonia. This inspired the Miller–Urey experiment [38], which in 1951 began the era of experimental prebiotic chemistry. They designed an experiment in a flask with heated water, and an "atmosphere" consisting of methane, ammonia, hydrogen and the circulating water vapor. After exposing this mixture to a continuous electrical discharge ("lightning"), they could detect aldehydes, carboxylic acid and amino acids (both D- and L-enantiomers). First analyses revealed glycine, alanine, α-amino-*n*-butyric acid, α-aminoisobutyric acid. Later, after more sophisticated analytic instrumentation become available, other canonical amino acids originating from the Miller–Urey experiment could be detected as well: Val, Ile, Leu, Pro, Asp, Glu, Ser and Thr [39]. However, the number of noncanonical amino acids was overwhelming, including norleucine, norvaline, α-aminobutyric acid, *N*-methylglycine (sarcosine), β-alanine, α-aminoisobutyric acid, α-amino-*n*-butyric acid, β-amino-*n*-butyric acid, γ-amino-*n*-butyric acid, isovaline, *N*-methylalanine, *N*-ethylglycine, β-aminoisobutyric, pipecolic acid and even diamino acids (Fig. 6.4). In 1961, Oro [40] succeeded in generating them by even simpler chemistry. For example, glycine, the most abundant amino acid generated in these experiments, can be formed from formaldehyde, cyanide and ammonia in a Strecker reaction. Another "hydrothermal" set of theories explains the origin of genetic monomers such as amino acids by nonenzymatic synthesis via a complex cycle of chemical interactions which are entirely geochemically and geobiologicaly controlled [41]. Hydrothermal vent environments are indicated as well-suited places for amino acid synthesis since they prove a milieu that permits their exergonic formation as shown by Amend and Shock [42]. They performed thermodynamic calculations which demonstrated that the autotrophic synthesis of all 20 amino acids is energetically favored in hot (100 °C), moderately reduced submarine hydrothermal solutions.

The delivery of extraterrestrial organic molecules to Earth by cometary dust, meteorites or interplanetary dust particles might be an equally important source of prebiotic amino acids on earth – a view known as "impact theory" (which is also based on doubts about the reductive character of the early atmosphere) [43]. Indeed, the analyses of the carbonaceous Murchison and Murray meteorites [44] showed that they contain a number of the same amino acids that Miller identified in roughly the same relative amounts [30]. However, the levorotatory amino acids were dominant in such "meteoritic" amino acids mixtures, indicating an extraterrestrial origin for homochirality, i.e. optically active amino acids were present in the early solar system before life began [45]. The hypothesis that circularly polar-

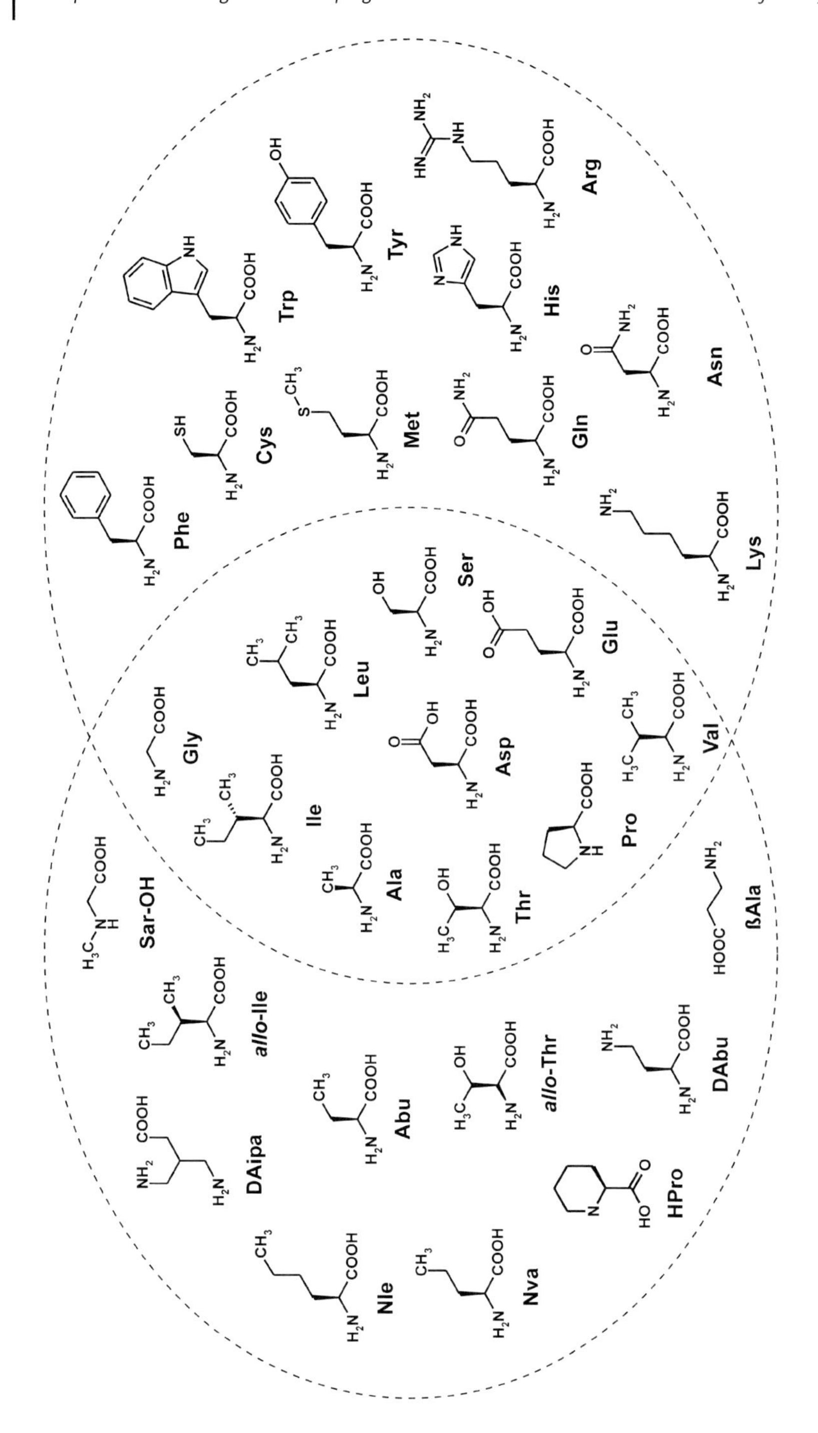
Phe
Cys
Trp
Tyr
Met
His
Arg
Gln
Asn
Lys
Ser
Leu
Glu
Gly
Asp
Val
Ile
Ala
Thr
Pro
Sar-OH
allo-Ile
ßAla
DAipa
Abu
allo-Thr
DAbu
HPro
Nle
Nva

ized ultraviolet light bathed the dusty cloud that condensed into our own solar system and preferentially destroyed the right-handed amino acids [46] correlates well with recent experiments performed under interstellar-like conditions [31, 47].

6.3.3
Metabolic Routes for Amino Acid Syntheses and Co-evolution Theory

From the canonical set of amino acids only a few chemically and structurally simple substances such as Ala and Gly are abundantly available in prebiotic synthesis; most of the other canonical amino acids are present in trace amounts. Such a partial overlap between the amino acids used in the universal code and those available from prebiotic sources as shown in Fig. 6.4 suggests that most of the remaining amino acids are generated metabolically. The connections between amino acid biosynthetic pathways and metabolism are indeed obvious. They share at least two common features: (i) the nitrogen atom of α-amino groups in amino acids originates from NH_4^+, and (ii) the sources of skeletal carbon atoms are intermediates of the tricarboxylic acid cycle and other metabolic pathways. These carbon atoms are assimilated in the form of CO_2 (inorganic carbon), a process which is in autotrophic microorganisms is directly linked to basic metabolic pathways [48].

The lack of production of half of the canonical amino acids in model prebiotic syntheses, along with the instability of amino acids such as glutamine and asparagine, was basis for Wong's proposal of the co-evolution theory [49]. This theory postulates that some canonical amino acids emerged as "members" of the genetic code repertoire only after their synthesis within the primordial biochemical system. This theory predicts the recruitment of novel amino acids into the code repertoire as soon as metabolism invents them, making it possible to reconstruct the history of the contemporary code by detailed analyses of synthetic pathways. In

Figure 6.4. Prebiotic (left circle) and genetically encoded (right circle) amino acids (L-enantiomers). Canonical amino acids generated (at least in traces) in classical Miller–Urey experiments, by an experimental setup that mimics the interstellar medium or found in carbonaceous meteorites are Pro, Thr, Ser, Glu and Asp (Gln and Asn are not stable under prebiotic conditions). Simple canonical aliphatic acids such as Val, Leu and Ile are more easily detected in such experiments; Gly and Ala are even abundantly present. Most recent analyses of the carbonaceous Murchison meteorite revealed the presence of more than 70 extraterrestrial (overwhelmingly noncanonical) amino acids and several other classes of compounds, including carboxylic acids, hydroxy carboxylic acids, sulphonic and phosphonic acids, aliphatic, aromatic and polar hydrocarbons, fullerenes, and heterocycles, as well as carbonyl compounds, alcohols, amines and amides. Typical noncanonical prebiotic amino acids are: (i) diastereomers of L-Thr (2*S*,3*R*) and L-Ile (2*S*,3*R*) called L-*allo*-threonine (*allo*-Thr, 2*S*,3*S*) and L-*allo*-isoleucine (*allo*-Ile, 2*S*,3*S*); (ii) simple aliphatic amino acids such as norvaline (Nva), norleucine (Nle) and α-aminobutyric acid (Abu) that are shown to be good candidates to "invade" proteomes of extant cells by substitution of canonical counterparts (see Section 5.1.4.1); and (iii) cyclic aliphatic amino acids such as pro-surrogate pipecolic acid (homoproline, HPro), *N*-methyl-amino acids like sarcosine (Sar-OH), β-amino acids (such as β-alanine, β-Ala) and even diamino acids (diaminobutyric acid, DAbu; diaminoisopentanoic acid, DAipa).

this way, the genetic code system would be a sort of an imprint of the prebiotic pathways of amino acid formation, since metabolically related amino acids are clustered together. In the frame of this theory the genetic code cannot freely optimize a particular function (e.g. polar requirement); however, amino acid clustering is also proposed to function to reduce damage due to the amino acid replacements arising for single base change [21, 50]. The co-evolution theory divides the time of code invasion into three phases (Fig. 6.5). In phase 1, "precursor" amino acids available in prebiotic syntheses (Gly, Ala, Ser, Asp, Glu, Val, Leu, Ile, Pro and Thr) invaded the early genetic code. In phase II, amino acids (Asn, Gln, Arg, His, Lys, Met, Cys, Trp, Phe and Tyr) were generated by tRNA-dependent modification, such as Gln and Asn which are formed from Glu and Gln in some microorganisms. On the other hand, the phase II amino acid might sufficiently chemically or sterically resemble a precursor, be accepted by a particular AARS and gain access for attachment to cognate tRNA in the competitive aminoacylation reaction. In both cases, novel amino acids took over a subset of codons from their metabolic precursors. In this way Wong identified eight precursor–product pairs: Asp → Asn, Glu → Gln, Gln → His, Ser → Trp, Ser → Cys, Val → Leu, Thr → Ile and Phe → Tyr. Finally, in phase III, the co-evolution theory predicts code repertoire expansion by use of preprogrammed context-dependent translation mechanisms in combination with the recruitment of intermediary metabolism (for these reasons the supporters of this theory believe that the genetic code is still evolving). For example, Sec is generated from Ser-tRNA^{Sec}, which undergoes successive rounds of modifications to yield phospho-Ser-tRNA^{Sec} and Ser-tRNA^{Sec} [21]. Interestingly, Söll and coworkers [51] found that *O*-phospohoserine (Sep) is charged to tRNA^{Cys} in *M. jannashi* by the class II-type enzyme *O*-phosphoseryl-tRNA synthetase (SepRS). Finally, the Sep-tRNA^{Cys} is enzymatically converted to Cys-tRNA^{Cys}. On the other hand, Pyl is directly translated from preprogrammed mRNA sequences by the natural 21st PylRS:tRNA^{Pyl} pair (see Section 3.10).

The validity of co-evolution theory was seriously undermined by Amirnovin [54] and Ronneberg and coworkers [55] who expressed doubts about the connection between the biosynthetic relationships between amino acids and the organization of the genetic code. First, when codon correlations between randomly generated codes and the universal genetic code are compared, many such correlations are also observable within random genetic codes. Second, some interconversions proposed by co-evolution theory are thermodynamically prohibitive since their inversion is mediated by ATP hydrolysis in extant metabolism. Third, some precursor–product assignments are wrong, such as Glu → Arg, which are separated by six enzymatic steps, whereas in the Asn → Arg interconversion there are only two steps from Asn to Arg. Finally, the statistical significance of codon pattern correlation to product–precursor pairs of amino acids is found to be vanishingly small.

Nonetheless, co-evolution theory is doubtless historically important, not only to the theory of code evolution, but also important food for thought for code engineers. The coupling of metabolic engineering with reprogrammed translation was principally shown by Mehl and coworkers [56] recently, and possibilities for linking direct and indirect metabolic pathways with reprogrammed translation are dis-

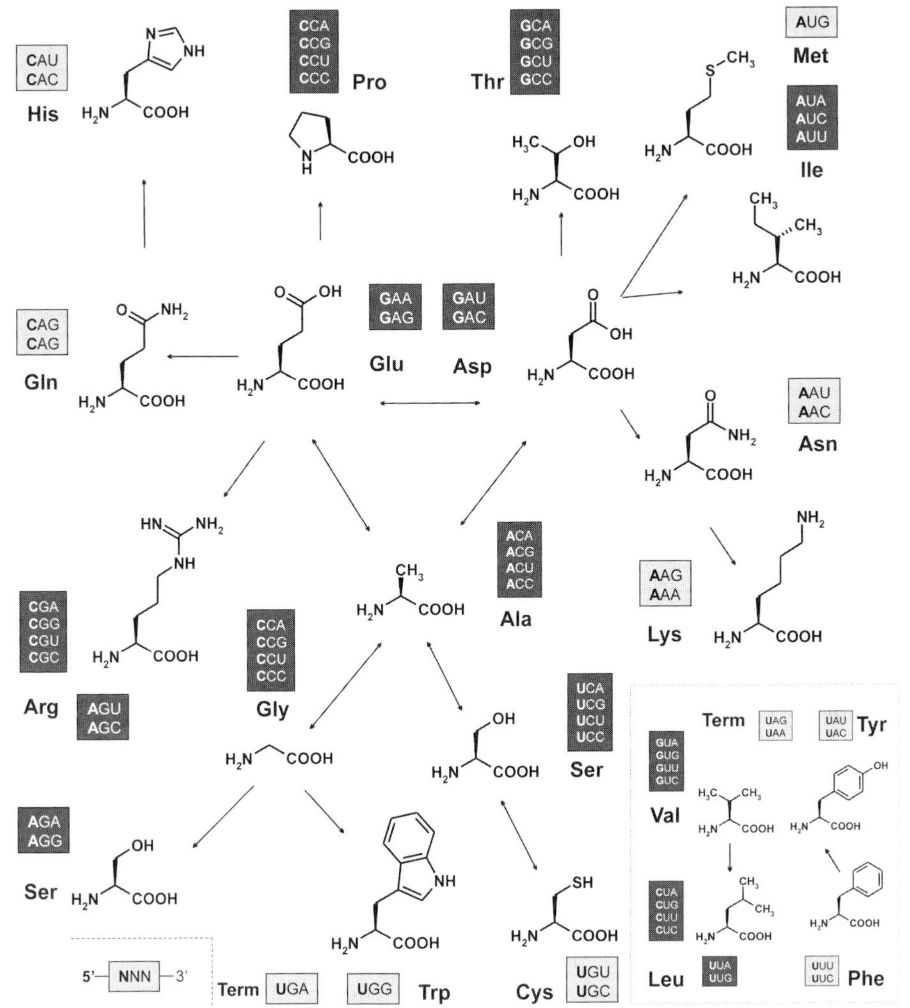

Figure 6.5. Flow chart map representing the connections of biosynthetic relationships between amino acids with the organization of the genetic code as proposed by co-evolution theory of the genetic code. This theory assumes that the amino acid repertoire cannot be freely expanded to maximize the efficiency of a single function (e.g. polar requirement), but must instead co-evolve with the expansion of other cellular metabolic and biosynthetic capacities. The early code used only a small subset of prebiotically generated amino acids (codons in dark boxes) termed here as "precursors" that were coded by an extremely degenerated code. Their metabolic products (grey boxes) were synthesized enzymatically and subsequently integrated into the genetic code repertoire. It should be kept in mind that such a distinction among amino acids (early/late; precursor/product) does not correlate with the two main structural classes of the AARS (see Fig. 3.4), which are believed to have no specific role in the evolution of the genetic code [52]. When compared to precursor codons, the codons for product amino acids are in most cases separated by only a single base change, like Cys/Trp or Ile/Met (they are sibling amino acids derived from a common precursor). Amino acids with the same first codon position tend to be metabolically related: amino acids assigned with GNN codons are prebiotic (Gly, Val, Asp, Glu and Thr), UNN codons usually encode amino acids derived from intermediates in glycolysis (Ser, Cys, Leu, Phe, Tyr and Trp), ANN coding triplets mainly encode amino acids derived from Asp (Asn, Arg and Lys), whereas those encoded by CNN triplets are mainly derivatives from Glu (Gln, Arg and His) [53].

cussed in Section 6.1.4. Since code engineers are less concerned with the patterns of biosynthetic relatedness in the modern code, their future work will certainly demonstrate that the design of a code with improved adaptive features is not necessary exclusive from the expansion of metabolic routes of translationally active amino acids.

6.4 Artificial Genetic Systems and Code Engineering

6.4.1 Cells with Chemically Ambiguous Proteomes – Codon Reassignment Issues

The enrichment of the translational machinery of living cells with artificial components (i.e. code engineering) has been a target for organic and protein chemists for more than a decade. These efforts recently attracted the attention of the scientific literature and the science press. For example, an engineered *E. coli* strain with recombinantly expressed extrachromosomal protein containing *p*-aminophenylalanine (pAF) through suppression of the amber terminal codon [56] (see Section 5.7.1.2) was described as the "first synthetic form of life" [57]. Although this system represents an excellent example of coupled metabolism/translation engineering experiment, it is still far from a "truly unnatural organism". In such and similar experiments, designed coding units (sense, nonsense and nontriplet) are temporally reassigned, whereas permanent reassignments or novel assignments of the coding units, i.e. codon capture (Section 5.7.2), are the first steps towards artificial organisms. In this particular experiment, a host *E. coli* strain would be at least in part a "truly unnatural organism" only if pAF would gain "full membership" in the host proteome. In other words, pAF as a "codon captor" of UAG should be regularly present in all host cellular proteins carrying UAG in their mRNAs, without endangering overall host viability.

The ultimate units of evolutionary selection are not translational components or single proteins, but rather whole cells. Thus, the generation of cells with chemically different genomes (e.g. novel coding units) or proteomes (with expanded chemistry) is the correct approach to generate synthetic life forms. Proteome-wide insertions of novel amino acids have not been reported so far; only the full replacement of a canonical → noncanonical amino acid (Met → SeMet) was first reported by Cowie and Cohen. Interestingly, *Chlorella vulgaris* is capable of growing in the presence of unusual isotopes such as D_2O, although such media inhibits growth of *E. coli* [25]. Recently, a statistical Val → α-aminobutyrate replacement (around 20%) in *E. coli* was also demonstrated [29]. The permissiveness of such substitutions is particularly promoted when "nonfunctional" aliphatic amino acids like Ala, Leu, Ile and Val with limited metabolic choices are globally replaced with chemically and sterically similar counterparts (i.e. analogs and surrogates) (see Section 5.1.4.1). Such experiments the provided basis for speculation about the possibility that early organisms with ambiguous codon assignments (see Section

4.4) were harboring statistical proteins. Chemically ambiguous proteomes might also confer a selective advantage for proto-cells since translational ambiguity might serve as a basic molecular mechanism that linked them with the primordial environment [5], i.e. proto-cells with ambiguous proteomes might more easily spread to new resource patches, serving as evolutionary intermediates towards more advanced cellular forms with a less ambiguous interpretation of the genetic code [4].

The experimental proof for the existence of cells with chemically ambiguous proteomes (i.e. statistical proteins) seems to favor the *ambiguous intermediate (decoding)* theory [58] of code invasion with novel amino acids. However, both theories, i.e. those of *ambiguous decoding* as well as *codon capture* [59] (see Section 4.8.2), successfully explain how it should be possible to insert a new amino acid into the existing repertoire despite a complex genome encoding thousands of highly evolved proteins. While the codon capture theory as a derivative of Kimura's neutral theory [60] resides on neutrality, i.e. codon reassignment without disruptive effects on the proteome, the *ambiguous intermediate* theory postulates a non-neutral mechanism allowing an alternative reading of particular codons with minimal phenotypic impact. Experiments in code engineering show that both theories are not mutually exclusive. Ambiguous reading is an important mechanism, especially for the suppression of missense, nonsense or frameshift mutations, and is also believed to be central for the experimental evolution of alternative genetic codes [61]. In addition, such global changes in proteome composition in highly evolved cells might function well only when rare codons are used as targets for reassignments (i.e. according to Osawa [62], to be reassigned, a codon must become very rare). Codon reassignments occur more or less via ambiguous intermediates in all currently available methods. The methodology used in experiments of pAF incorporation into target recombinant proteins [56] is the approach which is currently most close to codon capture (defined as permanent codon reassignment). As already discussed in Section 5.7.2, it is highly desirable to "cement" experimentally gained read-throughs in the form of permanent reassignment (i.e. codon capture) in order to achieve high fidelity in the translation of the desired noncanonical amino acid, in an adapted organism fully capable of utilizing its novel amino acid chemistry (see Fig. 5.27).

6.4.2
Is it Possible to Improve the Adaptive Features of the Genetic Code?

In a virtual computing experiment, by assuming each coding triplet is assigned to an amino acid independently, there are about 10^{84} possible codes [63]. However, the extant code structure is not random. Coding triplets (created by combination of 4 bases) are assigned to 20 canonical amino acids. Thus, redundancies in the code structure are inevitable. Indeed, most of the amino acids are assigned by more than one codon (i.e. the code is degenerative), with Met and Trp as exceptions. The central issue about the relation of this degeneracy to the physicochemical properties of amino acids is extensively elaborated in Section 4.5 (Fig. 4.3). The regular organization of the universal genetic code, where similar codons encode similar amino acids, is explained by its adaptive nature [64]. The structure of the canonical code was strongly influenced by natural selection for error minimization, i.e. an adaptive code was selected from a large pool of variants or was evolved within adaptive, error-minimizing constraints [65]. This means that at the level of an individual in a single generation, the genetic code structure is optimized by natural selection that keeps encoded proteins "fit" against the mutation and mistranslation [66].

Based on considerations of synonymous versus nonsynonymous substitutions as well as polar requirement measurements, these observations were already conceptualized in 1965 in the form of two similar models ("translation error/lethal mutation minimization") proposed by Sonneborn, [67] Woese [68], and Pauling and Zukerknadl [69]. Shortly after, Alf-Steinberger [70] performed Monte-Carlo simulations with the universal code against 200 alternative randomly generated codes (made by shuffling or partitioning amino acid properties among 64 codons and comparing the average error induced by single-base misreading at each position). She found that the standard code performed better error minimization relative to alternatives. The third and first positions in coding triplets seem to be highly optimized, while no evidence for optimization was found for the second base. Recently, Haig and Hurst [13] confirmed and further refined these results – it was found that out of 1000 virtual codes, only a few out-perform native ones with the respect to polar requirements. Indeed, this property is the best preserved (1 in 10 000) in the universal code. The second position of U in the coding triplets encodes hydrophobic amino acids, i.e. the genetic code efficiently conserves hydrophobicity against random nucleotide mutations (see Section 4.5; Fig. 4.3). Not surprisingly, this property is also an essential driving force in protein folding [71].

Does it mean that that genetic code is the "best of all possible codes"? Although in the context of the living cells the answer to this question might be positive, the fact that the genetic code is result of an interplay between chance and necessity during its evolution history [24, 72] clearly indicates that it is not optimal, as already discussed in Section 4.1. Indeed, one cannot imply that the code, as a product of optimization, is necessary optimal. By considering the generic properties of canonical amino acids represented by the Venn diagram in Fig. 4.2, it is clear that there are areas that define sets of properties not represented by the code's standard amino acids. These gaps can be filled with functional groups such as halides, carbonyls, phosphates, sulfonates, alkynes, azides or other peptide backbone chemistries based on thioester condensation, hydroxy, diamino and dicarboxylic acids or various esters, to mention but a few. The availability of such amino acids makes is possible to speculate about different avenues for code engineering. First, the code might be "improved" without repertoire expansion by permanent replacements of particular canonical amino acids, in a way that, for example, polar distances are minimized and redundancy patterns changed in a desired way. Second, the amino acid repertoire of the genetic code could be enriched and target-engineered code with a much higher degree of optimization than the natural one designed. Finally, the combination of both approaches is also conceivable. The arrangements of artificial code out of a million random trials that performed better than the natural code are known [64]. However, these calculations were based on the standard 20 amino acids, whereas useful theoretical models which include noncanonical amino acids ready to be verified experimentally are still not available.

Apart from these top-down approaches, bottom-up approaches based on the assumption that biogenesis moves from simplicity to complexity are also conceivable in code engineering. Bottom-up approaches might also yield designer organisms built from scratch with genetic codes having completely novel arrangements based on the use of different amino and nucleic acid chemistries in their design. A hypothetical genetic code designed from scratch might be used to remodel current life and progressively develop truly novel life forms in the laboratory. However, it would certainly have chemistries that at least partially overlap with those already represented in the extant code [73]. As already discussed in Section 4.4, the chemistry of basic metabolism [74] as well as protein functions [75] is necessary and deterministic. Therefore, any aqueous carbon-based life found anywhere in the universe would likely have many features in common with those already existing on earth.

6.4.3
Possibilities for *De Novo* Design of Organisms with their "Own" Genetic Codes

Von Neumann, the inventor of the "automata" concept, was among first to speculate about the ways to create artificial life. His "automata" are generally not much different from carbon-based life forms (cellular automata) and could principally develop the same complexity levels. Even at the current stage of development, the main difference between cells and cellular automata is the complexity level. Once such a level of complexity (i.e. artificial life) is generated by conscious human efforts, it would result in biological entities whose performance and functions would transcend those existing in nature. This is the ultimate goal of all efforts to generate artificial life from scratch. To be able to carry on the life functions of living creatures, such cellular automata require inert structures to encode information as well as conservative mechanisms for its high-fidelity transmission. Full transmission of such information would enable its use in order to construct appropriate dynamic configurations for such an evolving system.

Strictly speaking, the design of artificial life is very different from classical genetic methods that only modify biological material that is already alive and functionally optimized during evolution, rather then creating something completely new (i.e. from scratch). For example, the term "protein engineering and design" refers to the modification or redesign of preprogrammed protein modules; these are normally identified by pattern recognition or exchange (permutation) among the canonical 20 amino acids. Although this approach might be expected to be a promising avenue toward artificial life forms, numerous contemporary studies and experimental observations indicate that there are still no general rules for successful *de novo* design of protein of anthropogenic origin. Not surprisingly, attempts for *de novo* engineering and the design of well-structured, stable and functional proteins were described to be "surprisingly difficult" [76]. Therefore, the statement that "we are in the same position as those builders from the beginning of dark middle ages of Western culture which were borrowing foundations and building material from sophisticated Greek and Roman structures" [77] is not an exaggeration at all. For that reason, future developments will be marked by approaches that borrow strategies of natural evolution to a much larger extents as now.

Genetic code engineering, in the context of a complex research field of artificial life design, contributes mainly top-down approaches where existing file forms would be used as platforms for the redesign of life and even *de novo* design. Even then, one has always to keep in mind the immense complications in these experimental efforts. Currently available extensive descriptions of the complexity in biological systems show us that the components of a living cell interact in so many ways that a change in any key component requires compensatory changes in many others. For example, the consequences of a single gene deletion in a particular organism's genome are affected to a large extent by the topological position of its protein product in a complex hierarchical web of molecular interactions. The simultaneous accommodation of more mutations would require coordinated mutational changes throughout the genome. In the case of noncanonical amino acid incorporation, it would be desirable to determine which portions of the proteome would lead to the disturbance of cellular adaptations upon substitution experiments.

Attempts at code engineering would face even bigger challenges – the genetic code is the basic level in the developing scheme of life's complex structure, i.e. the oldest molecular fossil descending directly from the last common ancestor – showing the biochemical uniformity in all living organisms. Therefore, changes in code structure that would yield engineered cells cannot be made without significant and global effects in such a microbial host. For example, the introduction of the third base pair into the DNA helix (see Section 5.4.2) which would consist of six bases A, C, G, T, X and Y (or *isoC* and *isoG*) (see Fig. 5.22) would add novel letters to the genetic alphabet, instead of the contemporary 64 (see Fig. 4.1). This would yield hypothetical organisms whose code has 216 coding triplets. In this genetic code with a greatly expanded amino acid repertoire (i.e. expanded coding capacity), brand new codon–amino acid arrangements will enable the synthesis of novel, truly unnatural proteins that should build a hierarchical web of dynamic molecular interactions with different complexities. Thus, cells with such novel genome/proteome compositions as well as genotype–phenotype–environment interactions are in fact new life forms – taxonomic entities different from extant life forms.

However, these dreams can be realized only when DNA replication, transcription and editing mechanisms are adjusted and optimized to the presence of noncanonical base pairs in the genome. In addition, the design of novel tRNA genes and recognition interactions of novel bases and coding units that contain these bases needed to be introduced at all levels in genetic message transmission. Last, but not least, the phenotypic repercussions are largely unpredictable and in most cases detrimental to the organism itself or its environment. For example, one "rationally" designed AARS:tRNA orthogonal pair might introduce massive mutagenesis throughout the proteome and finally cell death, while cells able to generate "Teflon"-like proteins (see Section 7.7.2) might be toxic to the natural environment. Nonetheless, the benefits from such endeavors will certainly outweigh the potential dangers. Finally, it is reasonably to expect that such creatures would be unable to survive out of the laboratory. Interestingly, Knight and coworkers [78] recently speculated how "the evolutionary history of oxygen metabolism illustrates that even extreme toxicity is a relative and often transitory phenomenon over an evolutionary timescale!".

There is also broad consensus about the fact that the present-day code does not change since it has achieved sufficient scope and accuracy for cells, while any further dramatic change would have a deleterious impact. Unfortunately, earlier stages in the code development are not preserved, although they might represent an almost ideal basis to redirect protein evolution experimentally, and subsequently to establish artificial sources of biodiversity and possibly novel levels of complexity that should enrich life systems on Earth. In evolutionary advanced organisms with relatively large genomes, and corresponding proteomes, it is expectedly to be harder to achieve chemical "evolvability" by expanding the amino acid repertoire of the genetic code. On the other hand, the phenomenon of "genome minimization" or "genome economization" (see Section 4.8.1) observed in both mitochondria and *Mycoplasma* might be used for such a purpose. When compared with a larger proteome, a smaller one would have fewer functions and a smaller degree of interdependence, i.e. less complicated interaction networks. Such economized proteomes are in fact specialized for very efficient replication by reducing the costs of translation (redundant genes are simply discharged). This allowed the abolition of many coding restrictions inherent in the genetic code and helped to re-establish coding rules, i.e. it enabled deviations in the standard codon assignments (see Section 4.7). It remains to be seen whether such genome/proteome organization is amenable to manipulations by introducing novel translational components such as novel tRNA species and the corresponding DNA/RNA containing coding triplets with completely novel chemical meanings (see Section 5.1.4.1). Similarly, Ellington and Bacher [25] adapted small bacteriophage QB to grow in the presence of fluorinated Trp analogs in order to evolve a virus strain whose proteome is invaded with these amino acids.

6.4.4
Code Engineering and Society – Philosophical and Ethical Implications

Progress in natural sciences in general often cannot occur without having an impact on numerous traditional philosophical, religious and ethical issues. For example, the fundamental problem of biological chemistry to understand how translation of a series of instructions stored in DNA as "linear tape" is organized into structured and "smart three-dimensional matter" (i.e. foldable proteins capable of carrying out specific biological functions) [79] is equally fascinating for philosophers as well. The emergence of the protein function of such "smart three-dimensional matter" can be assigned as a teleological (i.e. purposive) process since functionally active protein is the end-directed entity. Conversely, natural selection is usually seen as a statistic process devoid of any suggestion that there are preordained goals (i.e. teleology). Namely, natural selection is a nonrandom process determined by the environment – a sort of statistical bias that increases the probability of otherwise extremely improbable, but successful, adaptive genetic and phenotypic combinations. In attempts to understand the general and intrinsic principles of the genetic code structure and organization by casual analysis, the teleological view is often not welcomed because teleology as a concept relies on noncasual explanations. It is therefore not surprising that a large body of recent and older literature dealing with different linguistic, theological or philosophical controversies related to this matter is available.

Nonetheless, it is quite difficult for an experimentalist to escape teleological language in the description of design. One can principally argue that the genetic code should be deterministic in its nature, therefore *per se* without purpose (i.e. nonteleological). Conversely, in the frame of the widely accepted Darwinian theory, the genetic code as an integral part of the cellular chemistry without doubts functions in the survival fitness of a living being in a given environment. To avoid misunderstandings, Mayr [80] established the term "natural teleonomy" which is related to the appearance of functions in living organisms during evolutionary history as end-directed processes shaped under natural selective pressures without conscious effort or design. Evolution via natural selection is a highly innovative process and teleonomic in its very nature, since the end result of it is improved functional performance in a given environment. In this context, the anthropogenic creation of novel protein functions and functional relationships in living beings that result from conscious human design and engineering efforts would represent an "anthropogenic teleonomy". In other words, no conscious effort is involved in the process of the emergence of biologically active natural proteins, while the engineering of novel functions of artificial proteins and cells with an expanded amino acid repertoire is an example of man-made conscious design.

In the near future a conceivable widespread need for tailor-made proteins and cells will be the main driving force for the engineering of the genetic code by expanding its amino acid repertoire. The emergence of designer organisms with unique genomes and proteomes intentionally designed for performance in user-defined environments probably marks the start of "post-proteome" era. It is difficult to imagine such progress not having an impact on numerous traditional philosophical, religious and ethical issues. Even if so, it would be desirable to refrain from the claim to describe current as well as and future progress as "revolutionary". As everywhere, the interplay of chance and necessity, and even serendipity, cannot be excluded as important routes for discoveries in this research field. Furthermore, it would be highly desirable to always be skeptical of revolutionary talk since the only context in which such developments could take place is with skilled and dedicated experimental labor. It remains to see whether (or to which extent) the progress anticipated in this way will cause or catalyze critical re-evaluation, disappearance of the obsolete and even the emergence of novel ideological, religious or ethical systems in particular societies. Nonetheless, it should not be regarded as a "paradigm shift". It is a child of the relatively young Mendelian and Darwinian tradition which has just started a novel epoch in the history of Western civilization. This epoch is already shaping a radically new picture of the man and his position in the universe, different from those based on the philosophical and literal traditions of Enlightenment and Romanticism of the 18th century.

6.4.5
Future Challenges, Chances and Risks

Apart from these abstract considerations, practical questions arise as to why molecular evolution, with its rather large number of different models, postulates, hypotheses and theories about the evolution of the genetic code, has contributed so little to its experimental engineering [81]. Some hypotheses on code evolution, like how it was shaped via various selective pressures [82], its capacity for translation-error minimization [67] and its flexibility [83], as discussed before, might indeed serve as good guidance or conceptual frameworks in experimental attempts at code engineering. However, the immense complexity of the biological systems and phenomena we are dealing with makes it almost impossible to generate widely accepted (and simple) concepts about the genetic code evolution. For that reason, we might forever remain agnostic about the exact historical sequence of events in code origin and establishment on Earth. The genetic code itself is product of Darwinian evolution and Darwinism *per se* is a historical concept [1]. Thus, work on code engineering will not reveal paths of its evolutionary establishment, but rather avenues for further evolution of the genetic code in laboratory. Darwinian evolution is based on the principle of *survival of the fittest* [84], which in the language of code engineers is an optimization problem – one has to design proteins or cells whose properties are optimally suited to solve a particular problem. Therefore, for code engineering, the experiment always remains the *ultima ratio* to analyze any prediction, hypothesis or expectation. In this context, a recent statement by Yarus [85] that "the immediate future of the code origin appears to be in hands of experimentalists" is easy to understand.

Progress in natural science, in general, as well as in genetic code engineering, in particular, would usually have an impact on society in terms novel technical advances and perspectives, as well as the risks that they bring. For example, building of the code structure based on different set of chemicals from the current one will yield tailor-made proteins and designer cells with novel teleonomic determinants; these will certainly bring about almost limitless technological promise. Such experimental efforts would bring easy conceivable benefits for the whole of mankind: environmental friendly biotechnologies, novel tools in diagnostics as well as in fighting infection diseases, to mention but a few. However, such developments might also bring potential risks as well. For example, it is conceivable to generate lethal viral/bacterial/fungal strains with their "own" genetic codes resistant to all their known defense repertoires. In addition, they might be capable of introducing lethal sense-to-sense or sense-to-nonsense codon reassignments into their hosts and subsequently seriously harm or even kill them either in a selective or nonselective manner. Such a detrimental codon reassignment [Leu(CUG)Ser] is already well documented in some pathogenic fungi such as *Candida albicans*. The reassignment of CUG in phylogenetically related *Saccharomyces cerevisiae* by introduction of the $tRNA_{CAG}$ gene from *C. albicans* yielded a *S. cerevisiae* strain with reduced viability due to the production of aberrant proteins [61]. However, future progress is not to be stopped; its direction will be determined by society itself, its capacity for consensus, its values and its political relations/constellations. These topics, although out of the scope of this book, have an important place in the agenda of social, political and other structures of nearly all contemporary human societies.

References

1 Budisa, N. (2004). Prolegomena to future experimental efforts on genetic code engineering by expanding its amino acid repertoire. *Angewandte Chemie International Edition* **43**, 3387–3428.

2 Copley, S. D. (2003). Enzymes with extra talents: moonlighting functions and catalytic promiscuity. *Current Opinion in Chemical Biology* **7**, 265–272.

3 Langeheim, J. H. (1994). Higher plant terpenpinds: a phytocentric overview of their ecological roles. *Journal of Chemical Ecology* **20**, 1223–1228.

4 Hendrickson, T. L., Crecy-Lagard, V. and Schimmel, P. (2004). Incorporation of nonnatural amino acids into proteins. *Annual Review of Biochemistry* **73**, 147–176.

5 Pezo, V., Metzgar, D., Hendrickson, T. L., Wass, W. F., Hazebrouck, S., Döring, V., Marliere, P., Schimmel, P. and de Crecy-Lagard, V. (2004). Artificially ambiguous genetic code confers growth yield advantage. *Proceedings of the National Academy of Sciences of the USA* **101**, 8593–8597.

6 Wang, P., Fichera, A., Kumar, K. and Tirrell, D. A. (2004). Alternative translations of a single RNA message: an identity switch of (2*S*,3*R*)-4,4,4-trifluorovaline between valine and isoleucine codons. *Angewandte Chemie International Edition* **43**, 3664–3666.

7 Putzer, H., Grunberg-Manago, M. and Springer, M. (1995). Bacterial aminoacyl-tRNA synthetases: genes and regulation of expression. In *tRNA: Structure, Biosynthesis, and Function* (Söll, D. and RajBhandary, U., eds), pp. 293–333. American Society for Microbiology, Washington, DC.

8 Meinel, T., Mechulam, Y. and Blanquet, S. (1995). Aminoacyl-tRNA synthetases: occurrence, structure and function. In *tRNA: Structure, Biosynthesis, and Function* (Söll, D. and RajBhandary, U., eds), pp. 251–292. American Society for Microbiology, Washington, DC.

9 Swanson, R., Hoben, P., Sumner-Smith, M., Uemura, H., Watson, L. and Söll, D. (1998). Accuracy of *in vivo* aminoacylation requires proper balance of tRNA and aminoacyl-tRNA synthetase. *Science* **242**, 1548–1551.

10 Tang, Y. and Tirrell, D. A. (2002). Attenuation of the editing activity of the *Escherichia coli* leucyl-tRNA synthetase allows incorporation of novel amino acids into proteins *in vivo*. *Biochemistry* **41**, 10635–10645.

11 Chlotia, C., Gough, J., Vogel, C. and Teichmann, S. A. (2003). Evolution of the protein repertoire. *Science* **300**, 1701–1703.

12 Lemeignan, B., Sonigo, P. and Marliere, P. (1993). Phenotypic suppression by incorporation of an alien amino acid. *Journal of Molecular Biology* **231**, 161–166.

13 Haig, D. and Hurst, L. D. (1991). A quantitative measure of error minimization in the genetic code.

Journal of Molecular Evolution **33**, 412–417.

14 Jones, D. T., Taylor, W. M. and Thornton, J. M. (1992). The rapid generation of mutation data matrices from protein sequences. *Bioinformatics* 8, 275–282.

15 Di Girolamo, M., Busiello, V., Di Girolamo, A., De Marco, C. and Cini, C. (1987). Degradation of thialysine- or selenalysine-containing abnormal proteins in CHO cells. *Biochemistry International* **15**, 971–980.

16 Budisa, N., Rubini, M., Bae, J. H., Weyher, E., Wenger, W., Golbik, R., Huber, R. and Moroder, L. (2002). Global replacement of tryptophan with aminotryptophans generates non-invasive protein-based optical pH sensors. *Angewandte Chemie International Edition* **41**, 4066–4069.

17 Bae, J. H., Rubini, M., Jung, G., Wiegand, G., Seifert, M. H. J., Azim, M. K., Kim, J. S., Zumbusch, A., Holak, T. A., Moroder, L., Huber, R. and Budisa, N. (2003). Expansion of the genetic code enables design of a novel "gold" class of green fluorescent proteins. *Journal of Molecular Biology* **328**, 1071–1081.

18 Budisa, N. (2004). Protein engineering with noncanonical amino acids. *Habilitation thesis.* TU München.

19 Renner, C., Alefelder, S., Bae, J. H., Budisa, N., Huber, R. and Moroder, L. (2001). Fluoroprolines as tools for protein design and engineering. *Angewandte Chemie International Edition* **40**, 923–925.

20 Francklyn, C. (2003). tRNA synthetase paralogs: evolutionary links in the transition from tRNA-dependent amino acid biosynthesis to *de novo* biosynthesis. *Proceedings of the National Academy of Sciences of the USA* **100**, 9650–9652.

21 Wong, T. T. F. (1998). Evolution of the genetic code. *Microbiological Sciences* **5**, 174–181.

22 Osawa, S., Jukes, T. H., Watanabe, K. and Muto, A. (1992). Recent evidence for evolution of the genetic code. *Microbiological Reviews* **56**, 229–264.

23 Tuite, M. F. and Santos, M. A. S. (1996). Codon reassignment in *Candida* species: an evolutionary conundrum. *Biochimie* **78**, 993–999.

24 Budisa, N., Moroder, L. and Huber, R. (1999). Structure and evolution of the genetic code viewed from the perspective of the experimentally expanded amino acid repertoire *in vivo*. *Cellular and Molecular Life Sciences* **55**, 1626–1635.

25 Bacher, J. M. and Ellington, A. D. (2003). The directed evolution of organismal chemistry: unnatural amino acid incorporation. In *Translation Mechanisms* (Lapointe, J. and Brakier-Gingras, L., eds), pp. 80–94. Landes Biosciences, Georgetown, TX.

26 Mortlock, R. P. (1982). Metabolic acquisitions through laboratory selection. *Annual Review of Microbiology* **36**, 259–284.

27 Kramer, F. R., Mills, D. R., Cole, P. E., Nishihara, T. and Spiegelman, S. (1974). Evolution *in vitro*: sequence and phenotype of a mutant RNA resistant to ethidium bromide. *Journal of Molecular Biology* **89**, 719–736.

28 Lenski, R. E. and Travisano, M. (2003). Dynamics of adaptation and diversification: a 10,000-generation experiment with bacterial populations. *Proceedings of the National Academy of Sciences of the USA* **100**, 6808–6814.

29 Doring, V., Mootz, H. D., Nangle, L. A., Hendrickson, T. L., de Crecy-Lagard, V., Schimmel, P. and Marliere, P. (2001). Enlarging the amino acid set of *Escherichia coli* by infiltration of the valine coding pathway. *Science* **292**, 501–504.

30 Kvenvolden, K. A., Lawless, J. G. and Ponnamperuma, C. (1971). Nonprotein amino acids in the Murchison meteorite. *Proceedings of the National Academy of Sciences of the USA* **68**, 486–490.

31 Munoz-Caro, G. M., Meierhenrich, U. J., Schutte, W. A., Barbier, B., Arcones-Segovia, A., Rosenbauer, H., Thiemann, W. H. P., Brack, A. and Greenberg, J. M. (2002). Amino acids from ultraviolet irradiation of

interstellar ice analogues. *Nature* **416**, 403–406.
32 Gilbert, W. (1986). The RNA world. *Nature* **319**, 618–619.
33 Crick, F. H. and Orgel, L. E. (1973). Directed panspermia. *Ikarus* **19**, 341–346.
34 Arrhenius, S. A. and Borns, H. (1908). *Worlds in the Making: The Evolution of the Universe.* Harper, New York.
35 Schuster, P. (2000). Taming combinatorial explosion. *Proceedings of the National Academy of Sciences of the USA* **97**, 7678–7680.
36 Prigogine, I. (1986). Life and physics – new perspectives. *Cell Biophysics* **9**, 217–224.
37 Miller, S. L., Schopf, J. W. and Lazcano, A. (1997). Oparin's "Origin of Life": sixty years later. *Journal of Molecular Evolution* **44**, 351–353.
38 Miller, S. L. and Urey, H. C. (1951). Production of some organic compounds under possible primitive Earth conditions. *Journal of the American Chemical Society* **77**, 2351–2357.
39 Ring, D., Wolman, Y., Friedmann, N. and Miller, S. L. (1972). Prebiotic synthesis of hydrophobic and protein amino acids. *Proceedings of the National Academy of Sciences of the USA* **69**, 765–768.
40 Oro, J. (1961). Comets and the formation of biochemical compounds on the primitive Earth. *Nature* **190**, 389–390.
41 Wächterhäuser, G. (2000). Origin of life. Life as we don't know it. *Science* **289**, 1307–1308.
42 Amend, J. P. and Shock, E. L. (1998). Energetics of amino acid synthesis in hydrothermal ecosystems. *Science* **281**, 1659–1662.
43 Kasting, J. F. (1993). Earth's early atmosphere. *Science* **259**, 920–926.
44 Hayatsu, R. and Anders, E. (1981). Organic compounds in meteorites and their origins. *Topics in Current Chemistry* **99**, 1–37.
45 Cronin, J. R. and Pizzarello, S. (1997). Enantiomeric excesses in meteoritic amino acids. *Science* **251**, 951–953.
46 Rubenstein, E., Bonner, W., Noyes, H. P. and Brown, G. S. (1983). Supernovae and life. *Nature* **308**, 118–119.
47 Bernstein, M. P., Dworkin, J. P., Sandford, S. A., Cooper, C. W. and Allamandola, L. J. (2002). Racemic amino acids from the ultraviolet photolysis of interstellar ice analogues. *Nature* **416**, 401–403.
48 Michal, G. (1999). *Biochemical Pathways: An Atlas of Biochemistry and Molecular Biology.* Wiley, New York.
49 Wong, T. T. F. (1975). A co-evolution theory of the genetic code. *Proceedings of the National Academy of Sciences of the USA* **72**, 1909–1912.
50 Woese, C. R., Dugre, D. H., Dugre, S. A., Kondo, M. and Saxinger, W. C. (1966). On fundamental nature and evolution of genetic code. *Cold Spring Harbor Symposia on Quantitative Biology* **31**, 723–736.
51 Sauerwald, A., Zhu, W., Major, T. A., Roy, H., Palioura, S., Jahn, D., Whitman, W. B., Yates, J. R., Ibba, M. and Söll, D. (2005). RNA-Dependent cysteine biosynthesis in Archaea. *Science* **307**, 1969–1972.
52 Brooks, D. J., Fresco, J. R., Lesk, A. M. and Singh, M. (2002). Evolution of amino acid frequencies in proteins over deep time: inferred order of introduction of amino acids into the genetic code. *Molecular Biology and Evolution* **19**, 1645–1655.
53 Taylor, F. J. R. and Coates, D. (1989). The code within codes. *BioSystems* **22**, 177–187.
54 Amirnovin, R. (1997). An analysis of the metabolic theory of the origin of the genetic code. *Journal of Molecular Evolution* **44**, 473–476.
55 Ronneberg, T. A., Landweber, L. F. and Freeland, J. S. (2000). Testing a biosynthetic theory of the genetic code: fact or artifact? *Proceedings of the National Academy of Sciences of the USA* **97**, 13690–13695.
56 Mehl, R. A., Anderson, J. C., Santoro, S. W., Wang, L., Martin, A. B., King, D. S., Horn, D. M. and Schultz, P. G. (2003). Generation of a bacterium with a 21 amino acid

genetic code. *Journal of the American Chemical Society* **125**, 935–939.

57 SERVICE, R. F. (2003). Metabolic engineering: Researchers create first autonomous synthetic life form. *Science* **299**, 640–640.

58 SCHULTZ, D. W. and YARUS, M. (1994). Transfer-RNA mutation and the malleability of the genetic code. *Journal of Molecular Biology* **235**, 1377–1380.

59 JUKES, T. H. and OSAWA, S. (1989). Codon reassignment (codon capture) in evolution. *Journal of Molecular Evolution* **28**, 271–278.

60 KIMURA, M. (1983). *The Neutral Theory of Molecular Evolution*. Cambridge University Press, Cambridge UK.

61 SANTOS, M. A. S., MOURA, G., MASSEY, S. E. and TUITE, M. F. (2004). Driving change: the evolution of alternative genetic codes. *Trends in Genetics* **20**, 95–102.

62 OSAWA, S. (1995). *Evolution of the Genetic Code*. Oxford University Press, Oxford.

63 JUNGCK, J. R. (1978). The genetic code as a periodic table. *Journal of Molecular Evolution* **11**, 211–224.

64 FREELAND, S. J. and HURST, L. D. (1998). The genetic code is one in a million. *Journal of Molecular Evolution* **47**, 238–248.

65 FREELAND, S. J., KNIGHT, R. D., LANDWEBER, L. F. and HURST, L. (2000). Early fixation of an optimal genetic code. *Molecular Biology and Evolution* **17**, 511–518.

66 ARDELL, D. H. and SELLA, G. (2002). The impact of message mutation on the fitness of a genetic code. *Journal of Molecular Evolution* **54**, 638–651.

67 SONNEBORN, T. M. (1965). Degeneracy of the genetic code: extent, nature and genetic implications. In *Evolving Genes and Proteins* (BRYSON, V. and VOGEL, H. J., eds), pp. 377–397. Academic Press, New York.

68 WOESE, C. R. (1965). On the evolution of the genetic code. *Proceedings of the National Academy of Sciences of the USA* **54**, 1546–1552.

69 ZUKERKNADL, E. and PAULING, L. (1965). Evolutionary divergence and convergence in proteins. In *Evolving Genes and Proteins* (BRYSON, V. and VOGEL, H. J., eds), pp. 97–166. Academic Press, New York.

70 ALF-STEINBERGER, C. (1969). The genetic code and error transmission. *Proceedings of the National Academy of Sciences of the USA* **64**, 584–591.

71 KAUZMANN, W. (1957). Physical chemistry of proteins. *Annual Review of Physical Chemistry* **8**, 413–438.

72 KNIGHT, R. D., FREELAND S. J. and LANDWEBER, L. F. (1999). Selection, history and chemistry: the three faces of the genetic code. *Trends in Biochemical Sciences* **24**, 241–247.

73 WEBER, A. L. and MILLER, S. L. (1981). Reasons for the occurrence of the 20 coded protein amino acids. *Journal of Molecular Evolution* **17**, 273–284.

74 MOROWITZ, H. J., KOSTELNIK, J. D., YANG, J. and CODY, G. D. (2000). The origin of intermediary metabolism. *Proceedings of the National Academy of Sciences of the USA* **97**, 7704–7708.

75 DENTON, M. and CRAIG, M. (2001). Laws of form revisited. *Nature* **410**, 417–417.

76 DEGRADO, W. F., SUMMA, C. M., PAVONE, V., NASTRI, F. and LOMBARDI, A. (1999). *De novo* design and structural characterization of proteins and metalloproteins. *Annual Review of Biochemistry* **68**, 779–819.

77 ROBSON, B. (1999). Beyond proteins. *Trends in Biochemical Sciences* **17**, 311–315.

78 KNIGHT, R. D., FREELAND S. J. and LANDWEBER, L. F. (2004). Adaptive evolution of the genetic code. In *The Genetic Code and the Origin of Life* (RIBAS DE POUPLANA, L., ed.), pp. 75–91. Landes Bioscience, Georgetown, TX.

79 GALAEV, I. Y. and MATTIASSON, B. (1999). "Smart" polymers and what they could do in biotechnology and medicine. *Trends in Biotechnology* **17**, 335–340.

80 MAYR, E. (2001). *What Evolution Is*. Perseus Books, New York.

81 CAVALCANTI, A. R. O. and LANDWEBER, L. F. (2003). Genetic code: what nature

missed. *Current Biology* **13**, R884–R885.

82 Judson, O. P. and Haydon, D. (1999). The genetic code: what is it good for? An Analysis of the effects of selection pressures on genetic codes. *Journal of Molecular Evolution* **49**, 539–550.

83 Maeshiro, T. and Kimura, M. (1998). The role of robustness and changeability on the origin and evolution of genetic codes. *Proceedings of the National Academy of Sciences of the USA* **95**, 5088–5093.

84 Darwin, C. (1869). *The Origin of Species by Means of Natural Selection, or Preservation of Favoured Races in the Struggle for Life*. Murray, London.

85 Yarus, M. (2000). RNA–ligand chemistry: a testable source for the genetic code. *RNA* **6**, 475–484.

7
Some Practical Potentials of Reprogrammed Cellular Translation

The point ... is to start with something so simple as not to seem worth starting, and to end with something so paradoxical that no one will believe it.

[Bertrand Russell]

7.1
Practical Choice of Methods and Some Controversies in the Field

The basic practical question regarding code-engineering methods presented in this book is that about their efficiency. Are they are mature enough to enable their rapid spread throughout biochemistry laboratories as routine methods? Table 7.1 summarizes the advantages and well as main drawbacks of two commonly used methods in the field. The general limits of both methods have already been discussed in Chapter 5. The most popular methods outside of research in the field are those based on the use of auxotrophic bacterial strains. This is not surprising since their basic advantage is the possibility for sense-codon-directed (residue-specific) substitutions, expression and purification of mutant/variant proteins at wild-type levels, and simple, easy and reproducible techniques. For that reason selective pressure-based methods are the methods of first choice for the production of labeled proteins on a larger scale. Examples include sample-intensive methods of structural biology like nuclear magnetic resonance (NMR) spectroscopy and X-ray crystallography or the production of therapeutic proteins on an industrial scale. The efficiency of most orthogonal pairs is still not high enough to be used in routine laboratory work. This is because most of the orthogonal aminoacyl-tRNA synthetases (AARS) are still characterized by low substrate selectivity and poor catalytic efficiency (see Section 5.1.5.1). To use them for incorporation attempts one need an experimental setup (i.e. fermentation calibration) identical to those for kinetic control incorporation experiments (see Section 5.1.4.3). Literally speaking, the whole field of code engineering is still in the "minimal medium" stage of development and a lot of work is necessary to generate AARS that would work efficiently in rich media (as is usually the case for routine gene expression experiments). There should be no doubt that combinations of these and others methods

Engineering the Genetic Code. Nediljko Budisa

ISBN: 3-527-31243-9

Table 7.1. Advantages versus drawbacks: comparison of two commonly used methods for the introduction of noncanonical amino acids into proteins.

Suppression-based methods	***Auxotroph-based methods***
Advantages	*Advantages*
site-specific incorporation	global incorporation mode
large number of translated noncanonical amino acids	yields comparable to wild-type protein
	simple, easily reproducible technology
Drawbacks	*Drawbacks*
generally low protein yields	limited number of noncanonical amino acids translationally active
expensive and difficult chemistry and biochemistry	
complicated working protocols, not easily reproducible	
read-through only at permissive sites (codon context)	
efficiency: 0–100% (never 100%!)	
Perspectives	*Perspectives*
possible practical applications after overcoming these drawbacks are practically limitless	potential production of tailored proteins on an industrial scale
	combination with other methods and approaches

in a unique approach is the most probable and most promising future avenue to achieve these goals.

"Site directedness" is a highly desirable feature for reprogrammed translation in the field of code engineering. It was often claimed that the suppression-based read-through methods for noncanonical amino acid incorporation are "site specific", because it is theoretically possible to insert a suppressible stop codon at any desired position in the protein sequence. This is most probably not fully correct due to the well-known fact that misacylated tRNA must compete with release factors, which often results not only in low suppression efficiency, but the complete absence of suppression. In this case, it would be necessary to screen for permissible positions that allow for high-level suppression (i.e. the lowest possible "leakage" of the system). However, it is hard to imagine this approach as "site directed" in the terms used in routine oligonucleotide-directed DNA mutagenesis experiments. The term "statistical" incorporation mode (labeling) is also not suitable for the following reasons. Whether the reported incorporation for particular amino acids by the use of read-through methods where suppression efficiency is 10% is "statistical" or "site directed"? Whether the 99% reassignment of the single UGG codon (i.e. Trp substitution) in one particular gene sequence is "statistical" or "multisite" or "site directed"? In fact, whether SeMet incorporation is "multisite", "statistical" or "site directed" depends upon the number of AUG codons that can be easily experimentally manipulated by recombinant DNA technology.

Obviously, in naturally occurring proteins the problem of "site directedness" is solved through evolutionarily optimized codon appearance (frequency) in the particular sequence. This is especially true for relatively rare amino acids like Trp, Cys and Met. In this context, the question whether incorporation of an amino acid is "site directed" or "multisite" is, in the first place, a matter of the codon composition of its coding template. In practical experiments with noncanonical amino acid incorporations there are no problems using codon-optimized gene sequences, i.e. target gene sequences with controlled codon composition. Furthermore, one can choose as replacement targets those amino acids which are naturally less abundant, such as Trp (1.1%) or Met (1.5%), but often play crucial roles in protein functionality. Finally, there should be numerous protein sequences where the problem of "multisite" replacements can be circumvented by a simple combination of site-directed mutagenesis and an expanded amino acid repertoire [1].

7.2 The Plasticity of the Translation Machinery, Amino Acid Generic Types and Protein Structure

Protein synthesis is a universal phenomenon achieved by broadly similar mechanisms across all life kingdoms. Despite the extreme and highly divergent environments to which all species have adapted, the basic translational functions and interactions are conserved. For example, the remarkable conservation of ribosomal components that bind mRNA and the anticodon stem–loops and monitor base pairing between the codon and anticodon, i.e. peptide bond formation, is understandable in the light of the universal structure of tRNAs. Furthermore, over the past decade it has become evident that ribosome-mediated synthesis of proteins allows for a vast numbers of amino acids to be translated into protein sequences (see Figs. 5.7 and 5.21). The remarkable tolerance of the ribosome towards many amino acids ranging from derivatives of biotin to polycyclic aromatic systems is now well documented (Fig. 7.1) [2]. While D- and β-amino acids are not translationally active, substrate analogs with backbone mutations such as α-hydroxy amino acids can be used as substitutes for normal α-amino acids in ribosomal synthesis [3]. Like the canonical imino acid Pro, the backbone ester generated in this way is not capable of forming α-hydrogen bonds in the α-helical structure. In addition, such esters can be easily hydrolyzed by treatment with aqueous base and consequently proteins containing a backbone ester will be cleaved under these conditions [4]. Similarly, many interesting fluorescent probes (e.g. 2-anthrylalanine, Fig. 5.21), electron-accepting and -donating amino acids (e.g. 2-anthraquinonylalanine, Fig. 5.21), and other amino acids whose side-chains possess special functions can be incorporated into protein probes only by using cell lysates. However, due to the cellular impermeability of many biotin-based constructs as well as fluorescent tags, a coupled transcription–translation system in which heterologous proteins are only available is the system of choice to test their translational activity (see Section 5.5.3).

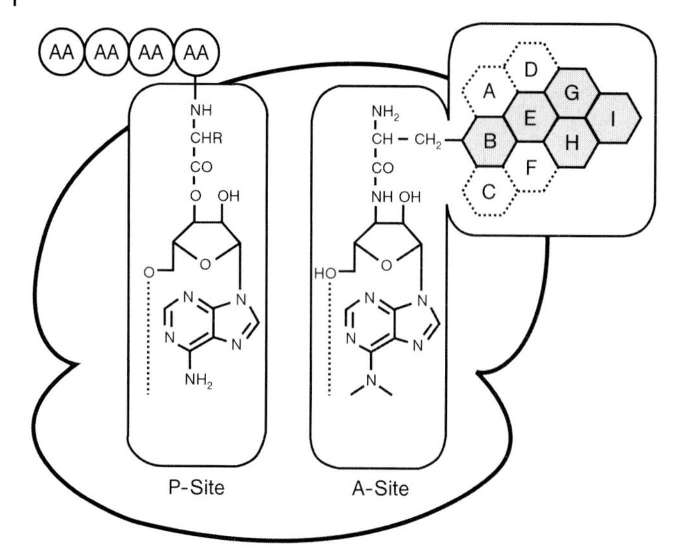

Figure 7.1. Adaptability of the ribosomal A site and structural criterion for efficient incorporation of β-aryl-alanines into proteins. When an aromatic amino acid has benzene rings outside the shaded regions, it will be not be accepted for translation by any *in vitro* coupled transcription–translation system based either on *E. coli* or rabbit reticulocyte lysates. The translational efficiency correlated directly with the adaptability of the ribosomal A site (studied by use of analogs of antibiotics such as puromycin). (Reprinted by permission of Federation of the European Biochemical Societies from Ref. [12]. Copyright 1993. See also Fig. 5.2.1.)

The only conceivable topological context for the incorporation of these large amino acids with chemically and sterically divergent side-chains into proteins is the protein surface. Their canonical counterparts such as Lys or Arg having coding triplets with a central A are polar side-chains (see Fig. 4.3). However, noncanonical amino acids with such features define sets of properties possessed by none of the canonical amino acids (e.g. spin probes, photoswitching, metal binding). In general, surface-exposed amino acids are characterized by a vast chemical and sterical diversity, and can be assigned generically as "divergent types". However, few noncanonical amino acids can be used for global substitutions of chemically relatively uniform, hydrophobic side-chains of amino acids that build the protein core [5]. Indeed, modifications to the protein sequence in the interior of the protein are likely to give rise to important and deleterious effects on protein function as compared to modifications to the protein sequence corresponding to surface locations. This is because the amino acid residues in the protein interior are packed very closely and any change in this sequence may lead to steric hindrance resulting in a change in the overall three-dimensional structure. In other words, core-building amino acids with hydrophobic side-chains are physicochemically very similar or even uniform ("convergent types"), leaving less possibilities for larger variations in their structure and chemistry. In fact, overpacking the core with larger side-chains causes a loss of native-like structure [6]. Ambivalent or amphipathic generic types among canonical amino acids are Met or Trp along with Cys, Thr, Tyr, Ala and the imino acid Pro. Many of their noncanonical analogs and surrogates are expected to be distributed in the protein core, at surfaces and in minicores in an identical manner. See Fig. 7.2. These issues are also extensively discussed in Chapters 4 and Section 6.2.2.

7.3 DNA Nucleotide Analogs: From Sequencing to Expanded Code and Therapy

The use of substrates analogs of endogenous molecules or generic building blocks is also widespread in other research areas and applications of the biosciences. In biomedicine, for example, an interesting strategy for the development of antiretroviral drugs is based on the very same principle as gene sequencing. The antiretroviral agent azidothymidine (AZT, zidovudine triphosphate), a thymidine analog in which the 3′-OH group is replaced by an azido group, inhibits HIV-1 reverse transcriptase activity (Fig. 7.3). It is achieved by competing with the natural substrate thymidine triphosphate and by causing DNA chain termination following its incorporation into viral DNA. In this way, viral DNA growth is terminated due to the lack of the 3′-OH group, which prevents DNA chain elongation. This is possible since HIV and other retroviruses do not encode specific enzymes required for the metabolism of purine or pyrimidine nucleosides. Therefore, many such compounds that exhibit antimetabolic effects or antiviral action that are currently used, or under advanced clinical trials, for the treatment of HIV infections belong to different classes of nucleoside/nucleotide analogs that act as reverse transcriptase inhibitors, like AZT, didanosine (ddI), zalcitabine (ddC) and stavudine (d4T) [8]. These nucleotide analogs work exactly like the dideoxynucleotide analogs (ddNTPs) used in Sanger sequencing of DNA. Namely, Sanger first recognized the possibility for enzymatic incorporation of ddNTPs into a growing DNA strand for DNA sequencing. Sanger and coworkers developed a method based on this experiment that is today at the heart of almost all current DNA-sequencing technologies [9, 10]. The modified nucleotides with detectable reporter groups such as digoxigenin or biotin, fluorochromes, or aliphatic side-chains covalently attached to the base are able to terminate DNA synthesis mediated by a number of DNA polymerases [11]. Finally, manipulation with the polymerase activity is also at the heart of the attempts to create the 65th codon–anticodon pair by using noncanonical nucleoside bases having nonstandard hydrogen-bonding, inter-base hydrophobic interactions patterns or some other chemistries (see Section 5.4.2).

7.4 Noncanonical Amino Acids in Material Science

Material science has recently succeeded in translating many natural concepts such as control of chain length, stereochemical purity or sequence composition into de-

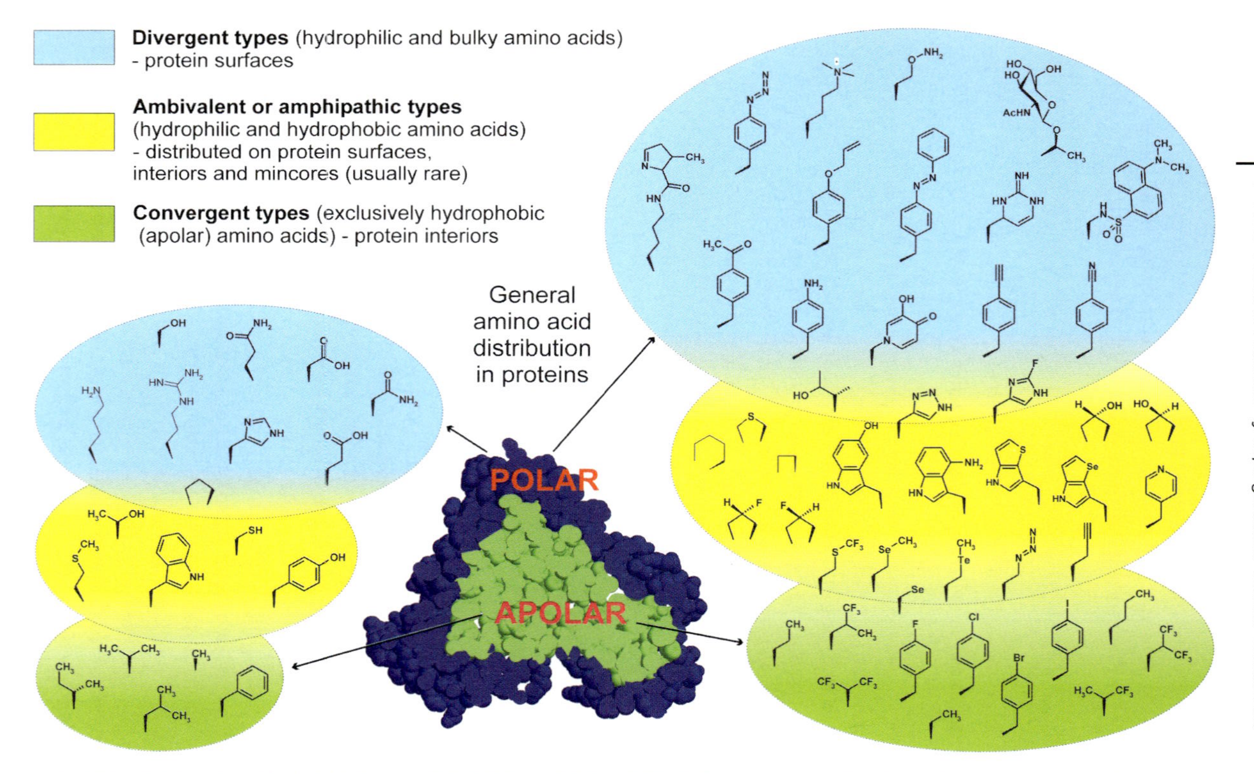

Figure 7.2. Generic types of canonical and noncanonical amino acids in the context of protein building and folding. Beyond the binary partition of polar versus nonpolar amino acids, other boundary sets and subsets such as aliphatic, aromatic, charged, positive, negative, small, etc., can also be defined (see Fig. 4.2). This binary partition of amino acid properties is also reflected in the general protein structure. Based on general residue distributions (polar in–apolar out), proteins are usually folded into tight particles that exhibit a characteristic architectural pattern with a hydrophobic core (interior) that is shielded from the surrounding solvent [7].

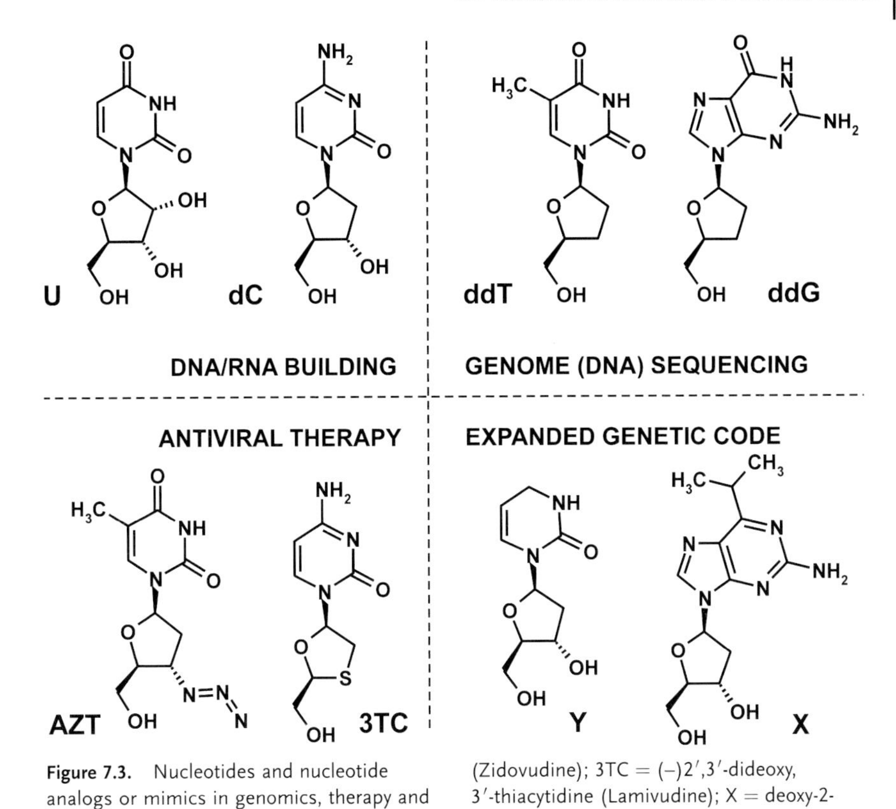

Figure 7.3. Nucleotides and nucleotide analogs or mimics in genomics, therapy and the expanded genetic code (U = uridine; dC = deoxycytidine; ddT = dideoxythymidine; ddG = deoxyguanosine; AZT = Azidothymidine (Zidovudine); 3TC = (–)2′,3′-dideoxy, 3′-thiacytidine (Lamivudine); X = deoxy-2-amino-6-(*N*,*N*-dimethylamino)purine; Y = deoxypyridin-2-one).

signer materials like synthetic silk, collagen, elastin, hydrogels, etc. Namely, for many specific functions, natural selection provided highly sophisticated protein-based structures during billions of years of evolution on Earth. Therefore, it is not difficult to conceive the future interest of material scientists in using reprogrammed translation for the preparation of high-molecular-weight materials with useful optical, electronic or liquid crystalline properties, in sufficient quantity for industrial applications [12]. For example, Kothakota and coworkers [13] reasoned it would be possible to "create conducting materials by incorporation of electroactive amino acid, which could subsequently be cross-linked (or grafted) to produce extended conjugated system". To achieve this goal, they expressed the gene-encoded repeat $[(\text{Ala–Gly})_3\text{Xxx–Gly}]_n$ with Xxx = Phe or thienyl-alanine (see Fig. 5.11) in a Phe auxotrophic strain and obtained recombinant material with more than 90% purity. Electron-rich selenium and sulfur heteroatoms have an unshared pair of electrons in a *p* orbital that can be conjugated with the C=C double bond in aromatic systems. In addition, unlike the atoms of carbon, nitrogen and oxygen,

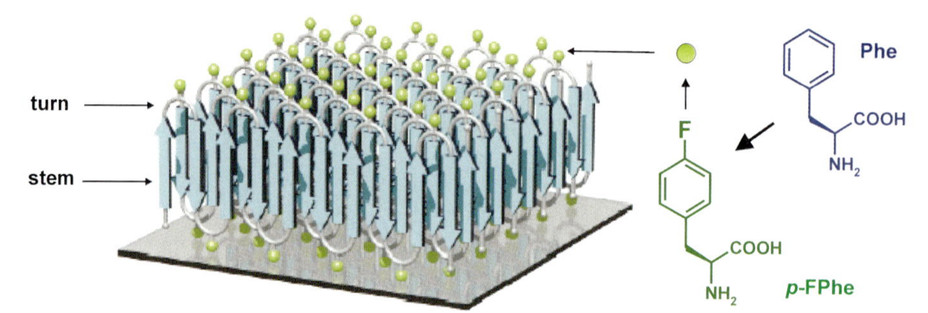

Figure 7.4. Natural and reprogrammed translation machinery provides precise structural control over synthesis of the desired material. A structural model of the lamellar crystalline phase generated by the β-sheet element repeat sequence consisting of stem and turn $[(\text{Ala–Gly})_3p\text{FPhe–Gly}]_n$ in which the *para* position of the phenyl ring is occupied by fluorine. In this way, the lamellar surface is provided with fluorine. It is well known that halogens like fluorine and chalcogens such as sulfur and selenium often have dramatic effects on the physical and chemical properties of materials. (Reproduced from [12] by permission of The Royal Society of Chemistry.)

selenium and sulfur atoms possesses vacant *d* orbitals in the outer shell and can thus act as electron acceptors. Such components readily interact with nearby charges and induce dipoles, additional dispersion forces, polarizability, resonance energy transfer or the formation of charge transfer complexes in the neighboring systems. As a result of these features, sulfur and selenium are capable of providing particular biomaterials with many useful properties like improved optical transparency or electroconductivity; good thermal stability also results from the inductive effects of the sulfur or selenium. In this context it is reasonable to expect that such unusual optical, chemical and thermodynamic properties will be also transmitted to proteins upon amino acid replacements [14]. Thus, it should not be surprising that thio- and seleno-based amino acids such as thienyl-alanine and SeMet are quite attractive targets in material science. Other chemical functionalities and reactive groups not present in the genetic code are also of great interest for material science as a tool to generate, for example, crystalline or liquid arrays with these groups displayed in a regular pattern along the peptide chain or on surfaces. For example, the Pro analog with an alkene group, 3,4-dehydroproline, and the conformationally restricted Pro surrogate azetidine-2-carboxylic acid (see Fig. 5.13), were incorporated into repetitive polypeptides [15, 16]. Similarly, *p*-fluoro-Phe (Fig. 7.4) was incorporated into genetically engineered polymers based on β-sheet element repeats in order to gain features of classical fluoropolymers such as low surface energy, low coefficient of friction, hydrolytic stability and solvent resistance [17] (e.g. Teflon). There is no doubt that future material sciences will profit from reprogrammed protein translation; by including other functionalities such as keto, cyano, azido, nitroso, nitro or silyl groups, alkenes or alkynes, it will become a *conditio sine qua non* in order to generate novel generations of biomaterials.

7.5 Isomorphous Replacement and Atomic Mutations in Structural Biology and Biophysics

7.5.1 Protein X-ray Crystallography

To solve the crystal structure of a novel protein it is necessary to have suitable diffraction markers ("heavy" or "anomalous" atoms). Since such atoms contain large numbers of electrons or exhibit special, anomalous scattering properties, their scattering signal is much stronger than that of the bulk of hundreds or thousands of lighter carbon, oxygen, nitrogen and hydrogen atoms in a protein structure. The traditional means of obtaining heavy-atom-containing proteins is to soak the protein crystal in solutions containing relatively small inorganic or organic molecules with heavy metals such as mercury, platinum, gold, uranium, etc. The heavy atom should diffuse into the crystal though solvent channels and bind to the protein at one or more sites without disturbing its conformation, i.e. the modification should be isomorphous [for that reason the method is termed multiple isomorphous replacement (MIR)]. The initial phases are usually deduced from differences between the diffraction measured from the native and the derivatized crystals. The localization of the heavy atoms in the crystal from the calculated scattering contribution is crucial for this step. These initial phases are then improved in an iterative fashion to gradually reveal the complete structure of the macromolecule in the crystal [18].

Biosynthetic incorporation of heavy-atom-containing amino acids at predefined positions in proteins would enable isomorphous protein derivatization before crystallization. This was the basic motivation behind the recent rediscovery of the classical Cowie and Cohen SeMet incorporation experiment [19] and its use in protein crystallography [20]. SeMet has become a valuable tool for structural biology since SeMet-containing protein crystals are reported to be isomorphous with the native protein crystals, with a difference of 18 electrons at each selenium atom. In addition, selenium-containing proteins proved to be capable of crystallizing under the same conditions as the parent protein and such crystals are normally stable in aqueous buffers under aerobic conditions. Using synchrotron facilities it is possible to measure X-ray diffraction at an optimally chosen wavelength; X-ray absorption edge of selenium (0.98 Å) is easily accessible with synchrotron radiation [21]. In fact, the main prerequisite for the multi-wavelength anomalous diffraction (MAD) method to directly solve the phase problem is the availability of an effective anomalous scatterer of X-rays, such as selenium. In this way, simple *in vivo* substitution experiments proved to be an important aid in evaluating phase angles of the diffracted X-ray waves. In fact, SeMet incorporation into proteins is part of the daily routine in nearly every laboratory dealing with protein structures. Other selenium-containing amino acids, such as Sec [22], TeMet [23, 24] and selenium-containing Trp surrogates (Fig. 7.5), were incorporated into proteins as well. For example,

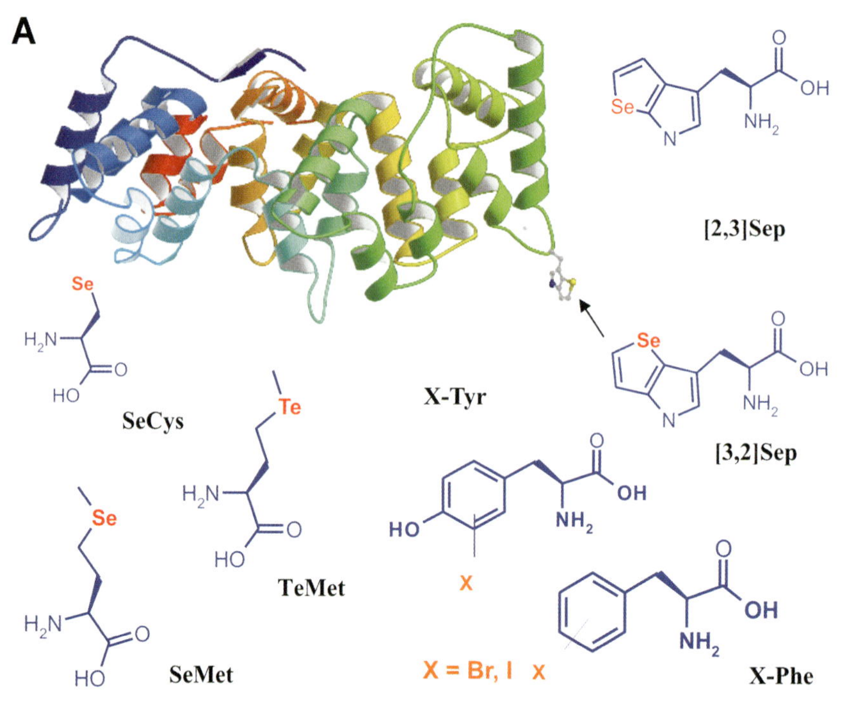

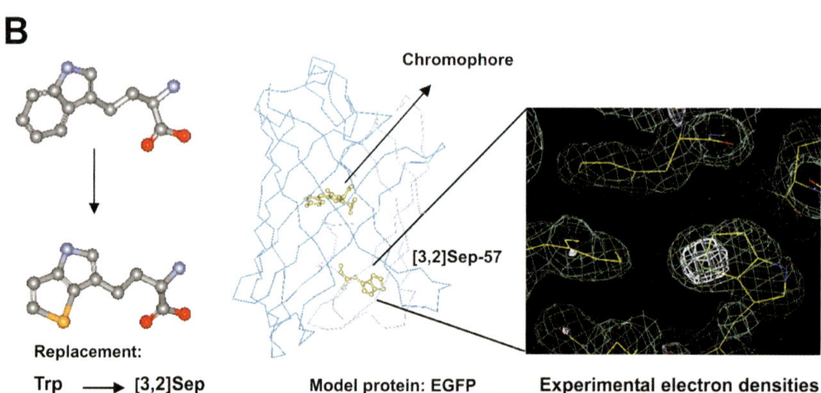

Figure 7.5. Expanded amino acid repertoire and protein X-ray crystallography. (A) Set of translationally active amino acids with "heavy" chalcogen or halogen atoms as suitable diffraction labels for protein X-ray crystallography, with a three-dimensional model of a human recombinant annexin-V containing [3,2]Sep at position 157 occupied by Trp in the native protein. Abbreviations: SeCys = selenocysteine, SeMet = selenomethionine, TeMet = telluro-methionine, [3,2]Sep = β-selenolo [3,2-*b*]pyrrolyl-L-alanine, [2,3]Sep = β-selenolo[2,3-*b*]pyrrolyl-L-alanine, X-Phe = 2-, 3- or 4-bromo(iodo)phenylalanine [alternative nomenclature: *o*-, *m*- or *p*-bromo(iodo)phenylalanine] and X-Tyr = 2- or 3-bromo(iodo)tyrosine [alternative nomenclature: *o*-, *m*- or *p*-bromo(iodo)tyrosine]. (B) The Trp → [3,2]Sep replacement in EGFP leads to isomorphous replacement of the single Trp57 with [3,2]Sep in its structure (middle). (Right) A portion of the unrefined difference electron density map ($F_o - F_c$) of EGFP between Trp57 and [3,2]Sep57 contoured at 3.5σ (white) superimposed on the $2F_o - F_c$ continuous electron density map countered at 1σ (green).

β-selenolo[2,3-*b*]pyrrolyl-L-alanine ([2,3]Sep) and β-selenolo [3,2-*b*]pyrrolyl-L-alanine ([3,2]Sep), that mimic Trp with the benzene ring of the indole moiety replaced by selenophene, were successfully incorporated into human annexin-V, barstar, enhanced green fluorescent protein (EGFP) and dihydrofolate reductase [25, 26]. Yokoyama, Budisa and coworkers [27, 28] reported highly efficient incorporation of SeMet and selenolo-pyrroles into a GFP and Ras protein by using *Escherichia coli* cell-free protein synthesis. In this way, the cell-free synthesis system could also become a powerful protein expression tool for the preparation of derivatized proteins for structure determination by X-ray crystallography.

In classical protein X-ray crystallography it is exceptionally possible to solve a structure using single-wavelength anomalous diffraction (SAD) or single-wavelength isomorphous diffraction (SIR) instead of using the MIR or MAD approaches [29]. The SAD method allows us to calculate the crystallographic phases based on collected diffractions with only one wavelength and one derivative, requires fewer data, and does not necessary need synchrotron radiation. The position-specific incorporation of heavy-atom-containing amino acids might offer a promising tool for this methodology. For example, Schultz and coworkers incorporated *p*-iodo-Phe [30] into proteins in response to an amber stop codon. They expressed a Phe153 → *p*-iodo-Phe mutant of bacteriophage T4 lysozyme and solved its structure by SAD experiments on in-house X-ray sources. The presence of single bulky atoms such as iodine in the tightly packed hydrophobic core of the protein is generally expected to be tolerated due to the well-known phenomena of protein plasticity [31, 32]. Incorporation of selenium, tellurium, iodine and bromine atoms into recombinant proteins can be useful for other emerging approaches for phase calculations in protein X-ray crystallography such as radiation damage-induced phasing [33]. Namely, irradiation of protein crystals by X-rays causes specific changes in their structures such as hydrolyses of disulfide bonds or halide-substituted aromatic rings. In this context, gradually "disappearing" heavy atoms such as bromine, iodine or even tellurium would generate isomorphous changes that might be used to calculate crystal phases.

Code engineering might represent an ideal approach for the production of heavy-atom-containing protein derivatives for phase calculations in protein X-ray crystallography. Preparation of various mutants that contain suitable diffraction labels is further facilitated by the fact that site-directed mutagenesis at the DNA level allows us to manipulate the number of heavy atoms either by introducing or eliminating of rare (e.g. AUG and UGG) codons from gene sequences or by the introduction of suppressible nonsense or quadruplet codons at permissive sites in the target sequences.

7.5.2
Atomic Mutations and Predictable Perturbations in the Protein Structure

The standard amino acid repertoire offers rather limited possibilities for isosteric replacements in protein structures, such as Ser/Ala/Cys, Thr/Val, Glu/Gln and Tyr/Phe. For example, it was recently shown that two isosteric mutations

(Thr203Val and Glu222Gln) in GFP selectively increase the energy of both the anionic and the neutral chromophore form, resulting in the stabilization of an intermediate form [34]. However, there are no such possibilities for canonical amino acids such as Trp, Pro or His, and there is also a limited number of amino acid analogs and surrogates to study, e.g. in terms of the hydrophobicity of the protein core. In fact, the conformational specificity of the protein core is a stereochemical code that discriminates between the native and other conceivable chain folds [35]. One of the possible ways to crack this code might be to use isosteric analogs of core-building amino acids. Unlike site-directed mutagenesis, where several sets of interactions are usually affected, altered protein properties arise solely from exchanges of single atoms, i.e. in the favorable case one can study the importance of a single interaction.

Budisa and coworkers [36] studied annexin-V and its variants with all Met sulfur atoms replaced by selenium and tellurium in order to "analyze distinct contributions of various local interactions to the protein fold, such as hydrogen bonding, van der Waals and hydrophobic interactions, thus simplifying the interpretation of the experimental data". They found that crystals of annexin-V and its atomic SeMet and TeMet variants are isomorphous, whereas in buffered aqueous solution they have secondary structures of typical of α-helical proteins. However, thermally induced denaturing profilesclearly indicate the decrease in the values of melting points and changes in cooperativity of unfolding. The authors argued that altered protein thermodynamic properties arise solely from specific atom exchanges. For example, (i) the electronegativity of sulfur (2.58) is more like that of selenium (2.55) than tellurium (2.0), (ii) the series from sulfur to tellurium includes larger electrovalent and coordination spheres, and (iii) the C–X (X = S, Se, and Te) bond lengths are 1.8, 2.1 and 2.4 Å, respectively [37]. Folding parameters correlated well with the van der Waals radii increase in order: S $\rightarrow$ Se $\rightarrow$ Te. In general, the varied polarities, specific volumes, electronegativities or hydrophobicities in the protein structure yielded folding parameters which correlated well with the physical properties (e.g. solubility) of the free amino acid Met and its chalcogen analogs. Later, Minks and coworkers reported the H $\rightarrow$ F "atomic mutation" approach on the single Trp residue (Trp187) of human recombinant annexin-V [38]. Interestingly, Matthews and coworkers found a significant increase of the stability of phage T4 lysozymes whose 14 Met residues were fully replaced by SeMet [39]. See Fig. 7.6.

More than two decades ago Wilson and Hatfield [41] noted that analogs of canonical amino acids are able to "induce predictable perturbation in protein structure". In the context of protein X-ray crystallography, it is equivalent to the concept of isomorphous replacement [18]. This concept assumes no gross changes in the conformational properties of protein molecules in crystal upon analog incorporation, but an influence in a predictable way on diffraction patterns, enabling phase calculation and subsequent structure elucidation, i.e. atomic coordinate assignment (Section 7.5.1). For biophysicist such replacements at the level of single atoms like H/F and CH_2/S/Se/Te represent "atomic mutations" [36]. They cause minimal changes in the local geometry often far beyond the resolution of X-ray and nuclear magnetic resonance (NMR) protein analyses, but might induce signif-

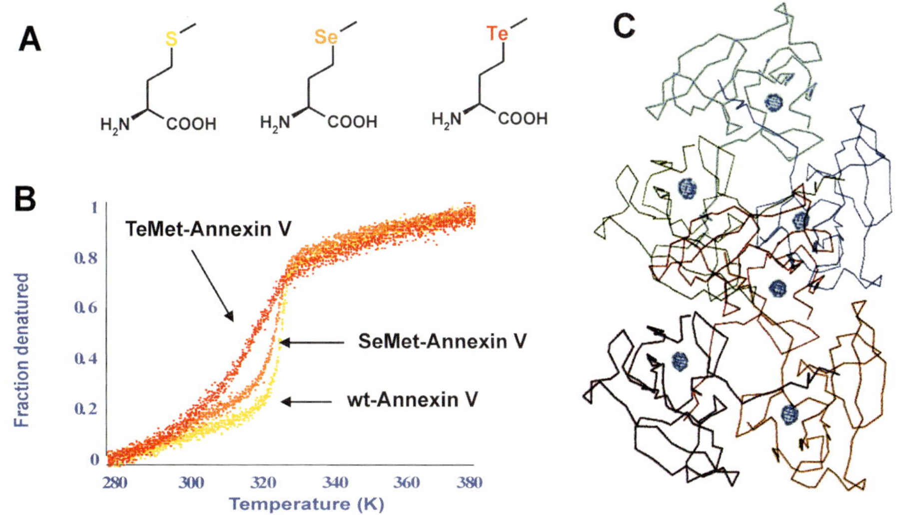

Figure 7.6. The concept of atomic mutations in protein folding. (A) The chalcogen series of Met/SeMet/TeMet amino acids can be seen as a series of atomic mutants with S → Se → Te in their side-chains. (B) Full replacement of Met with SeMet and TeMet in human annexin-V yielded changes in the thermally induced unfolding profiles of SeMet and TeMet variants when compared with wild-type annexin-V. Note that denaturation occurs as a two-state transition from a more or less compact and α-helical ground state to a denatured state. Related variants of human annexin-V have unfolding profiles whose denaturation midpoints are shifted toward lower values with an increase of the atomic number (S → Se → Te). (C) The isomorphous presence of the metallic tellurium in the N-terminal domain of tailspike protein (TSP) consisting of six monomers (one Met residue per monomer). The protein backbone is shown as the C^{α} representation, whereas tellurium atoms are visualized by the calculated difference Fourier map which is contoured at 6σ [40].

icant changes in the context of marginally stable protein structures. In fact, they produce novel interactions in proteins, like enhanced or decreased polarity or electronegativity, and offer the unique possibility to study and understand how these properties are integrated, propagated and modulated into the cooperatively folded protein structure. Although the concept of isomorphous replacement (i.e. predictable perturbations or "atomic mutations") is usually not adequately appreciated in the current literature, its importance is enormous – it can be generally extended from structural biology and biophysics to biomedicine and biological chemistry.

7.5.3
Proteins Enriched with Chalcogen, Hydroxyl and Aza Analogs and Surrogates of Trp

Although Trp as the main chromophore in proteins has the unique advantage of being an intrinsic probe, in some cases it may be less suitable for investigating protein–protein or protein–nucleic acid interactions. Namely, the absorption spectra of nucleic acids overlap that of Trp and thus prevent the assignment of the sole spectral contribution of Trp residues to the total signal output [42]. Similar difficulties arise when protein–protein interactions are investigated since many interacting proteins in multiprotein complexes contain Trp residues, and thus their absorption and fluorescence signals are more or less indistinguishable. The introduction of the so-called "silent fluorophores" such as (4-F)Trp [43] and (7-F)Trp [44] might offer efficient tools to study these issues. The repertoire of "silent" fluorophores as protein building blocks for *in vivo* translation was recently greatly expanded by Budisa's group by introducing sulfur- and selenium-containing Trp surrogates (see Fig. 5.8, Section 5.2.1.1). In these surrogates, the benzene ring in the indole moiety is replaced by thiophene/selenophene, making them mutually isosteric. Sulfur or selenium as heteroatoms cause physicochemical changes in these Trp surrogates, endowing them with completely new properties not found in the native Trp. Related protein variants expectedly retain the secondary and tertiary structure of the native proteins, but exhibit significantly changed optical and thermodynamic properties (Fig. 7.7). These surrogates provide proteins with at least two notable intrinsic spectroscopic properties: (i) altered absorption profiles, and (ii) a fluorescence switch-off due to the efficient static fluorescence quenching by sulfur and selenium atoms.

Proteins tailored in this way are expected to serve as useful tools for investigating protein–DNA interactions or multiprotein assemblies. It is reasonable to expect that the UV spectra in many protein chalcogen variants would not be dramatically changed when there is dominant contribution of Tyr residues to the overall protein absorbance (e.g. in human annexin-V which has only one Trp residue). Conversely, the UV spectral properties of β-(thienopyrrolyl)alanines in solution are fully reflected in the spectra of proteins, such as in barstar where the Trp contribution is dominant. Interestingly, the Trp $\rightarrow$ [3,2]Tpa replacements in human [3,2]Tpa-annexin-V and barstar yield dramatically different circular dichroism (CD) spectra as shown in Fig. 7.7, whereas [2,3]Tpa-containing protein is similar to the native one. Such effects are usually explained by a coupled-oscillator interaction between the aromatic side-chains [45]. An additional explanation for the observed effects might be that these result solely from the intrinsic properties of [3,2]Tpa [46].

The thermal unfolding profiles of chalcogen-containing barstar variants are significantly changed when compared with the native ones. The main reason is that the Trp residue at position 53 is completely buried in the protein interior and plays a central role in protein stability. Around this residue, a clustering of aromatic residues occurs since it is "sandwiched" between Phe56 and Phe74. Such clusters in proteins are often found to form a network of three or more interacting side-chains, and these interactions are supposed to serve as nucleation sites in protein-folding pathways and as the main stabilizing forces of the tertiary structures. The unfolding profiles of [3,2]Tpa-barstar show not only lower T_m values, but the cooperativity of its unfolding process (seen as decreased steepness of the transition profile) is also affected. In contrast, the incorporation of [2,3]Tpa into barstar leads to enhanced folding cooperativity (by almost 15%) when compared with the parent protein (Fig. 7.7). Similar observations are made for human recombinant

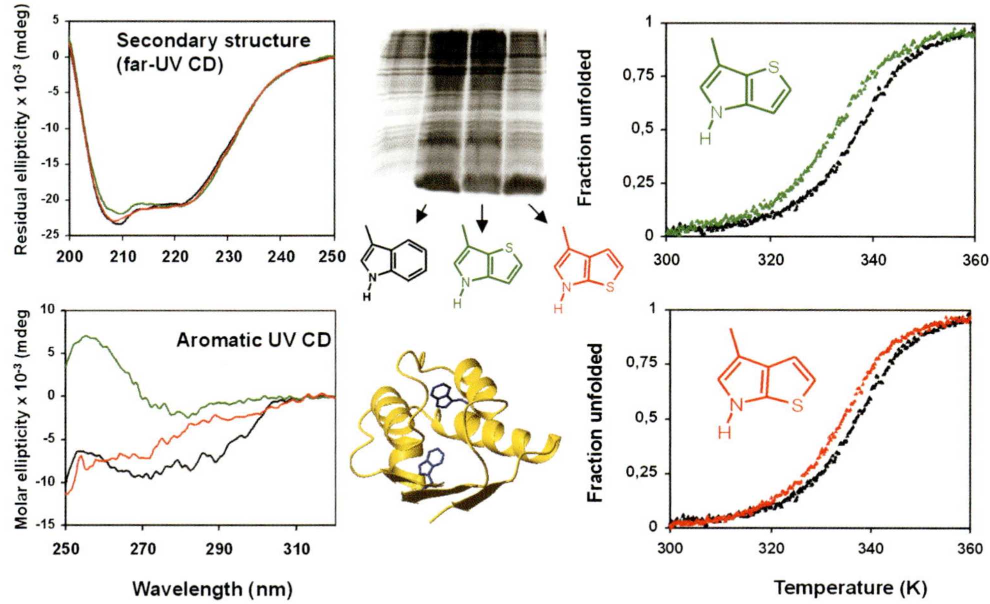

Figure 7.7. Expression profile, CD and thermodynamic analyses of chalcogen variants of barstar mutants with only two Trp residues: Trp44 being partially buried and Trp53 being completely buried in the protein hydrophobic core. There is no difference in expression efficiency between native and variant proteins. The far-UV CD spectra indicate that all three proteins have almost identical secondary structure profiles. The spectral profiles in the aromatic region of native (black) and [2,3]Tpa-containing protein (red) are similar. Conversely, [3,2]Tpa-barstar (green) exhibits a dramatically different CD profile and intensity of the aromatic UV CD spectra. The unfolding profiles of [3,2]Tpa-barstar show not only the lower T_m value, but also that the cooperativity of its unfolding process (seen as decreased steepness of the transition profile) is affected. In contrast, the incorporation of [2,3]Tpa (red) into barstar leads to enhanced folding cooperativity (by almost 15%) when compared with the parent protein.

annexin-V. It is generally observed that isosteric replacements of Met and Trp are accompanied by rather large differences in van't Hoff enthalpies despite the moderate shifts of the T_m values [36].

What is the biophysical basis for such unique behavior of chalcogen-containing proteins? It is well known that all benzene-based amino acids (Phe, Tyr and Trp) exhibit a strong quadrupole moment because of their nonspherical charge distribution [47]. The replacement of the benzene moiety in indole with five-membered hydrocarbon rings containing sulfur or selenium as heteroatoms further complicates the already complex indole photophysics. Selenium and sulfur heteroatoms in their aromatic five-membered rings have an unshared pair of electrons in a p orbital conjugated with the C=C double bond. In addition, unlike the atoms of carbon, nitrogen and oxygen, selenium and sulfur atoms possess vacant d orbitals in the outer shell and can thus act as electron acceptors. Thus, such systems might be more able to interact with nearby charges that might induce dipoles in the system, additional dispersion forces, polarizability, exciplex formation, resonance energy transfer or the formation of charge transfer complexes. Other properties that differentiate them from benzene include enhanced hyperpolarizability and differences in aromatic delocalization energies (for a further discussion, see [46] and references therein).

The Trp analogs (7-Aza)Trp and (5-OH)Trp are particularly interesting as incorporation targets in proteins due to their unique absorption and emission properties. Both have a red-shifted absorption spectrum, i.e. their excitation values are 20 nm red shifted when compared with that of Trp. The insertion of either of these analogs into recombinant proteins makes it possible for selective excitation in the range 315–320 nm [48]. These unique features have been exploited in studying protein–DNA interactions, hirudin–thrombin recognition, integral membrane protein–ligand interactions, phosphorescence of α_2 RNA polymerase, interactions of bacteriophage λcI repressor and toxin-elongation factor 2 interactions, to name but a few (see, e.g. [49]). Interestingly, the incorporation of the 5-OH Trp in annexin-V does not affect the quantum yield dramatically, but induces a red shift of about 14 nm in the fluorescence emission spectrum. The (7-Aza)Trp-containing annexin-V exhibits identical behavior to the (4-F)Trp-protein in aqueous solution at neutral pH [50]. Spectral behavior as well as protein expression with these analogs have been discussed in great detail in a series of excellent reviews and papers written in the last decade by Szabo, Ross and others [42, 51].

The hydroxy group at position 5 of indole in (5-OH)Trp has a pronounced spectral shoulder at 310 nm in the UV absorbance profile of related protein variants. The reasons for such a behavior lie in the nature of the UV transition as well as the fluorescence of indole. They are composed of the two overlapping transitions ^1L_a and ^1L_b that are nearly orthogonal in polarization [52]. It is well known that the hydroxyl group at position 5 of indole leads to a significant enhancement of the ^1L_b transition band and thus to a pronounced spectral shoulder in the absorbance with a maximum at 310 nm [53].

7.6 Translationally Active Amino-Trp Analogs: Novel Spectral Windows and Protein Sensors

7.6.1 Providing Proteins with pH Sensitivity

Motivated by earlier reports on the spectroscopic properties of amino-substituted indoles [53, 54], Budisa and coworkers [55] recently reported the global replacement of Trp residues in proteins with amino-Trp analogs in proteins. Namely, in the repertoire for the Trp-coding triplet (UGG) that has recently been expanded with various novel translationally active chromophores (see Fig. 5.8), only (7-Aza)Trp was interesting with respect to its pH-dependent protonation/deprotonation [48]. The presence of a nitrogen atom at position 7 in the indolyl

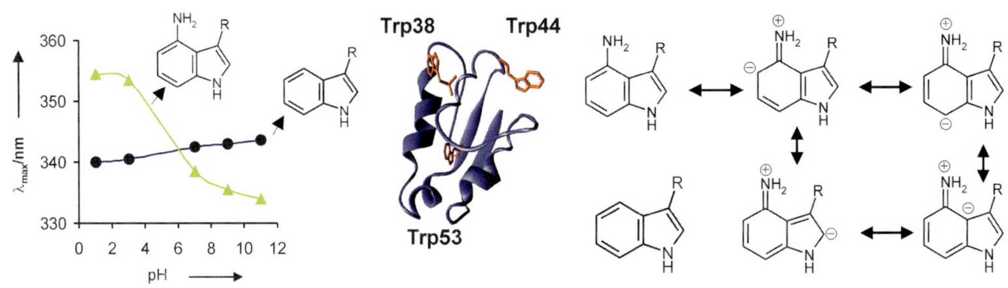

Figure 7.8. The design of protein-based molecular sensors: charge transfer and pH sensitivity in 4-aminotryptophan-containing barstar. Upon full replacement of three Trp residues in barstar (middle) with (4-NH_2)Trp the pH-insensitive native (parent) protein was converted into a pH-sensitive fluorescent protein [i.e. an increase in pH induces a spectral shift in the emission maxima of the substituted protein (left)]. (Right) Canonical structures representing the intramolecular charge transfer in (4-NH_2)Trp. The augmentation of the indole ring with an amino group as a good electron donor leads to the contribution of an additional four mesomeric structures. (Taken from [56].)

nucleus results in a red shift of 10 nm in absorption and 46 nm in emission of (7-Aza)Trp when compared with Trp [48]. In addition, the fluorescence emission maximum and quantum yield of (7-Aza)Trp are extremely sensitive to the local environment, and normally strongly quenched in water [56]. In this context, the substitution of indole with electron-donating amino groups would lead to enhanced intramolecular charge transfer which should be even more sensitive to pH changes. This was confirmed by replacing all Trp residues of barstar [55] as well as the single Trp of human annexin-V [44] by amino analogs (5-NH_2)Trp and (4-NH_2)Trp; the fluorescence of both proteins becomes pH sensitive upon substitution. In other words, the pH sensitivity of substituted proteins designed in this way is possible due to the capacity of charge transfer found uniquely in amino-Trp analogs.

The amino substituents of aromatic ring systems are less basic in the excited than in the ground state; the protonation level in the excited state affects the resonant integration of their free electrons in conjugated ring systems causing charge migration (Fig. 7.8) seen as blue- or red-shifted spectroscopic bands. Therefore, the pH-sensitive fluorescence of the aminoindole side-chains of Trp analogs such as (5-NH_2)Trp and (4-NH_2)Trp is expected to be fully transferred to the target protein, allowing for a pH shuttle of its optical properties [55]. Interestingly, the presence of electron-donating amino and hydroxy groups in different positions of the indole moiety leads to a dual fluorescence with one peak centered around 350 nm and the other at around 520 nm in an apolar solvent. Interactions such as hydrogen bonding of water molecules to the aminoindole in the excited state cause enhancement of radiationless processes and explain the absence of the second fluorescence band in the related protein variants in aqueous buffered solutions [54]. In a particular milieu such as membrane environments or apolar solutions, such proteins might be conceived as dual-fluorescence protein-based pH sensors.

The pH sensitivity, both in terms of fluorescence emission maximum and intensity as the consequence of intramolecular charge migration, originates solely from cation-to-anion transitions of the aminoindoles. Thus, the conversion of any native protein from a pH-insensitive to a pH-sensitive one in terms of fluorescence is generally possible since these properties are a result of the intrinsic properties of the amino-Trp analogs translated into the protein structure. Because of the multiple functions of Trp residues in proteins it is indeed conceivable that proteins "tailored" with amino- or other indole-like side-chains may represent useful non-invasive tools for *in vivo* or *in vitro* monitoring of protein–membrane, protein–protein, protein–ligand and enzyme–substrate interactions.

7.6.2
Novel "Golden" Class of Autofluorescent Proteins

A further step in utilizing aminoindoles is their integration into the chromophore of GFP from the jellyfish *Aequorea victoria* which is currently a standard reporter in cellular and molecular biology as well as a model for studying chromophore/protein photophysics [57]. The GFP chromophore a 4-(*p*-hydroxybenzilidene)-imidazolid-5-one, completely encoded in the amino acid sequence, is autocatalytically formed in the posttranslational reaction of amino acid residues Ser65, Tyr66 and Gly67. Chromophore-building residues of GFP are ideal models for such substitutions since the effects produced are immediately observable in terms of the optical properties. The standard amino acid repertoire limits the possibilities to gain altered spectral windows by direct redesign of the GFP chromophore since only four aromatic rings are available (i.e. Phe, His, Tyr and Trp), and the mutants Tyr66 → Phe, His and Trp show a characteristic blue emission since the excited state proton transfer responsible for the characteristic green fluorescence in native GFP is prohibited [58]. The expansion of the spectral window into the red region of the spectrum has been achieved by mutations in the protein-binding pocket which affect the hydrogen bonding network. However, no significantly red-shifted variants of *A. victoria* GFP were found (i.e. not beyond 530 nm). Nature, however, extended the chromophore π system by using different principles. First, the conjugated system of the red fluorescent protein from *Discosotoma striata* (*dsRed*) is expanded by integration of an additional amino acid into its chromophore ($\lambda_{max} = 573$ nm). Second, the chromophore of chromoprotein *asFP595* ($\lambda_{max} = 595$ nm) from *Anemone sulcata* was extended by augmentation of its π system with an imino group [59]. Finally, a new photophysics principle in red shifting the fluorescence gained by incorporation of aminoindole into related chromophores was developed by Budisa's group. It is based on protein relaxation as the response of a strong change in dipole moment in the chromophore upon excitation [44].

Extensive variations of the chromophore itself as well as its protein matrix by classical protein engineering approaches (site-directed mutagenesis, guided evolution, random mutagenesis coupled with the screening, etc.) based on the 20 canonical amino acids were performed to develop novel classes of the proteins as well as to expand the biological applications [60]. Seven major classes of *A. victoria*

GFP proteins were derived on the basis of their spectral characteristics (for review, see [61]). The mutation Tyr66Trp represents a class V with a characteristic blue/green emission known as "enhanced cyan fluorescent protein" ECFP; (Ser65Thr/Tyr66Trp) has an indole ring as an integral part of the chromophore. It is important to note that the placement of any aromatic amino acid in position 66 is crucial for the emergence of a fluorescently active GFP chromophore. The basic characteristic of class V is the presence of the indole moiety in the chromophore with other compensatory mutations in the closer or distant environment [60]. The replacement of the chromophore indole with aminoindole, which contains a free electron pair, is expected to produce larger red shifts upon integration either in the indolyl moiety or the chromophore of ECFP.

Such expectations were completely fulfilled with the introduction of the electron-donating (4-NH_2)Trp into ECFP, resulting in a "gold" fluorescent protein (GdFP) which is, with its Stokes shift of about 100 nm, the most red-shifted *Aequorea* GFP variant known to date ($\lambda_{max} = 574$ nm). It has a characteristic golden fluorescence with an emission maximum similar to that of *dsRed*, which is undoubtedly a consequence of the intramolecular charge migration. It has not only a different color in comparison with the other GFP mutants from *Aequorea*, but also dramatically increased thermostability and stability in the monomeric state as well as having expression yields similar to that of the parent ECFP (around 30 mg L^{-1} of culture). Apart from its special spectral properties, GdFP is especially interesting with respect to the longevity of the monomeric state. It is well known that native *Aequorea* GFPs and their various mutants tend to aggregate upon storage. For example, fresh samples of ECFP are overwhelmingly in the monomeric state (around 90%) and to a lesser extent in the dimeric state (around 10%). Their storage in buffered aqueous solutions (pH 7.0–8.0) at 4 °C for 6 weeks resulted in a dramatic decrease of the monomeric species by almost 50%. Conversely, GdFP samples are relatively stable for over longer periods of time (up to 1 year) at room temperature as well as at 4 °C as determined by using sedimentation velocity measurements. After 6 months under identical conditions, all ECFP samples almost completely aggregated [62]. Finally, the investigation of GdFP in both prokaryotic and eukaryotic cells indicated that it is also intracellularly stable, and gives fluorescent signals strong enough for various investigations such as in apoptosis (as a fusion partner of annexin-V). The cells from such cultures can be unambiguously identified under confocal microscopy, as shown in Fig. 7.9 where two populations of Trp auxotrophic *E. coli* cells expressing "cyan" and "gold" fluorescent proteins can be easily distinguished upon excitation at a single wavelength.

The high-resolution crystal structures of ECFP and GdFP were determined as well; their comparison offers a framework for possible models that correlate the chromophore's environmental configuration with its spectroscopic function. While the structures are almost identical, the most prominent novel interactions that can be identified from crystallographic distance considerations are that the amino-chromophore of GdFP is slightly shifted toward Phe165. In this case possible amino–aromatic interactions cannot be excluded since the two residues are 3.2–4.5 Å apart. These amino–aromatic interactions might be promoted by a more

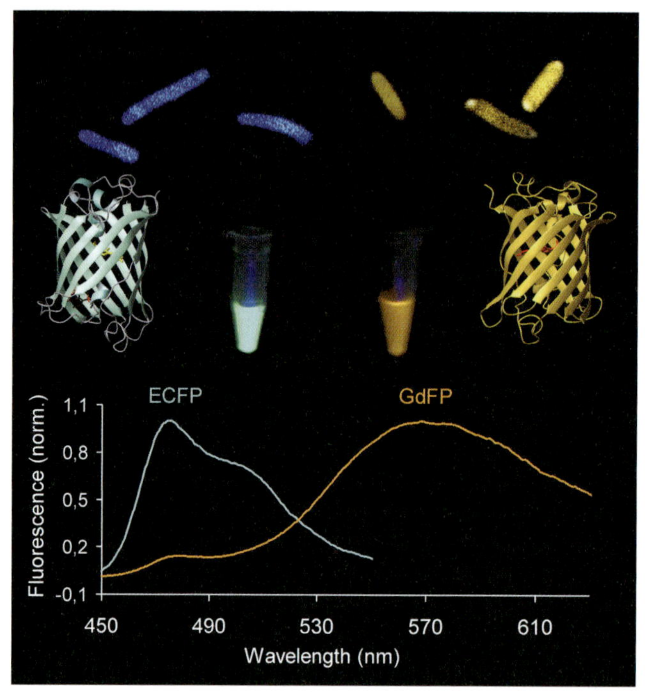

Figure 7.9. Designing novel classes of autofluorescent proteins: turning "cyan" into "gold" fluorescent proteins by (4-NH_2)Trp incorporation. *E. coli* Trp auxotrophic bacterial cells expressing ECFP and GdFP have been imaged on a standard confocal microscope by using different emission filters. Bacterial cells, simultaneously excited at a single wavelength (457 nm), are clearly distinguishable, demonstrating possibilities for multiple labeling applications. Lower part: Comparison of normalized fluorescence emission spectra of ECFP ($\lambda_{max} = 504$ nm) with its substituted derivative GdFP ($\lambda_{max} = 574$ nm). (Taken from [56].)

effective intramolecular charge transfer in the excited state in comparison to the ground state of the chromophore and might be responsible for the large Stokes shift. However, this large red shift of GdFP is always accompanied by a loss in quantum yield in the GdFP when compared to the native protein ECFP. This is most probably due to the enhanced rates of radiationless processes.

Interestingly, Wang and coworkers reported selective substitution of Tyr66 of GFP with *p*-amino-Phe, *p*-methoxy-Phe, *p*-iodo-Phe, *p*-bromo-Phe and naphthyl-Ala. They found that the absorbance (range 375–435 nm) and fluorescence (range 428–498 nm) maxima and quantum yields correlate with the structural and electronic properties of the substituents on the amino acids [63]. However, most of the currently available GFP variants are exclusively produced using classical protein engineering procedures. It is widely believed that the desired effects for GFP mutants such as high extinction coefficients, quantum yields and efficient folding could be mainly achieved via multiple mutations in their target sequences [59].

In contrast, GdFP clearly shows that a simple exchange of atomic groups in the chromophore might yield structurally identical, but spectroscopically and thermodynamically different, GFP structures. This suggests the possibility for a rational strategy in designing GFPs with improved fluorescence and increased stability of the monomeric state. One might expect that future experiments on substitution with amino-Trp analogs on known red-emitting fluorescent proteins like *dsRed* will result in further red-shifted mutants. Furthermore, novel generations of red-shifted autofluorescent proteins are also conceivable by the introduction of novel substituents like cyano, nitro, nitroso, etc., in chromophores formed by indole, imidazole, phenol and other aromatic rings. Such substances, when used as substrates for protein synthesis (i.e. those being armed with an expanded amino acid repertoire), in combination with classical random and combinatorial mutational approaches, might well pave the way towards novel families of tailor-made autofluorescent proteins that could be used as multiple labels, reporters and resonance energy transfer acceptors.

7.7 Fluorinated Amino Acids in Protein Engineering and Design

7.7.1 Monofluorinated Amino Acids in Protein Studies, Engineering and Design

Fluorine, which in the form of fluoride minerals is one of the most abundant halogens in the Earth's crust, appears only in 12 naturally occurring organofluorine compounds. Conversely, there is a large number and range of commercially generated fluorinated compounds – all of them being of anthropogenic origin. Fluorine chemistry was not evolved in living cells possible due to its low availability. It is well known that the available fluoride is mostly insoluble, e.g. sea water contains 1.3 p.p.m. fluorine and 19 000 p.p.m. chloride [64]. The rarity of naturally fluorinated products contrasts with the existence of more than 3500 naturally occurring halogenated compounds. Naturally occurring posttransitionally halogenated amino acids are usually derivatives of His, Trp and Tyr, and proved to be interesting targets for biomedical research [65]. For example, some antibiotics from various marine algae and some higher plants are in fact chlorinated derivatives of aliphatic or aromatic amino acids [66]. Examples of posttranslational iodinations, brominations and chlorinations of Tyr were first described in vertebrate tyreoglobulins and invertebrate scleroproteins, while they were recently found in a variety of marine organisms [67]. Many organisms use them for modification of their amino acids into pharmacologically active substances for different purposes, usually for chemical defenses from predators. This is especially pronounced in a marine ambient with a higher natural abundance of iodine, bromine and chlorine ions [68]. Nonetheless, in the context of protein translation, fluorinated and chlorinated amino acids are "xeno" compounds, and consequently both fluorine and chlorine are "xeno" elements. Therefore, it should not be surprising that these different properties will produce different effects upon their integration into cooperatively folded protein structures.

Among all halogen-containing amino acids, the most interesting properties were obtained when fluorine was used for modification, for several reasons. Fluorine generally causes minimal structural perturbations during CH $\rightarrow$ CF replacements and such changes can be easily accommodated even in the protein core. Although chlorine and fluorine are neighbors in the group of halogens in the periodic system of elements, they are markedly different in their electrochemical and stereochemical properties. As the most electronegative element known, fluorine exerts a much stronger electronegativity (3.98) in its local environment than chlorine does (3.16 according to the Pauling scale). Both produce a strong electron-withdrawing inductive effect ($-$I effect) and induce a moderate electron-donating resonance effect on the whole aromatic ring (+M effect). They also possess nonshared electron pairs and carbon-bound fluorine is known to be a weak hydrogen bond acceptor. Their dipole moments are opposite to that of a C–H bond (C–F = 1.41 and C–Cl = 1.46). However, the most dramatic differences between these halogens lie in their steric properties. Chlorine is more bulky than fluorine (van der Waals radius of F is 1.35 Å and Cl is 1.8 Å) and is much more polarizable. Thus, chlorine as a bulkier atom (C–H = 1.09 Å, C–F = 1.39 Å and C–Cl = 1.77 Å) should produce more dramatic steric changes in the protein core than fluorine does [69].

Monofluorinated analogs of mainly aromatic amino acids were extensively used in early incorporation experiments as well as in more recent years for different purposes. For example, fluorination of the catalytic residues might lead to changes of enzyme pK_a and, subsequently, in a novel pH range of catalytic activity. Ring and Huber [70] probed the role of catalytic Tyr in the active center of β-galactosidase by the global replacement of all Tyr side-chains with (3-F)Tyr, whereas Brooks and coworkers [71] incorporated in addition 2-fluorotyrosine and 2,3-difluorotyrosine in order to study the catalytic mechanism of ketosteroid isomerase from *Pseudomonas testosteroni*. The fluorination of catalytic His residues in enzymes might induce a drop of the pK_a value by about 5 pH units; this dramatically alters the activity and properties of the fluorohistidine-containing proteins [72]. More recently, Budisa and coworkers [73, 74] reported interesting spectral and dynamic properties of GFP upon global substitution of its Tyr residues with *o*- and *m*-fluorotyrosines (see Section 5.2.1.2; Fig. 5.9).

Fluorination of aromatic amino acids might dramatically change their spectral properties. For example, Phe residues normally do not contribute significantly to the overall protein absorption. However, the absorbance intensity of Phe might be dramatically enhanced upon fluorination of its benzene ring at the *ortho*, *meta* and *para* positions. In addition, fluorinated Phe analogs alone and when integrated into proteins exhibit two characteristic and prominent shoulders ("fingers") in the UV spectrum in the range of 260–270 nm, which do not overlap with the contributions of Tyr and Trp residues in the protein UV spectra (Fig. 7.10). Thus, the presence of such "fluoro-Phe fingers" ("FF fingers") opens a new spectral window to identify the labeled target protein among other nonlabeled cellular proteins in preparative work by simple UV spectroscopy [75]. The novelty of such and similar approaches

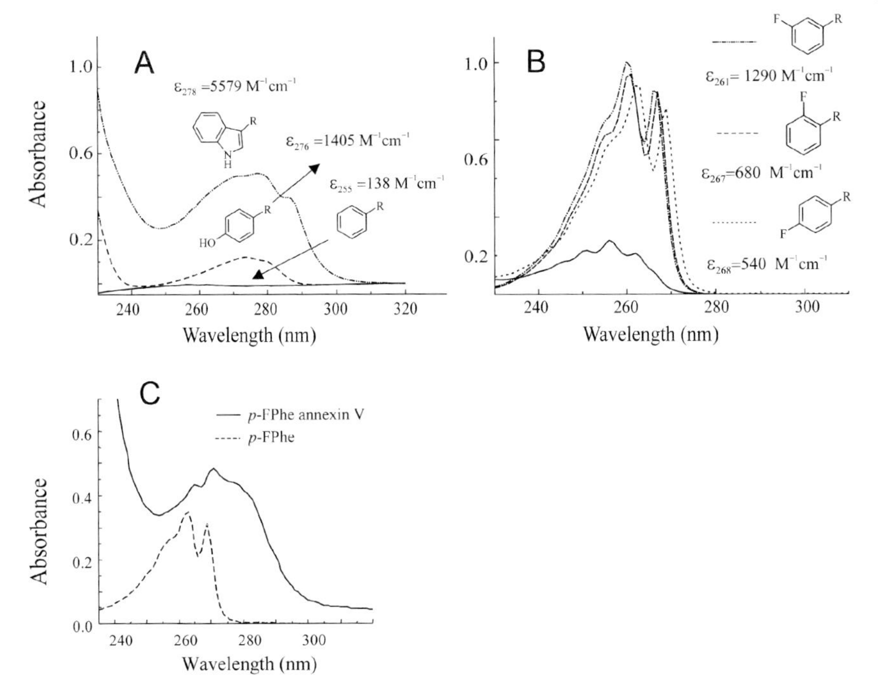

Figure 7.10. Spectral enhancement of the Phe UV absorption spectra (modified from [75]). (A) UV spectra of 100 μM Phe, Tyr and Trp measured in the range 230–320 nm in neutral aqueous buffer. (B) UV spectra of 1 mM Phe, *o*-FPhe, *m*-FPhe and *p*-FPhe measured in the range 230–320 nm. (C) UV spectra of 50 μM *p*-FPhe and 50 μM *p*-FPhe annexin-V under the same conditions. The extinction coefficients of the absorbance maxima are given in M^{-1} cm^{-1}. For more details, see [75].

lies in the possibilities to offer a reliable, cheap and easily reproducible methodology that might have a great potential for speeding up the identification and characterization of target molecules in the total protein output from the genomes of a variety of organisms.

In recent decades, fluorinated Trp analogs along with 7-azatryptophan and 5-hydroxytryptophan were often used for *in vivo* and *in vitro* translation in order to obtain "spectrally enhanced proteins" [76]. The dominance of Trp in protein spectral properties makes it an almost ideal model for replacement studies since the introduced modifications are easily observable. The incorporation of (5-F)Trp into annexin-V results in an enhanced quantum yield and a red-shifted emission maxima (334 nm) by about 14 nm (Fig. 7.11). In contrast, the incorporation of (4-F)Trp leads to a total loss in the Trp fluorescence and reveals a spectral contribution of the Tyr residues. (4-F)Trp is well known as a nonfluorescent analog ("silent fluorophore") that allows the identification of spectroscopic contributions of "hidden" chromophores like Tyr or Phe in proteins. This property of (4-F)Trp is well documented in the literature [43].

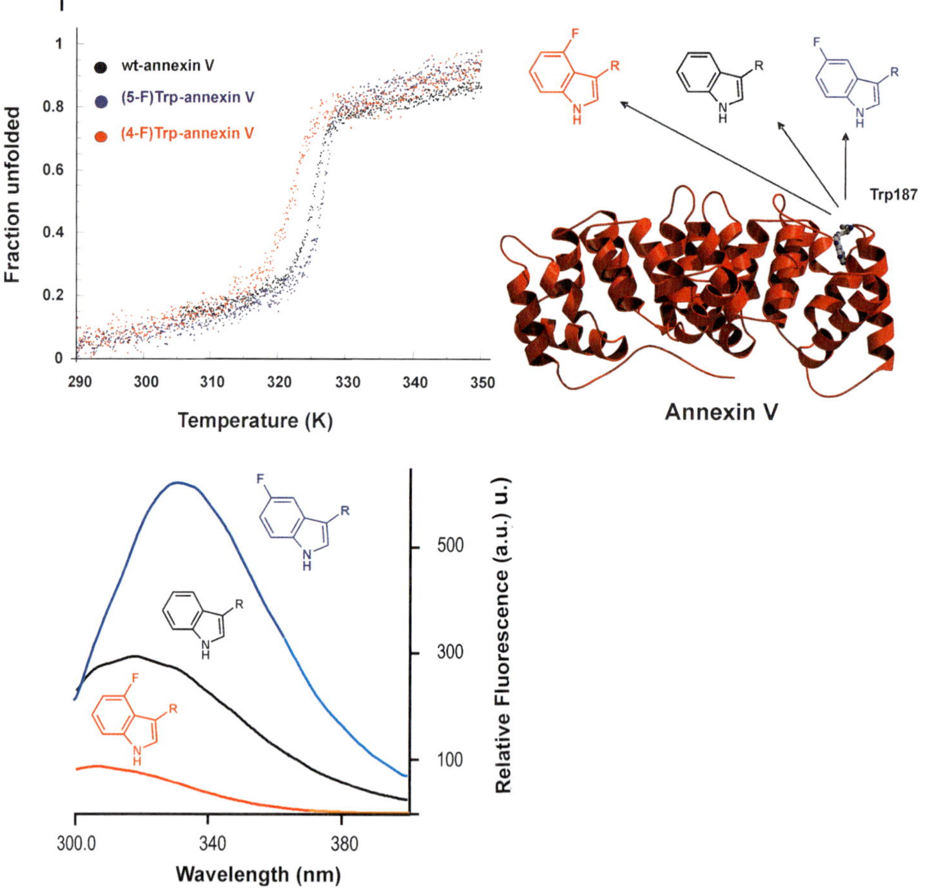

Figure 7.11. Thermodynamic and spectroscopic effects of fluorination of the single Trp187 in human recombinant annexin-V. The side view (middle) of the three-dimensional structure of human recombinant annexin-V with single Trp187 marked as "balls and sticks". In this particular experiment, human recombinant annexin-V, containing 320 residues and 5328 atoms, was replaced with two different isosteric Trp analogs with H → F substitutions at positions 4 and 5 in the indole ring. The thermal denaturation experiment (left) clearly indicates that the melting point of (5-F)Trp-annexin-V is higher than native and (4-F)Trp-containing proteins [38]. On the other hand, fluorescence emission of (5-F)Trp-annexin-V is enhanced and whereas Trp fluorescence in (4-F)Trp-annexin-V is fully suppressed (the emission signal with a maximum at 307 nm is due to the contributions of Tyr side-chains from the protein structure).

Fluorinated Trp analogs have also been the subject of numerous studies and reviews dealing with different aspects of their chemistry, photophysics in protein structures, translational activity and physiological stability (reviewed in [42, 51, 56]). In many cases, such analogs are excellent tools with the potential to supplement or even to replace classical mutagenesis approaches. Trp residues are usually mutated to Phe in an attempt to minimize structural perturbations by replacing one aromatic planar moiety with another. However, this strategy is often limited due to the fact that some Trp residues might be essential for the structural integrity and functionality of proteins, and therefore cannot be replaced by any of the remaining 19 canonical amino acids. In this context, fluorinated Trp analogs bring about the lowest possible level of structural alterations, i.e. single atom exchanges H $\rightarrow$ F, thus providing "atomic mutations" for studying protein folding, activities, dynamics, spectroscopy and stability as discussed earlier (Section 7.5.2). For example, the replacement of the single Trp187 with its fluorinated analogs was demonstrated to change not only the intrinsic spectroscopic properties of annexin-V, but also its folding properties (Fig. 7.11). Thermodynamic analyses using CD spectroscopy clearly indicate that (5-F)Trp stabilizes annexin-V when present in its structure. The effect of analogs varies from protein to protein, whereby the position of the Trp residues in the context of the protein structure (buried/exposed) plays an important role. It is generally accepted that these rather large global effects, resulting from minimal local changes, have to be attributed either to the relatively strong changes in polar interactions of the indole ring, to differences in the van der Waals radii or to a combination of both factors [38]. A series of fluorinated Trp analogs was incorporated into proteins by *in vitro* suppression methods in order to study the significance and additive effects of the introduced fluorine atoms on the cation–π interactions [4].

Many efforts have been made in order to develop optimal incorporation protocols which use fluorinated analogs of Trp and Tyr as suitable fluorine nucleus donors for protein NMR [77]. In the last three decades ^{19}F-NMR has proven to be a powerful technique which allows us to probe subtle conformational changes and to reveal intrinsic dynamic features of proteins. In fact, ^{19}F-NMR studies have taken advantage of the extreme sensitivity of ^{19}F chemical shifts to changes in a target residue microenvironment, including van der Waals packing interactions and local electrostatic fields without background signals. In most cases, the structure of the fluorinated proteins is indistinguishable from that of the wild-type form, which is confirmed by the available crystallographic studies [73]. Relatively rigid aromatic amino acids are especially suitable targets for H $\rightarrow$ F replacement, since they are usually less likely to be involved in internal motion than smaller amino acids, which facilitate chemical shift calculations [78]. A recent example includes the ^{19}F-NMR study on "cyan" fluorescent protein (ECFP) from *A. victoria* with fluorinated Trp side-chains, which enabled the detection of the slow molecular motions in the vicinity of the chromophore of this protein [79]. In this particular case, the combination of dynamic NMR and ^{19}F relaxation measurements at different temperatures revealed the existence of a slow exchange process in the range 1.2–1.4 ms between two different conformational states in the environment of ECFP. Later, X-ray analysis of native ECFP crystals indeed confirmed the presence of two configurations for residues Tyr145 and His148 around the rigid ECFP chromophore (Fig. 7.12).

Moroder, Raines and coworkers [80, 81] recently demonstrated the utility of fluorinated derivatives of Pro as general vehicles for a rational redesign of natural structural scaffolds in order to gain structures endowed with predetermined and

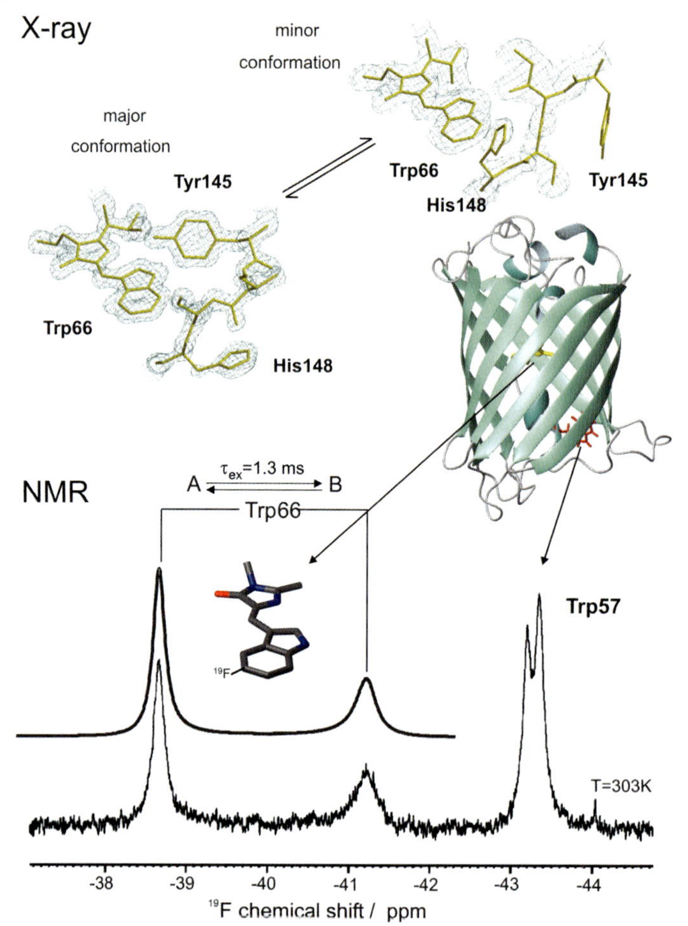

Figure 7.12. Fluorination as a tool for the interpretation of dynamic features of proteins. ECFP contains two Trp residues, Trp57 and Trp66, being an integral part of the chromophore. Both chromophore-bound and fluorinated Trp66 exhibits two peaks in NMR spectra with the mean lifetimes about 1.2–1.4 ms [79] (lower part). However, X-ray analyses of native ECFP revealed two crystal forms with "minor" and "major" conformations related to the relative positions of Tyr145 and His148 in the vicinity of the chromophore (upper part) [62]. Combined together, X-ray and ^{19}F-NMR studies of fluorinated ECFP provide valuable information about the dynamic behavior of its chromophore in the context of its molecular microenvironment, i.e. that Tyr145 and His148 exchange between these two conformations. For more details, see [79].

unusual functions. Pro-rich sequences are often found in proteins that are known to display helical structures like the triple helix of collagen. Since collagen is an important biomaterial for the design of various biodegradable structures such as artificial heart valves, the biosynthetic production of collagens with improved properties is an attractive research goal [82]. The important breakthrough in collagen

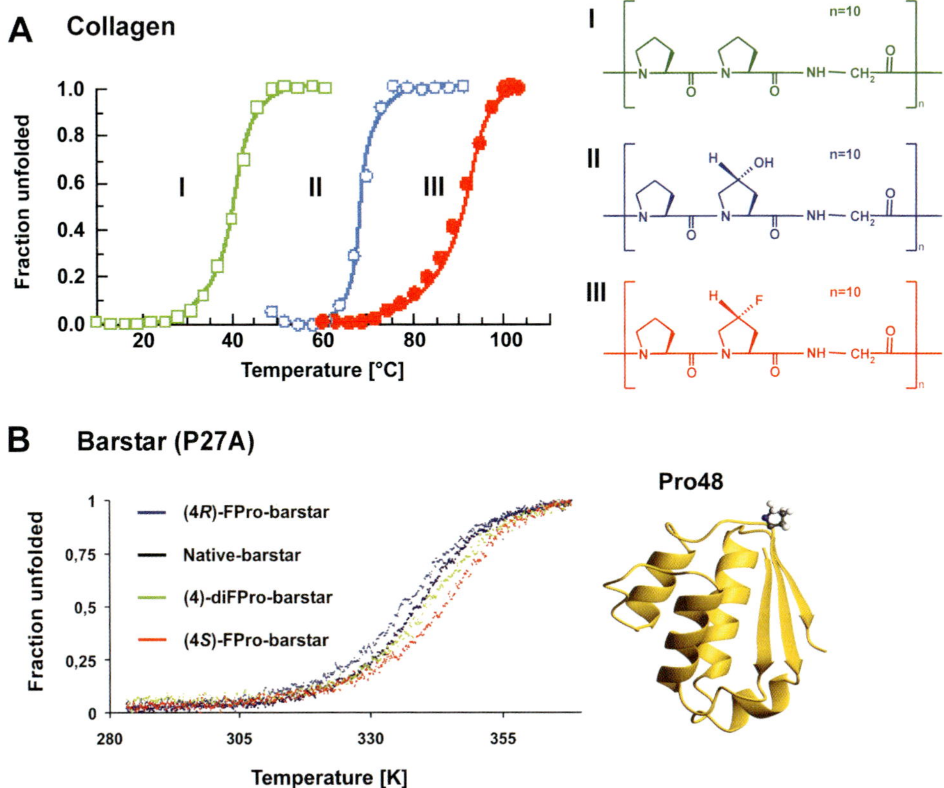

Figure 7.13. Material design and protein engineering with fluorinated Pro analogs. (A) The synthetic repeat $(Pro–Y–Gly)_{10}$ can be designed as a relatively unstable "unnatural" form (**I**, Y = Pro), as a "natural" hydroxylated form [**II**, Y = 4(*R*)-OHPro] or as a hyperstable fluorinated triple helix [**III**, Y = 4(*R*)-FPro]. Values of T_m were measured in dilute solutions of aqueous acetic acid. Note that *trans*-hydroxylation of Pro(Y) leads to an increase in T_m of almost 30 °C, whereas its *trans*-fluorination elevates the melting temperature by 50 °C. Reproduced from Ref. [81]. Copyright 1999, with permission from Elsevier. (B) Thermally induced unfolding profiles of barstar and its fluorinated variants determined by the changes of protein dichroic intensities at 222 nm. The derived T_m value for native barstar is 64 °C. As expected, (4*S*)-FPro-barstar has an increased T_m value (69 °C), (4*R*)-FPro-barstar has a decreased T_m (61.7 °C) and diFPro-barstar has an almost unchanged T_m (65.2 °C). In the structure of native barstar (right), Pro48 is placed at a loop near β-sheet 2 and is highly exposed to solvent. (Modified from [44].)

design was the discovery of a dramatic increase in triple-helical stability upon the replacement of 4(*R*)-hydroxyproline [4(*R*)-OHPro] residues by 4(*R*)-fluoroproline [4(*R*)-FPro] [81] (Fig. 7.13). Three types of collagen were generated by the peptide synthesis of a 10 tripeptide unit: (i) "unnatural" $(Pro–Pro–Gly)_{10}$ collagen exhibits the lowest melting point (T_m = 41 °C), (ii) the "natural" $(Pro–4(R)\text{-}OHPro–Gly)_{10}$ sequence is much more stable (T_m = 69 °C), whereas (iii) the fluorine-containing

(Pro–4(*R*)-FPro–Gly)$_{10}$ triple helix is hyperstable ($T_m = 91$ °C). The hyperstability of the fluorinated collagen triple helix was attributed to the inductive effects introduced by fluorine atoms in its structure [82]. This is indeed plausible, since fluorine exerts a strong inductive effect that is sufficient to affect properties such as the polarity or binding capacity of functional groups that are even distantly positioned in the structure.

The crucial role of Pro in protein folding is also reflected in the fact that the rate-limiting step in the slow folding reactions of many proteins involves *cis*/*trans* isomerization about peptidyl-prolyl amide bonds. In folded proteins peptide bonds not proceeding Pro are almost always *trans*, but about 7% of all Xaa–Pro peptide bonds (Xaa = any amino acid) show the *cis* conformation in proteins with known three-dimensional structures [83]. In this way, prolyl residues restrict the conformational space for the peptide chain, since the rotation barrier of *cis*/*trans* isomerization is high ($\Delta G = 85 \pm 10$ kJ mol^{-1}) [84]. The *cis*/*trans* isomerization of a Xaa–Pro bond has been implicated in critical roles in the biochemistry of Pro-containing peptides and proteins. This is especially well studied in barstar, which has two Pro residues – Pro27 being in *trans* and Pro48 in *cis* in the native state [85]. Its Pro27 can be replaced with Ala (Pro27Ala) using classical site-directed mutagenesis, without any deleterious effects on protein structure. Barstar tailored in this way has only one Pro residue, Pro48, that is exclusively in the *cis* conformation in the native state, serving as an almost ideal model for studying Pro *cis*/*trans* isomerization in the course of folding. The main folding phase of barstar proceeds much faster than Pro48 *cis*/*trans* isomerization, i.e. Pro48 isomerization it is the rate-limiting step in this process.

Using NMR and IR spectroscopy, Renner and coworkers [80] provided an excellent account on the thermodynamics of the model compounds *N*-acetyl-Pro-methylesters of their respective (4)-fluoroproline derivatives. They demonstrated that depending upon whether the *cis* or *trans* isomer was used, the *cis* or *trans* conformation can be favored or disfavored, even in proteins. Accordingly, barstar with the single Pro40 being in the *cis* conformation in the native state is expected to be stabilized upon fluorination at *cis*, and even destabilized upon fluorination at the *trans* position of the Pro48 ring, while difluoroproline is expected to produce no significant effect. In other words, the more the preference of the Pro analog for adopting the *cis* conformation, the more stable the corresponding barstar variant is, because only this conformation is realized in the native state. This was fully confirmed in the thermal unfolding experiments of barstar and its related fluoro-variants: (4*S*)-FPro-barstar has an elevated melting point, while (4*R*)-FPro-barstar has a lowered melting point. As expected, 4-difluoro-Pro-barstar is without marked changes in thermostability expressed in terms of T_m values (Fig. 7.13). In this way it was demonstrated how the thermodynamic properties of fluoroproline-containing proteins can effectively be predicted based on studies of model fluorinated substances. Therefore, the bio-incorporation of these amino acids with predefined properties into well-characterized model proteins might offer possibilities to rationally manipulate the kinetics, folding rates, stability and reactivity of substituted proteins, i.e. act as useful tools for protein design and engineering.

7.7.2
Nonsticking Eggs and Bio-Teflon – Trifluorinated Amino Acids in Protein Engineering and Design

In the context of monofluorinated amino acids, the fluorine atom is not a sterically demanding substituent, although the C–F bond is 0.4 Å longer than the C–H bond. Therefore, its introduction generally causes minimal structural perturbations. However, the $CH_3 \rightarrow CF_3$ substitution increases the steric bulk by more than 12 Å^3 per single replacement. Nonetheless, materials modified by multiple fluorinated carbons posses elevated hydrophobicity, and often have profound and unexpected properties. The best known example is the perfluorinated polymer Teflon. The most interesting property of the trifluoromethyl group is its higher hydrophobicity (it is almost as twice as hydrophobic as the methyl group). This feature is exploited to suppress metabolic detoxification, to increase the delivery bioavailability of many pharmaceuticals or to enhance drug activity [86].

Perfluorocarbons are described as fluorophilic, and not hydrophobic or hydrophilic, i.e. such molecules are expected not to interact either with hydrophilic or hydrophobic molecules. Therefore, the hydrophobic effect (i.e. exclusion of apolar residues from the aqueous phase) as the driving force of the protein folding process should also hold for fluorous proteins in both aqueous and organic phases. In other words, such proteins should be capable of folding in aqueous media and to resist denaturation by organic solvents. Such resistance is achieved by exclusion of fluorocarbon side-chains from the organic phase and formation of a compact interior or core that is shielded from the surrounding solvent (Fig. 7.14). This property can be gained by, for example, Leu → HFL substitution in the hydrophobic core in order to design programmed self-sorting coiled-coils peptides as demonstrated by Kumar and coworkers [88]. Similarly, proteins engineered in this way should posses a similar "fluorous effect", i.e. the same effect responsible for the nonsticking properties of Teflon, offering a vision of a whole class of proteins with unknown properties.

The most conceivable way to endow protein structures with such unusual properties, is to replace the "core" amino acids Leu, Val, Ile or Met with their trifluorinated counterparts (Fig. 7.15). Sequence comparisons usually assign similar hydrophobicities to Met, Val, Leu and Ile residues in proteins [89]. The fact that many of their tri- and hexafluorinated counterparts are translationally active can be used to manipulate protein stability or protein–protein interactions. It is well known that naturally occurring protein or peptide surfaces that mediate protein–protein interactions are usually equipped with specifically positioned hydrogen bonds or salt bridges, or with hydrophobic amino acid patches that maximize van der Waals interactions. In this context, trifluorinated amino acids offer the possibility to rationally modify the interface of proteins or peptides making it both more hydrophobic and more lipophobic. In addition, the enhanced hydrophobic character of trifluorinated amino acids could also be utilized in protein engineering in order to prepare peptides and proteins with enhanced structural stability. For example, Leu zipper and coiled-coil polypeptides in which the Leu residues have been replaced with

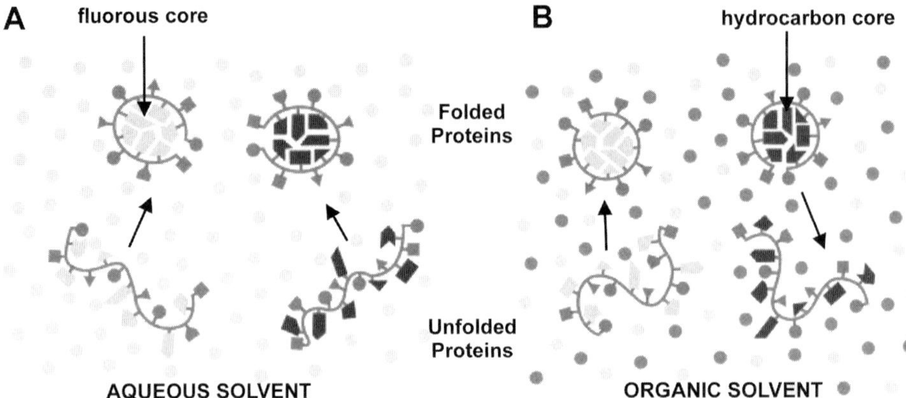

Figure 7.14. Fluorous effect: hydrocarbon versus fluorous protein core (faint spots = aqueous solvent; dark spots = organic solvent). (A) In aqueous solvents the hydrophobic effect would drive the folding of polypeptide sequences containing hydrocarbon (marked with black color) or fluorous amino acids (marked with grey color) into stable proteins whose three-dimensional structure would have either a fluorous or hydrocarbon core. (B) Most natural proteins with a hydrocarbon core are not resistant to denaturation induced by organic solvents. In contrast, fluorous proteins should be resistant to denaturation by organic solvents because the fluorocarbon side-chains will partition away from the organic phase. (Reproduced from [87] with permission from Elsevier, copyright 2000.)

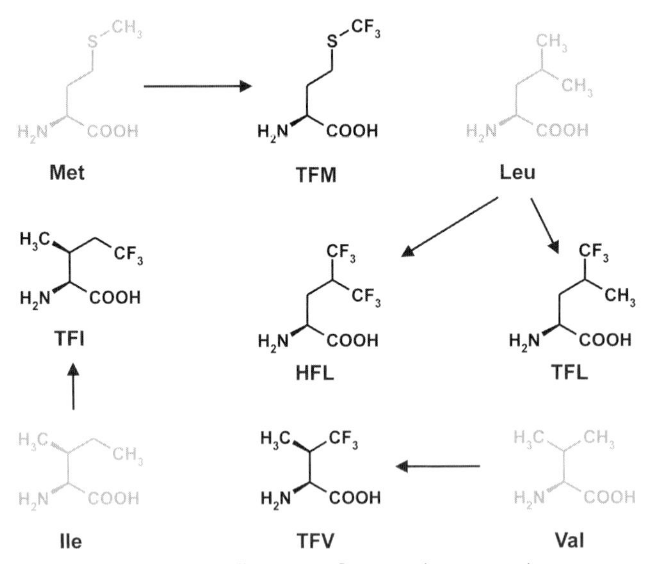

Figure 7.15. Translationally active trifluorinated amino acids (black) and their natural counterparts (grey). Abbreviations: TFM = 6,6,6-trifluoromethionine, TFL = 5,5,5-trifluoroleucine, HFL = 5,5,5,5′,5′,5′-hexafluoroleucine, TFV = (2*S*,3*R*)-4,4,4-trifluorovaline, TFI = 5,5,5-trifluoroisoleucine.

TFL and HFL are endowed with increased thermal and chemical stability compared to their corresponding wild-type natural proteins. Tang and Tirrell [90] demonstrated that the melting temperature of the Leu zipper protein A1 can be increased by about 10 °C by replacing Leu residues with TFL, whereas Leu → HFL replacement elevates the melting temperature by more then 20 °C (Fig. 7.16). HFL-A1 is also remarkably resistant to denaturation by urea.

Met residues, which are rare in proteins and significantly more flexible than those of both Leu and Ile with their branched and more rigid side-chains, are also substituted by di- and trifluorinated counterparts. Honek, Budisa and coworkers [91, 92] provided experimental reports about the possibility of TFM and DFM incorporation into phage lysozyme and engineered GFP that contains only two Met residues (one buried and one surface exposed). However the Met → TFM replacements were never quantitative. The reason for such behavior is most probably the increased steric bulk (about 15%) of noncanonical TFM (84.4 Å^3) when compared with that of canonical Met (71.2 Å^3) [93]. However, even lower levels of replacements can be useful for sensitive methods such as ^{19}F-NMR spectroscopy [78]. For example, TFM incorporation into GFP provides a direct insight into solvent exposure, i.e. dynamic features of both TFM residues in the structure of this protein as shown in Fig. 7.16.

Although the experiments with trifluorinated amino acid incorporation into smaller proteins and peptides such as Leu zipper A1 might indicate a general

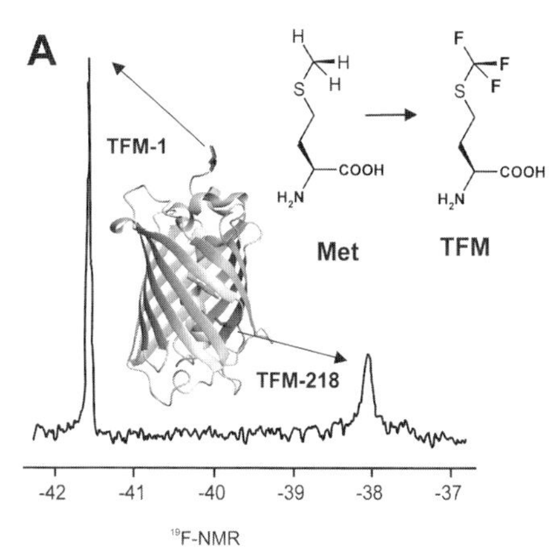

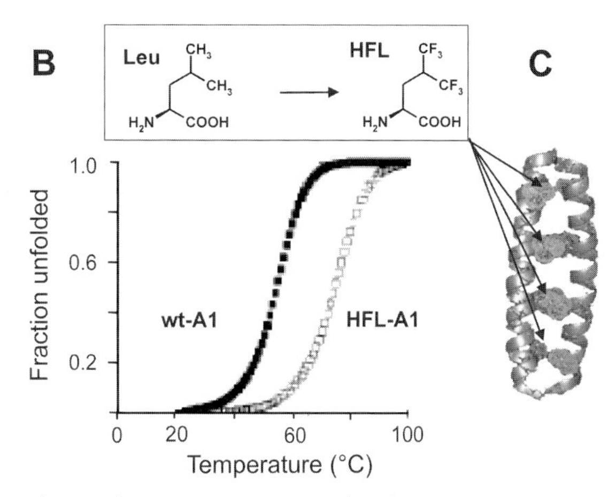

Figure 7.16. Biophysical studies of proteins labeled with trifluorinated amino acids. (A) Solvent exposure of TFM residues in GFP containing only two Met residues as determined by one-dimensional ^{19}F-NMR. The profile of the ^{19}F chemical shift reveals different environments for TFM residues in GFP: N-terminal and flexible TFM1, and buried and relatively rigid TFM218. This experiment provided a direct insight into solvent exposure, i.e. the dynamic features of both TFM residues in the GFP structure. Reproduced from Ref. [92], copyright 2004, with permission from Elsevier. (B) Dramatic stabilization of the coiled-coil peptide synthetic Leu zipper A1 results from introduction of HFL. The stability increase in terms of measured T_m values was about 22 °C ($T_{m(wt\text{-}A1)} = 54$ °C and $T_{m(HFL\text{-}A1)} = 76$ °C). (Reproduced from [90, 92] with permission.) (C) General ribbon model of the coiled-coil peptide with marked Leu residues (space-filling mode).

method to improve the stability of proteins, one has to be cautious. It should be always kept in mind that H → F replacements are not fully isosteric: the C–F bond of 1.39 Å is significantly longer than that of the C–H bond (1.09 Å), although the van der Waals radius of fluorine (1.35 Å) is only 0.15 Å larger than that of hydrogen. In fact, the volume of the trifluoromethyl group was estimated to be closer to that of the isopropyl group [93]. Therefore, such increased steric bulk of the trifluoromethyl group relative to the nonfluorinated methyl group would lead to unfavorable interactions with the surrounding residues in a particular protein interior. Thus, it is conceivable that the $CH_3 \rightarrow CF_3$ replacement might be tolerated at single sites, while global substitution of all Leu/Ile/Val side-chains in larger proteins would significantly enlarge their fluorous hydrophobic core. Namely, numerous slight changes in side-chain volumes and local geometry are accumulated, and these induce considerable local reorganization and disturbance in the side-chain packing of the protein core. In fact, larger proteins such as annexin-V (35 kDa) or GFP (27 kDa) TFL reproducibly led to only marginal replacements of Leu residues (Fig. 7.17) [92].

Thus, the major deterrent for full integration of trifluorinated amino acids into protein structures may be their steric bulkiness and subsequent difficulty in ac-

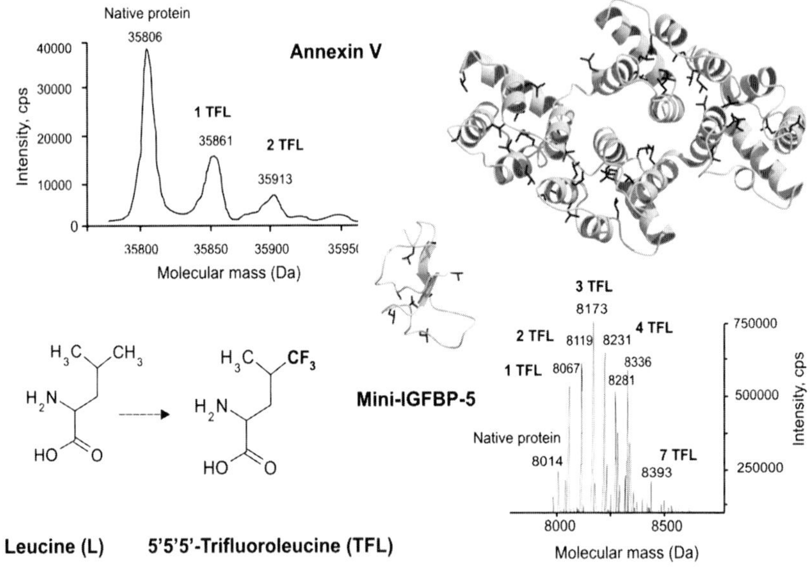

Figure 7.17. Global Leu → TFL substitution in annexin-V (31 Leu residues) and insulin-like growth factor binding protein-5 (mini-IGFBP-5, seven Leu side-chains) as determined by electrospray mass spectrometry. Large proteins like annexin-V with a highly cooperatively packed hydrophobic core that contains many leucyl side-chains are quite sensitive to the accumulation of numerous slight changes in side-chain volumes and local geometry brought by trifluoro moieties. In contrast, a relatively smaller and more flexible (i.e. less cooperative) scaffold like mini-IGFBP-5 is more likely to be more permissive for such substitutions. Even then, full substitution of all leucyl residues was never possible.

commodating them into compact protein cores. Natural scaffold themselves are also not suitable for the accommodation of these amino acids as protein building blocks. Namely, the natural structural frameworks with hydrocarbon cores have been shaped and optimized by billions of years of evolution [94], and are, as such, unsuitable for accommodating a large number of fluorine atoms. Furthermore, the balance of the forces that fold a protein into a unique compact native state is encoded within its amino acid sequence [7]. It is well known that any given sequence will form basic secondary structure formats: helices, sheets, turns and loops. The basic question in this context is what factors determine which of the four available formats will be adopted by an arbitrary chosen chain segment. Thus, the challenge to find, design or select the amino acid sequence to specify a functional fluorous fold is in fact the challenge of *de novo* design of proteins. The dream of the generation of "Teflon" or "nonsticking" proteins, capable of folding and functioning in both organic and aqueous solvents, will be realized by those who first learn how to enable novel repacking of the fluorous protein core, i.e. *de novo* design [95–97].

7.8 Protein Processing, Bioorthogonality and Protein Surface Diversifications

7.8.1 Chemoenzymatical Control of Protein Processing and Posttranslational Modifications

Chemical control of posttranslational modifications (e.g. processing of complex oligosaccharides on glycoproteins) was also attempted in the early experiments with noncanonical amino acid analogs in order to study protein turnover and processing of viral precursors, as well as precursors of other proteins such as insulin, albumin, collagen, etc. [98]. For example, an analog of Asn, β-fluoroasparagine, was described as a potent inhibitor of Asn-linked glycosylation [99]. The inhibitory effects were attributed to the strong electron-withdrawing effect of fluorine on the sugar acceptor site. Another example of an analog used for protein processing studies includes Leu $\rightarrow$ β-hydroxyleucine replacement that makes some secretory proteins unable to transverse the membranes of the endoplasmic reticulum. The mechanism of inhibition includes prevention of the binding of signal recognition particle to ribosomes, when nascent secretory proteins are labeled in this way [100]. Other examples of processing control including analogs such as β-hydroxynorvaline or processes like sulfating Tyr residues are extensively reviewed by Hortin and Boime [98, 99].

It is believed that 1% of the mammalian genome is allocated to encode the most widespread and complex form of posttranslational modification, i.e. glycosylation [101]. For that reason, many efforts were focused on mimicking these modifications in the hope of elucidate their precise biological role as well as to develop suitable pharmaceuticals. One of the basic features of posttranslationally modified proteins is their heterogeneity, i.e. an organism produces almost combinatorial libraries of proteins modified to different extents. This makes it difficult to purify or to have such conjugates as therapeutic proteins for commercial use (since the creation of homogeneous protein-based drugs is the major request). Thus, methods for creating artificial posttranslational modifications steadily gaining at importance. The most common targets for classical protein labeling are primary amines, which are found mainly on Lys residues. They are abundant, widely distributed and easily modified because of their reactivity and their location on the surface of proteins. The second most common target is sulfhydryls, which exist in proteins under reducing conditions, but more often are present in the oxidized form as disulfide bonds. As sulfhydryls are less abundant than primary amines, targeting them results in more specific conjugates.

Thus, naturally occurring Cys as well as Cys inserted by site-directed mutagenesis are especially attractive for specific labeling. In a simple approach, the site-selective formation of a disulphide link by the reaction of synthetic (e.g. glycosylation) or biotinylation reagents with a Cys reside introduced at the protein surface can be achieved [102]. Similarly, by the introduction of the noncanonical amino acid *p*-acetyl-Phe (i.e. ketone functionality) at surfaces of proteins, Schultz and coworkers demonstrated that proteins can be derivatized by aminooxy saccharides in order to generate homogeneous neoglycoproteins [103]. The same group reported the most advanced approach to generate tailor-made glycoproteins in *E. coli* cytoplasm (which lacks the posttranslational machinery for glycosylations) by direct incorporation of the core sugar into the target protein. This approach utilized evolved orthogonal AARS that can preferentially recognize the glycosylamino acid *N*-acetyl glycosamine-*β*-serine, charge it onto cognate tRNA and subsequently incorporate it at the surface of the target protein [104]. After that, glycoprotein can easily be isolated and sugar chains further elongated by using the enzyme glycosyltransferase *in vitro*. Emerging applications for the technologies that might be derived from these approaches include analyses of complex cell–surface interactions, engineering of antigen-presenting cells, development of cancer vaccines and possibly protection against graft rejection.

7.8.2
Staudinger–Bertozzi Ligation and "Click" Chemistry on Proteins

The chemical labeling of proteins continues to be an important tool for the study of their function and cellular fate. Highly desirable *in vivo* labeling requires biocompatible synthetic organic chemistry capable of generating highly selective ligation reactions in an aqueous physiological milieu. In recent years, several different two-step labeling strategies have emerged. They normally rely on the introduction of a bioorthogonal attachment site into a macromolecule followed by ligation of a reporter molecule to this site using bioorthogonal organic chemistry. However, protein labeling in living cells imposes some important restrictions that should be taken into account during the design of the chemical tags. For example, many useful biotinylated constructs, radiolabels and fluorescent tags are not suitable for *in vivo* labeling due to their poor bioavailability (e.g. cell impermeability). To circumvent these difficulties it is necessary to (i) introduce a bioorthogonal attachment site onto the target protein in a living cell and (ii) to attach the desired tag by addition of the reporter moiety, through chemoselective ligation [105].

The alkyne and azide moieties introduced recently as part of aliphatic or aromatic side-chains of noncanonical amino acids became extremely important for the following reasons. (i) They are absent in almost all naturally occurring compounds (i.e. bioorthogonal) [106]. (ii) Despite their high intrinsic reactivity they undergo selective ligation with a very limited number of reaction partners. (iii) They can be delivered into cells in an "amino acid format" and incorporated into target proteins via reprogrammed protein translation. In this context, the rediscovery of the Staudinger ligation [107] (i.e. the smooth reaction of azide with triarylphosphanes) by Saxon and Bertozzi [108] represented a great breakthrough in the field. They improved the method by generating an electrophilic trap, such as an ester moiety, within the structure of the phosphane that would capture the nucleophilic aza-ylide by intramolecular cyclization. This process would ultimately produce a stable amide bond before the competing aza-ylide hydrolysis could take place (Fig. 7.18A). Shortly after, Bertozzi, Raines and coworkers [109, 110] reported a "traceless Staudinger ligation", in which the phosphane oxide moiety is cleaved during the hydrolysis step. Finally, Tirrell and coworkers [111] used murine dihydropholate reductase labeled with the Met surrogate azidohomoalanine as a target for chemoselective modifications of proteins by Staudinger ligation (the protein was labeled with FLAG peptide) (see also Section 1.3, Fig. 1.1).

Another ligation type that can be executed very specifically under aqueous conditions is Huisgen-type [3 + 2] cycloaddition, i.e. a reaction between an azide and a terminal alkyne [112, 113]. These two reactive components are largely inert in physiological environments and thus suitable for such "click chemistry" reactions that additionally require the presence of a copper catalysts. Tirrell and coworkers [114, 115] shown that azides at cell surfaces can be covalently biotinylated via copper-catalyzed triazole formation and subsequently stained with fluorescent streptavidin, facilitating flow cytometric separation of labeled from unlabeled cells. Deiters and coworkers [116, 117] incorporated both *O*-propargyltyrosine and *p*-azidophenylalanine into proteins expressed in yeast. After affinity purification of the target proteins, they were selectively ligated by their azide- or alkyne-functionalized dansyl, fluorescein or PEG derivatives in a Huisgen-type cycloaddition. This allowed for in-gel visualization of the modified target proteins as well as selective PEGylation of proteins for therapeutic applications. In general, these approaches should enable chemical control and a better understanding of the widespread types of posttranslational modifications such as glycosylation, phosphorylation, ubiquitination and methylation, as well as other less common types like prenylation, sulfation or hydroxylation.

7.8.3 Tagging, Caging, Crosslinking and Photoswitching at the Protein Surface

Both classical and genetically encoded protein modifications have as the main goal to prepare conjugates with suitable reporter groups, usually at the protein surface,

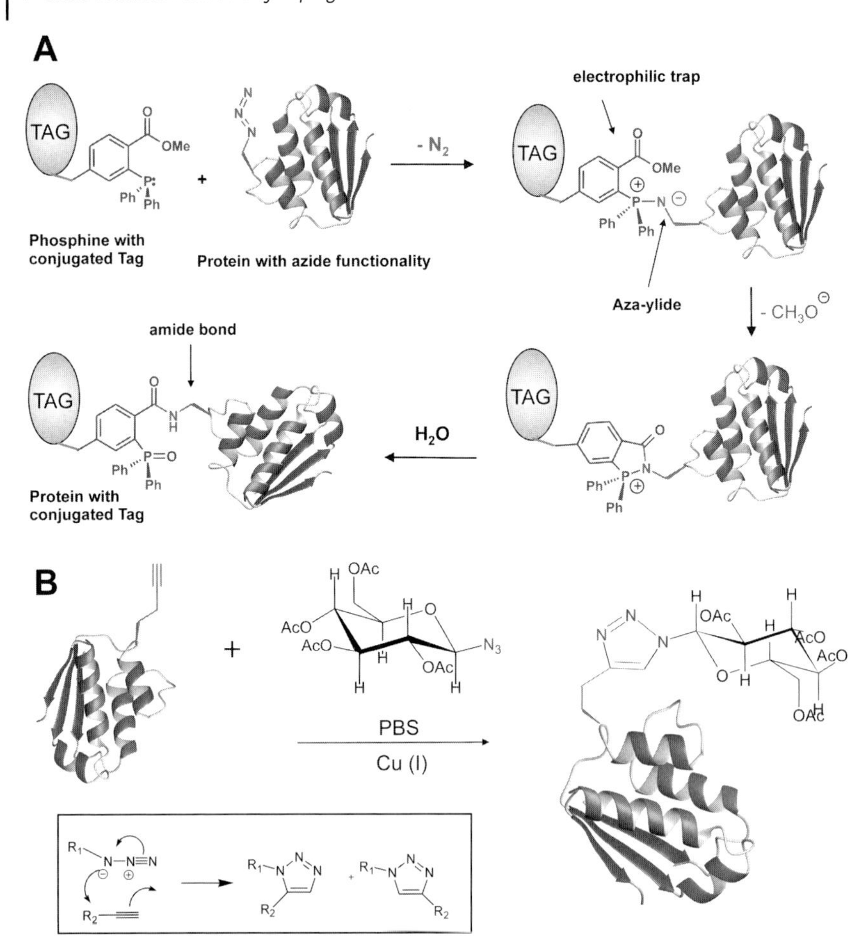

Figure 7.18. (A) Protein with surface decorated with an azide as bio-orthogonal probe can be specifically labeled with the desired tag via Staudinger–Bertozzi ligation. (B) "Clicking" azido-sugar at the protein surface provided with alkyne bioorthogonal functionality via homopropargyl-glycine (red) incorporation. (Insert) The mechanism of copper-catalyzed Huisgen-type cycloaddition.

such as a dye, spin label, affinity tag, recognition motif, reactive group, photo- or thermosensitive group, radioactive isotope, etc. Widely used tags are short oligopeptide stretches that can be selectively isolated through immunoprecipitation (such as the FLAG tag) or affinity chromatography (e.g. Ni-binding His-tag). However, it is highly desirable to introduce fluorescent amino acids, electron acceptors and donors, spin-probe photoswitching, metal binding or amino acids with other special properties at a predesigned site. A special class of noncanonical amino acids with side-chains that are highly reactive under photolytic conditions has been under investigation for decades [118]. Instead of performing classical modification chemistry, codon reassignment experiments with such amino acids can be applied on the proteins of the cell surface. This is a potentially powerful technology through which the surface protein composition of cells can be manipulated without gross changes in the nucleotide compositions of the gene.

Most of the photoreactive amino acids contain an aryl azide group that is chemically inert until it is exposed to UV, causing the formation of short-lived intermediates, normally carbenes or nitrenes, that react with the surrounding functional groups, including amide linkages. Until recently, site-specific incorporation of desired amino acids could only be achieved in *in vitro* translation systems (reviewed extensively in [4, 119–121]). A step further in these strategies is certainly the *in vivo* translation system as a platform for efficient incorporation of such functionalities at protein surfaces. For example, Schultz and coworkers [122] succeeded in incorporating azidophenylalanine at the binding surface of glutathione-S-transferase which upon irradiation with UV light enables *in vitro* crosslinking to another glutathione-S-transferase molecule in dimeric form. These results were successfully reproduced *in vivo* as well by using a different photocrosslinking amino acid, *p*-benzoyl-L-Phe [123]. This certainly opens a perspective for a general *in vivo* translation technique for probing protein–protein interactions in living cells.

7.9 Pharmacologically Active Amino Acids

7.9.1 Bioisosteric Compounds, Antagonists, Agonists and Antimetabolites

Bioisosteric compounds are substances which posses nearly equal molecular shapes and volumes, and a similar distribution of electrons and local geometry. They might affect biochemically associated systems as agonists or antagonists and thereby produce biological properties that are related to each other. For example, many noncanonical amino acids act as antimetabolites. In recent decades their strong antimicrobial, antimycotic and antitumor activities have been extensively studied and demonstrated [124]. It is therefore surprising how little attention is being paid to these interesting pharmacological properties that certainly have a great potential in biomedicine. In general, the roles of noncanonical amino acids, other than participation in protein synthesis, are (i) specific interactions with metabolic and catabolic enzymes (inhibition, activation and modulation), and (ii) interference in amino acid biosynthesis, turnover, transport and storage [65].

It should always be kept in mind that most of the noncanonical amino acids are toxic. This toxicity is often generated by their conversion into toxic substances by a relatively complex metabolic route. Microorganisms and fungi produce many unusual amino acid-containing toxic substances that are physiologically highly active in mammals or other microorganisms, usually in a highly deleterious manner (e.g. cephalosporins or penicillins) [66]. Plants are the most excellent chemists among living organisms because of their remarkable capacity to produce a wide variety of secondary metabolites such as alkaloids, terpenes and tannins to protect themselves from predators, parasites and infection by viruses. Many novel plant amino acids are structurally very similar to canonical ones and have an impressive potential to deleteriously affect the metabolism of other organisms. For example, the amino acid analog mimosine causes loss of hair and wool in cattle and sheep grazing on *Leucaena leucocephala* or *Mimosa pudica*, while numerous selenium analogs of sulfur-containing amino acids are found in a variety of plants that have wide-ranging toxic effects on grazing animals [125].

The particularly well-documented toxicity of 3-fluoro-Tyr, 3-fluoro-Phe and 5-fluoro-Trp is due to the formation of fluoroacetate formed via the dominant Tyr metabolic pathway discovered over 30 years ago [126]. The toxic fluoroacetate is the most ubiquitous among the organofluorine compounds, and has been identified in more than 40 tropical and subtropical plant species [66]. It is also produced by some microorganisms when grown on media containing fluoride. Can such principles from nature be "borrowed" in a way that they enable us to kill tumor cells and spare normal ones by the use of cytotoxic amino acids? Indeed, in drug design for chemotherapy the best way to achieve specificity of action and selectivity of delivery is to use proteins as delivery vehicles (protein shuttles). For example, it is known that lethal metabolic intermediates such as fluorocitrate and fluoroacetate are formed from 3-fluoro-Tyr in mammalian tissues [65]. The same must hold true for tumor cells, if this cytotoxic amino acid were to be specifically delivered to them. When the target tissue is reached and the substituted protein delivered, the cytotoxic amino acid is expected to be set free during internalization and protein turnover. The free cytotoxic amino acid than has two choices: (i) re-incorporation into other proteins (i.e. re-entering protein translation) or (ii) entry into the cellular metabolism. The cytotoxic action most probably follows the general principle formulated by Pattison that "any compound which can form fluoroacetic acid by some simple biochemical process is toxic" [127].

The rationale behind this proposal is the possibility for noninvasive delivery of the substances covalently integrated into a protein (prodrug form) orthogonal in their physiological action to their natural (canonical) counterparts (Fig. 7.19). Substitution of putative shuttle proteins with drug-like amino acids should introduce as few as possible perturbations in terms of structural, functional and immunogenic properties. In other words, the basic advantage of this system should be non-invasiveness, i.e. the cytotoxic amino acid is covalently integrated into the polypeptide in the inactive prodrug form and exerts no toxicity during delivery. This is indeed plausible for isosteric analogs or surrogates since they resemble canonical counterparts in terms of shape, size and chemical properties. Protein structures substituted with such substances are normally indistinguishable from the parent molecules [62, 73]. This approach is expected to introduce as few as possible perturbations in the structural, functional and immunogenic properties of protein carrier, i.e. to provide full noninvasiveness [128]. In other words, with the substitutions of suitable "protein shuttle" candidates like antibodies, cytokines, growth factors or other tissue-specific proteins, there is an opportunity to intervene in a pathological process with a high degree of specificity and minimal perturbation of normal physiological processes (Fig. 7.20). Similarly, a variety of synthetically sub-

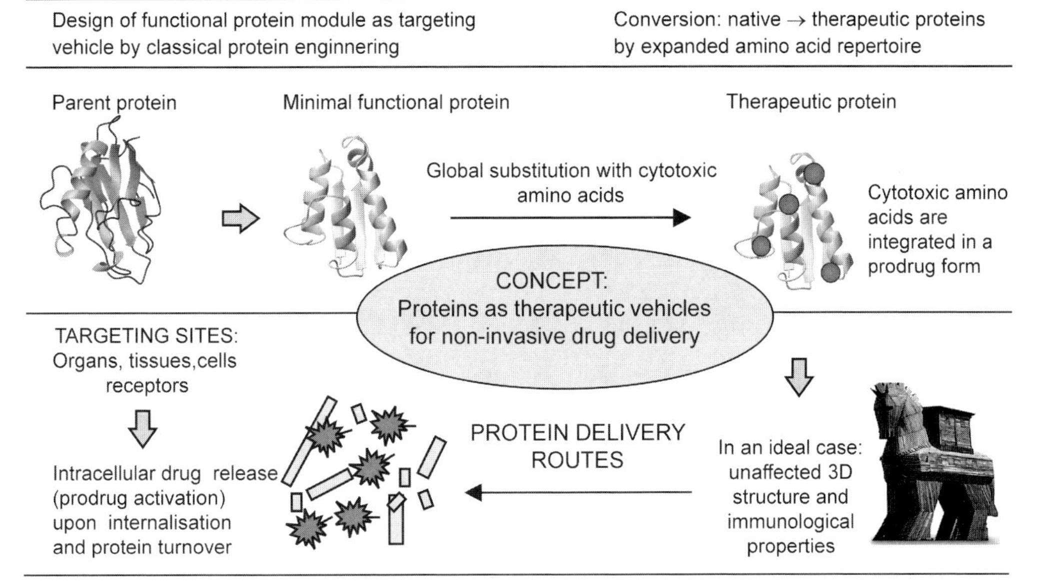

Figure 7.19. Schematic representation of the concept of use of proteins "armed" with cytotoxic amino acids as possible noninvasive carriers for drug delivery. Antibiotic amino acids such as penicillamine, azaleucine, azatyrosine, thiaproline and furanomycin, or strong antimetabolites like fluorinated amino acids or canavanine can be incorporated into proteins. Recombinant proteins that contain such pharmaceutically active amino acids are expected to act as specific delivery vehicle "shuttles", "Trojan horses" or even "magic bullets" due to their potential ability to selectively deliver and target in the human body.

stituted small peptides prepared under different names like "smugglins" or "portable transporters" were used as carriers for some antibacterial amino acids [130]. Taking into account the nearly uniform metabolic roles of amino acids in different tissues or cell types, the delivery of cytotoxic amino acid analogs into cytoplasm might lead to the inhibition of the cellular growth, most probably by blocking cellular signaling, leading eventually in apoptosis or necrosis. Future research should show whether such a concept might be converted into a practical approach to less-toxic cancer therapy.

7.9.2
Neuroactive Amino Acids and their Derivatives

Neuroregulatory amino acids and their derivatives play critical biochemical roles not only in the brain, but also in the whole nervous system. Numerous neurologically active substances overwhelmingly have amino acids as precursors. Glutamic acid is involved in neurotransmission as an excitatory amino acid, while its decarboxylation product γ-amino-butyrate (GABA) is a neuroinhibitory substance, unique

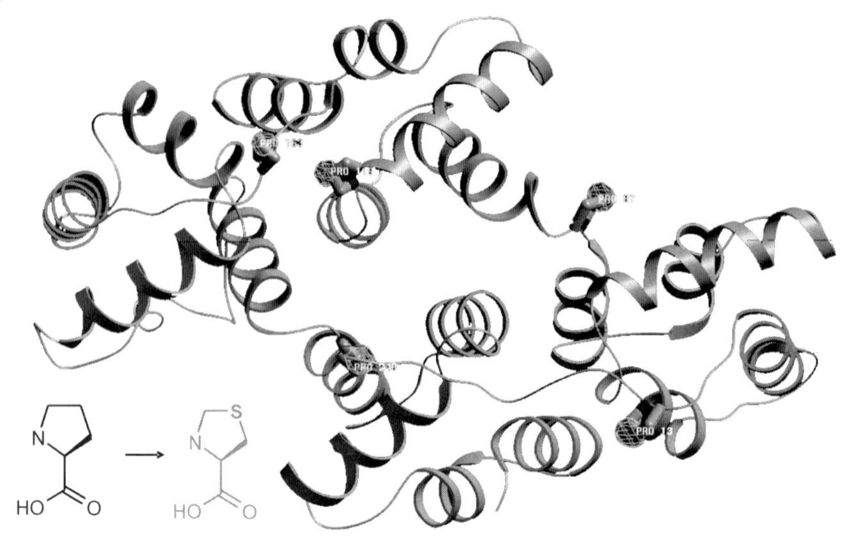

Figure 7.20. Human recombinant annexin-V isomorphously substituted with the pharmacologically active noncanonical amino acid thiaproline [129]. Pharmacologically active, cytotoxic or "diagnostic" amino acids are covalently integrated into the polypeptide in the inactive prodrug form and should exert no toxicity during delivery.

to the brain and central nervous system. Tyrosine is an important factor in the regulation of the synthesis of the catecholamines, dopamine and adrenaline. Thus, the analogs and surrogates of these and other amino acids might exert interesting effects in terms of selective modulations of neurological reactions [131].

The uptake of amino acids into the brain has considerable physiological importance in the control of brain function. The composition of the free amino acid pools of the brain differ largely from those of other tissues. Availability (i.e. uptake) and utilization of such amino acids are often the rate-limiting steps in the neurological actions of their derivatives. For instance, studies of the role of Trp in brain metabolism revealed that the administration of large amounts of Trp in rats led to an increase in the synthesis of the neurohormone 5-hydroxytryptamine (serotonin) [132].

The neurological activity of some noncanonical amino acid analogs, surrogates and their derivatives has been known for a long time. Koe and Weissman [133] discovered that 4-chlorophenylalanine specifically depletes brain serotonin. Stark and Fuller [134] reported the great potency of 3-chlorotyrosine to induce various behavioral effects on experimental dogs. More recently, there have been attempts for systematic structure–function studies on integral membrane proteins such as ion channels and neuroreceptors that mediate synaptic transmission by use of caged special noncanonical amino acids [135]. Interesting Trp surrogates, thienopyrroles, whose translational activity was demonstrated recently [46], have side-chains that act as hallucinogenic substances and serotonin antagonists [136]. The delivery of such a surrogate alone or as a part of a peptide or protein carrier that passes brain–body barriers might provide the brain with a novel substance of anthropogenic origin that exhibits interesting and useful properties.

It could be expected that the brain enzymatic machinery exhibits substrate promiscuity [137] in a similar manner as for enzymes from the translational apparatus or metabolic circuits (see Section 6.2.1). Indeed, such substrate promiscuity is well documented in some systems. For example, well-known enzymes with rather "relaxed" substrate specificity are (i) the aromatic amino acid decarboxylase which is responsible for the conversion of Tyr derivatives into analogs of norepinefrine and (ii) dopamine-β-hydroxylase which converts 5-hydroxytryptophan into serotonin [138]. Furthermore, the storage mechanism between processed (e.g. decarboxylated) canonical amino acids and their amino acid surrogates is also nonselective. This feature of brain enzymes might be used to modulate synaptic activity because noncanonical amino acids may enter some of the metabolic pathways of their canonical counterparts and form metabolic derivatives that might displace natural neurotransmitters. Thus, the arrival of an impulse at the nerve ending or at a synapse triggers the release of a neurotransmitter surrogate stored in vesicle, resulting in different mode of synaptic activity that might also be modulated by such a surrogate neurotransmitter with possible long-term effects on the central or peripheral nervous system.

Since amino acid-based neurohormones are responsible for many aspects of neural and brain function, it is conceivable that alternations in their balance caused by surrogates and analogs released in the brain might even be responsible for different consciousness states. In other words, approaching the chemistry of the brain by amino acid surrogates should provide an efficient and valuable tool not only to study the structural and physicochemical determinants that shape the molecules of thought, memory and sensory perception, but might also be able to act as an effective medicine [139].

References

1 Budisa, N. (2004). Prolegomena to future experimental efforts on genetic code engineering by expanding its amino acid repertoire. *Angewandte Chemie International Edition* **43**, 3387–3428.

2 Hohsaka, T., Sato, K., Sisido, M., Takai, K. and Yokoyama, S. (1993). Adaptability of nonnatural aromatic amino acids to the active center of the *Escherichia coli* ribosomal A-site. *FEBS Letters* **335**, 47–50.

3 Koh, J. T., Cornish, V. W. and Schultz, P. G. (1997). An experimental approach to evaluating the role of backbone interactions in proteins using unnatural amino acid mutagenesis. *Biochemistry* **36**, 11314–11322.

4 Dougherty, D. A. (2000). Unnatural amino acids as probes of protein structure and function. *Current Opinion in Chemical Biology* **4**, 645–652.

5 Munson, M., Balasubramanian, S., Fleming, K. G., Nagi, A. D., O'Brien, R., Sturtevant, J. M. and Regan, L. (1996). What makes a protein a protein? Hydrophobic core designs that specify stability and

structural properties. *Protein Science* **5**, 1584–1593.

6 Bowie, J. U., Reidhaar-Olson, J. F., Lim, W. A. and Sauer, R. T. (1990). Deciphering the message in protein sequences: tolerance to amino acid substitutions. *Science* **247**, 1306–1310.

7 Anfinsen, C. (1973). Principles that govern the folding of protein chains. *Science* **181**, 223–230.

8 Hoetelmans, R. M. W. (1999). Pharmacology of antiretroviral drugs. *Antiviral Therapy* **4**, 29–41.

9 Hattori, M. and Sakaki, Y. (1986). Dideoxy sequencing method using denatured plasmid templates. *Analytical Biochemistry* **152**, 232–238.

10 Zimmermann, J., Voss, H., Schwager, C., Stegemann, J. and Ansorge, W. (1988). Automated Sanger dideoxy sequencing reaction protocol. *FEBS Letters* **233**, 432–436.

11 Prober, J. M., Trainor, G. L., Dam, R. J., Hobbs, F. W., Robertson, C. W., Zagursky, R. J., Cocuzza, A. J., Jensen, M. A. and Baumeister, K. (1987). A system for rapid DNA sequencing with fluorescent chain-terminating dideoxynucleotides. *Science* **238**, 336–341.

12 van Hest, J. C. M. and Tirrell, D. A. (2001). Protein-based materials, toward a new level of structural control. *Chemical Communications*, 1897–1904.

13 Kothakota, S., Mason, T. L., Tirrell, D. A. and Fournier, M. J. (1995). Biosynthesis of a periodic protein containing 3-thienylalanine – a step toward genetically engineered conducting polymers. *Journal of the American Chemical Society* **117**, 536–537.

14 Dougherty, M. J., Kothakota, S., Krejchi, M. T., Zhang, G. H., Mason, T. L., Tirrell, D. A. and Fournier, M. J. (1992). Biosynthesis of new polymers of controlled molecular structure. *Makromolekulare Chemie Macromolecular Symposia* **62**, 225–229.

15 Deming, T. J., Fournier, M. J., Mason, T. L. and Tirrell, D. A. (1996). Structural modification of a periodic polypeptide through biosynthetic replacement of proline with azetidine-2-carboxylic acid. *Macromolecules* **29**, 1442–1444.

16 Deming, T. J., Fournier, M. J., Mason, T. L. and Tirrell, D. A. (1997). Biosynthetic incorporation and chemical modification of alkene functionality in genetically engineered polymers. *Journal of Macromolecular Science Pure and Applied Chemistry* A**34**, 2143–2150.

17 Yoshikawa, E., Fournier, M. J., Mason, T. L. and Tirrell, D. A. (1994). Genetically engineered fluoropolymers – synthesis of repetitive polypeptides containing *p*-fluorophenylalanine residues. *Macromolecules* **27**, 5471–5475.

18 Blundell, T. L. and Johnson, L. N. (1976). *Protein Crystallography.* Academic Press, New York.

19 Cowie, D. B. and Cohen, G. N. (1957). Biosynthesis by *Escherichia coli* of active altered proteins containing selenium instead of sulfur. *Biochimica et Biophysica Acta* **26**, 252–261.

20 Hendrickson, W. A. and Ogata, C. M. (1997). Phase determination from multiwavelength anomalous diffraction measurements. *Macromolecular Crystallography A* **276**, 494–523.

21 Hendrickson, W. A. (2000). Synchrotron crystallography. *Trends in Biochemical Sciences* **25**, 637–643.

22 Bock, A., Thanbichler, M., Rother, M. and Resch, A. (2003). Selenocysteine. In *Aminoacyl-tRNA Synthetases* (Ibba, M., Francklyn, C. and Cusack, S., eds), pp. 320–328. Landes Bioscience, Georgetown, TX.

23 Boles, J. O., Lewinski, K., Kunkle, M., Odom, J. D., Dunlap, R. B., Lebioda, L. and Hatada, M. (1994). Bio-incorporation of telluromethionine into buried residues of dihydrofolate reductase. *Nature Structural Biology* **1**, 283–284.

24 Budisa, N., Steipe, B., Demange, P., Eckerskorn, C., Kellermann, J. and Huber, R. (1995). High-level bio-synthetic substitution of methionine in proteins by its analogs 2-amino-

hexanoic acid, selenomethionine, telluromethionine and ethionine in *Escherichia coli*. *European Journal of Biochemistry* **230**, 788–796.

25 Bae, J. H., Alefelder, S., Kaiser, J. T., Friedrich, R., Moroder, L., Huber, R. and Budisa, N. (2001). Incorporation of beta-selenolo[3,2-*b*]pyrrolyl-alanine into proteins for phase determination in protein X-ray crystallography. *Journal of Molecular Biology* **309**, 925–936.

26 Boles, J. O., Henderson, J., Hatch, D. and Silks, L. A. (2002). Synthesis and incorporation of 6,7-selenatryptophan into dihydrofolate reductase. *Biochemical and Biophysical Research Communications* **298**, 257–261.

27 Kigawa, T., Yamaguchi-Nunokawa, E., Kodama, K., Matsuda, T., Yabuki, T., Matsuda, N., Ishitani, R., Nureki, O. and Yokoyama, S. (2002). Selenomethionine incorporation into a protein by cell-free synthesis. *Journal of Structural and Functional Genomics* **2**, 29–35.

28 Budisa, N. (2003). Expression of "tailor-made" proteins via incorporation of synthetic amino acids by using cell-free protein synthesis. In *Cell Free Protein Expression* (Swartz, J. R., ed.), pp. 89–98. Springer Verlag, Berlin.

29 Drenth, J. (ed.) (1999). *Principles of Protein X-Ray Crystallography*, 2nd edn. Springer Verlag, New York.

30 Xie, J., Wang, L., Wu, N., Brock, A., Spraggon, G. and Schultz, P. G. (2004). The site-specific incorporation of *p*-iodo-L-phenylalanine into proteins for structure determination. *Nature Biotechnology* **22**, 1297–1301.

31 Cordes, M. H. J., Walsh, N. P., McKnight, C. J. and Sauer, R. T. (1999). Evolution of a protein fold *in vitro*. *Science* **284**, 325–327.

32 Koehl, P. and Levitt, M. (1999). *De novo* protein design: plasticity in sequence space. *Journal of Molecular Biology* **293**, 1183–1193.

33 Ravelli, R. B. and McSweeney, S. M. (2000). The "fingerprint" that X-rays can leave on structures. *Structure Folding and Design* **8**, 315–328.

34 Wiehler, J., Jung, G., Seebacher, C., Zumbusch, A. and Steipe, B. (2003). Mutagenic Stabilization of the photocycle intermediate of green fluorescent protein (GFP). *ChemBioChem* **4**, 1164–1171.

35 Rose, G. D. and Wolfenden, R. (1993). Hydrogen-bonding, hydrophobicity, packing, and protein-folding. *Annual Review of Biophysics and Biomolecular Structure* **22**, 381–415.

36 Budisa, N., Huber, R., Golbik, R., Minks, C., Weyher, E. and Moroder, L. (1998). Atomic mutations in annexin-V – thermodynamic studies of isomorphous protein variants. *European Journal of Biochemistry* **253**, 1–9.

37 Moroder, L. (2005). Isosteric replacement of sulfur with other chalcogens in peptides and proteins. *Journal of Peptide Science* **11**, 187–214.

38 Minks, C., Huber, R., Moroder, L. and Budisa, N. (1999). Atomic mutations at the single tryptophan residue of human recombinant annexin-V: effects on structure, stability, and activity. *Biochemistry* **38**, 10649–10659.

39 Gassner, N. C., Baase, W. A., Hausrath, A. C. and Matthews, B. W. (1999). Substitution with selenomethionine can enhance the stability of methionine-rich proteins. *Journal of Molecular Biology* **294**, 17–20.

40 Steinbacher, S., Miller, S., Baxa, U., Budisa, N., Weintraub, A., Seckler, R. and Huber, R. (1997). Phage P22 tailspike protein: crystal structure of the head binding domain at 2.3 Å, fully refined structure of the endorhamnosidase at 1.56 Å resolution, and the molecular basis of O-antigen recognition and cleavage. *Journal of Molecular Biology* **267**, 865–880.

41 Wilson, M. J. and Hatfield, D. L. (1984). Incorporation of modified amino acids into proteins *in vivo*. *Biochimica et Biophysica Acta* **781**, 205–215.

42 Ross, J. B. A., Rusinova, E., Luck, L. A. and Rousslang, K. W. (2000).

Spectral enhancement of proteins by *in vivo* incorporation of tryptophan analogues. In *Trends in Fluorescence Spectroscopy* (Lakowitz, J. R., ed.), vol. 6, pp. 17–42. Plenum, New York.

43 Bronskill, P. M. and Wong, J. T. F. (1988). Suppression of fluorescence of tryptophan residues in proteins by replacement with 4-fluorotryptophan. *Biochemical Journal* **249**, 305–308.

44 Budisa, N. (2004). Protein engineering and design with an expanded amino acid repertoire. *Habilitation thesis*. TU München.

45 Woody, R. W. and Dunker, K. (1996). Aromatic and cysteine side-chain circular dichroism in proteins. In *Circular Dichroism and the Conformational Analysis of Biomolecules* (Fasman, G. D., ed.), pp. 109–157. Plenum Press, New York.

46 Budisa, N., Alefelder, S., Bae, J. H., Golbik, R., Minks, C., Huber, R. and Moroder, L. (2001). Proteins with beta-(thienopyrrolyl)alanines as alternative chromophores and pharmaceutically active amino acids. *Protein Science* **10**, 1281–1292.

47 Dougherty, D. A. (1996). Cation–pi interactions in chemistry and biology: a new view of benzene, Phe, Tyr, and Trp. *Science* **271**, 163–168.

48 Ross, J. B. A., Szabo, A. G. and Hogue, C. W. V. (1997). Enhancement of protein spectra with tryptophan analogs: fluorescence spectroscopy of protein–protein and protein–nucleic acid interactions. *Fluorescence Spectroscopy* **278**, 151–190.

49 De Filippis, V., De Boni, S., De Dea, E., Dalzoppo, D., Grandi, C. and Fontana, A. (2004). Incorporation of the fluorescent amino acid 7-azatryptophan into the core domain 1–47 of hirudin as a probe of hirudin folding and thrombin recognition. *Protein Science* **13**, 1489–1502.

50 Minks, C. (1999). *In vivo* Einbau nicht-natürlicher Aminosäuren in rekombinante Proteine. *PhD thesis*. TU München.

51 Senear, D. F., Mendelson, R. A., Stone, D. B., Luck, L. A., Rusinova, E. and Ross, J. B. A. (2002). Quantitative analysis of tryptophan analogue incorporation in recombinant proteins. *Analytical Biochemistry* **300**, 77–86.

52 Lakowitz, J. R. (1999). *Protein Fluorescence*, 2nd edn. Kluwer, New York.

53 Kishi, T., Tanaka, M. and Tanaka, J. (1977). Electronic absorption and fluorescence-spectra of 5-hydroxytryptamine (serotonin) – protonation in excited-state. *Bulletin of the Chemical Society of Japan* **50**, 1267–1271.

54 Sinha, H. K., Dogra, S. K. and Krishnamurthy, M. (1987). Excited-state and ground-state proton-transfer reactions in 5-aminoindole. *Bulletin of the Chemical Society of Japan* **60**, 4401–4407.

55 Budisa, N., Rubini, M., Bae, J. H., Weyher, E., Wenger, W., Golbik, R., Huber, R. and Moroder, L. (2002). Global replacement of tryptophan with aminotryptophans generates non-invasive protein-based optical pH sensors. *Angewandte Chemie International Edition* **41**, 4066–4069.

56 Pal, P. P. and Budisa, N. (2004). Designing novel spectral classes of proteins with tryptophan-expanded genetic code. *Biological Chemistry* **385**, 893–904.

57 Cubitt, A. B., Heim, R., Adams, S. R., Boyd, A. E., Gross, L. A. and Tsien, R. Y. (1995). Understanding, improving and using green fluorescent proteins. *Trends in Biochemical Sciences* **20**, 448–455.

58 Kummer, A. D., Kompa, C., Lossau, H., Pollinger-Dammer, F., Michel-Beyerle, M. E., Silva, C. M., Bylina, E. J., Coleman, W. J., Yang, M. M. and Youvan, D. C. (1998). Dramatic reduction in fluorescence quantum yield in mutants of green fluorescent protein due to fast internal conversion. *Chemical Physics* **237**, 183–193.

59 Zimmer, M. (2002). Green fluorescent protein (GFP): applications, structure, and related photophysical behavior. *Chemical Reviews* **102**, 759–781.

60 Palm, G. J. and Wlodawer, A. (1999). Spectral variants of green fluorescent

protein. *Methods in Enzymology* **302**, 378–394.

61 Tsien, R. Y. (1998). The green fluorescent protein. *Annual Review of Biochemistry* **67**, 509–544.

62 Bae, J. H., Rubini, M., Jung, G., Wiegand, G., Seifert, M. H. J., Azim, M. K., Kim, J. S., Zumbusch, A., Holak, T. A., Moroder, L., Huber, R. and Budisa, N. (2003). Expansion of the genetic code enables design of a novel "gold" class of green fluorescent proteins. *Journal of Molecular Biology* **328**, 1071–1081.

63 Wang, L., Xie, J. M., Deniz, A. A. and Schultz, P. G. (2003). Unnatural amino acid mutagenesis of green fluorescent protein. *Journal of Organic Chemistry* **68**, 174–176.

64 O'Hagan, D., Schaffrath, C., Cobb, S. L., Hamilton, J. T. G. and Murphy, C. D. (2002). Biosynthesis of an organofluorine molecule – a fluorinase enzyme has been discovered that catalyses carbon–fluorine bond formation. *Nature* **416**, 279–279.

65 Kirk, K. L. (1991). *Biochemistry of Halogenated Organic Compounds.* Plenum, New York.

66 Rosenthal, G. (1982). *Plant Nonprotein Amino and Imino Acids. Biological, Biochemical and Toxicological Properties.* Academic Press, New York.

67 Jimenez, E. C., Craig, A. G., Watkins, M., Hillyard, D. R., Gray, W. R., Gulyas, J., Rivier, J. E., Cruz, L. J. and Olivera, B. M. (1997). Bromocontryphan: post-translational bromination of tryptophan. *Journal of Biological Chemistry* **36**, 989–994.

68 Craig, A. G., Jimenez, E. C., Dykert, J., Nielsen, D. B., Gulyas, J., Abogadie, F. C., Porter, J., Rivier, J. E., Cruz, L. J., Olivera, B. M. and McIntosh, J. M. (1997). A novel post translational modification involving bromination of tryptophan – identification of the residue, L-6-bromotryptophan, in peptides from *Conus imperialis* and *Conus radiatus* venom. *Journal of Biological Chemistry* **272**, 4689–4698.

69 Dunitz, J. D. (2004). Organic fluorine: odd man out. *ChemBioChem* **5**, 614–621.

70 Ring, M. and Huber, R. E. (1993). The properties of beta-galactosidases (*Escherichia coli*) with halogenated tyrosines. *Biochemistry and Cell Biology Biochimie et Biologie Cellulaire* **71**, 127–132.

71 Brooks, B., Phillips, R. S. and Benisek, W. F. (1998). High-efficiency incorporation *in vivo* of tyrosine analogues with altered hydroxyl acidity in place of the catalytic tyrosine-14 of Delta(5)-3-ketosteroid isomerase of *Comamonas (Pseudomonas) testosteroni*: effects of the modifications on isomerase kinetics. *Biochemistry* **37**, 9738–9742.

72 Jackson, D. Y., Burnier, J., Quan, C., Stanley, M., Tom, J. and Wells, J. A. (1994). A designed peptide ligase for total synthesis of ribonuclease-A with unnatural catalytic residues. *Science* **266**, 243–247.

73 Bae, J. H., Pal, P. P., Moroder, L., Huber, R. and Budisa, N. (2004). Crystallographic evidence for isomeric chromophores in 3-fluorotyrosyl-green fluorescent protein. *ChemBioChem* **5**, 720–722.

74 Pal, P. P., Bae, J. H., Azim, M. K., Hess, P., Friedrich, R., Huber, R., Moroder, L. and Budisa, N. (2005). Structural and spectral response of *Aequorea victoria* green fluorescent proteins to chromophore fluorination. *Biochemistry* **44**, 3663–3672.

75 Minks, C., Huber, R., Moroder, L. and Budisa, N. (2000). Noninvasive tracing of recombinant proteins with "fluorophenylalanine-fingers". *Analytical Biochemistry* **284**, 29–34.

76 Wong, C. Y. and Eftink, M. R. (1998). Incorporation of tryptophan analogues into staphylococcal nuclease, its V66W mutant, and Delta 137–149 fragment: Spectroscopic studies. *Biochemistry* **37**, 8938–8946.

77 Pratt, E. A. and Ho, C. (1974). Incorporation of fluorotryptophans into protein in *Escherichia coli*, and their effect on induction of beta-galactosidase and lactose permease. *Federation Proceedings* **33**, 1463–1463.

78 Rehm, T., Huber, R. and Holak, T. A. (2002). Application of NMR in structural proteomics. *Structure* **10**, 1613–1618.

79 Seifert, M. H., Ksiazek, D., Azim, M. K., Smialowski, P., Budisa, N. and Holak, T. A. (2002). Slow exchange in the chromophore of a green fluorescent protein variant. *Journal of the American Chemical Society* **124**, 7932–7942.

80 Renner, C., Alefelder, S., Bae, J. H., Budisa, N., Huber, R. and Moroder, L. (2001). Fluoroprolines as tools for protein design and engineering. *Angewandte Chemie International Edition* **40**, 923–925.

81 Holmgren, S. K., Bretscher, L. E., Taylor, K. M. and Raines, R. T. (1999). A hyperstable collagen mimic. *Chemistry and Biology* **6**, 63–70.

82 Jenkins, C. L. and Raines, R. T. (2002). Insights on the conformational stability of collagen. *Natural Product Reports* **19**, 49–59.

83 Schiene, C., Reimer, U., Schutkowski, M. and Fischer, G. (1998). Mapping the stereospecificity of peptidyl prolyl *cis/trans* isomerases. *FEBS Letters* **432**, 202–206.

84 Dugave, C. and Demange, L. (2003). *Cis–trans* isomerization of organic molecules and biomolecules: implications and applications. *Chemical Reviews* **103**, 2475–2532.

85 Golbik, R., Fischer, G. and Fersht, A. R. (1999). Folding of barstar C40A/C82A/P27A and catalysis of the peptidyl-prolyl *cis/trans* isomerization by human cytosolic cyclophilin (Cyp18). *Protein Science* **8**, 1505–1514.

86 Yoder, N. C. and Kumar, K. (2002). Fluorinated amino acid in protein design and engineering. *Chemical Society Reviews* **31**, 335–341.

87 Marsh, E. N. G. and Neil, E. (2000). Towards the nonstick egg: designing fluorous proteins. *Chemistry and Biology* **7**, R153–R157.

88 Bilgicer, B., Xing, X. and Kumar, K. (2001). Programmed self-sorting of coiled coils with leucine and hexafluoroleucine cores. *Journal of the American Chemical Society* **123**, 11815–11816.

89 Dayhoff, M. O. (1972). *Atlas of Protein Sequence and Structure 5*. National Biomedical Research Foundation, Washington, DC.

90 Tang, Y. and Tirrell, D. A. (2001). Biosynthesis of a highly stable coiled-coil protein containing hexafluoroleucine in an engineered bacterial host. *Journal of the American Chemical Society* **123**, 11089–11090.

91 Duewel, H., Daub, E., Robinson, V. and Honek, J. F. (1997). Incorporation of trifluoromethionine into a phage lysozyme: implications and a new marker for use in protein F-19 NMR. *Biochemistry* **36**, 3404–3416.

92 Budisa, N., Pipitone, O., Slivanowiz, I., Rubini, M., Pal, P. P., Holak, T. A., Huber, R. and L., M. (2004). Efforts toward the design of "Teflon" proteins: *in vivo* translation with trifluorinated leucine and methionine analogues. *Chemistry and Biodiversity* **1**, 1465–1475.

93 Vaughan, M. D., Cleve, P., Robinson, V., Duewel, H. S. and Honek, J. F. (1999). Difluoromethionine as a novel F-19 NMR structural probe for internal amino acid packing in proteins. *Journal of the American Chemical Society* **121**, 8475–8478.

94 Klimov, D. K. and Thirumalai, D. (1996). Factors governing the foldability of proteins. *Proteins: Structure Function and Genetics* **26**, 411–441.

95 Regan, L. and DeGrado, W. F. (1988). Characterization of a helical protein designed from first principles. *Science* **241**, 976–978.

96 Dalal, S., Balasubramanian, S. and Regan, L. (1997). Protein alchemy: changing beta-sheet into alpha-helix. *Nature Structural Biology* **4**, 548–552.

97 Koehl, P. and Levitt, M. (1999). *De novo* protein design: in search of stability and specificity. *Journal of Molecular Biology* **293**, 1161–1181.

98 Hortin, G. and Boime, I. (1983). Applications of amino acid analogs for studying co-translational and

posttranslational modifications of proteins. *Methods in Enzymology* **96**, 777–784.

99 Hortin, G. and Boime, I. (1983). Markers for processing sites in eukaryotic proteins: characterization with amino acid analogs. *Trends in Biochemical Sciences* **8**, 320–323.

100 Blobel, G. (2000). Protein targeting (Nobel Lecture). *ChemBioChem* **1**, 86–102.

101 Lowe, J. B. and Marth, J. D. (2003). A genetic approach to mammalian glycan function. *Annual Review of Biochemistry* **72**, 643–691.

102 Swanwick, R. S., Daines, A. M., Flitsch, S. L. and Allemann, R. K. (2005). Synthesis of homogenous site-selectively glycosylated proteins. *Organic and Biomolecular Chemistry* **3**, 572–574.

103 Liu, H., Wang, L., Brock, A., Wong, C. H. and Schultz, P. G. (2003). A method for the generation of glycoprotein mimetics. *Journal of the American Chemical Society* **125**, 1702–1703.

104 Zhang, Z., Gildersleeve, J., Yang, Y. Y., Xu, R., Loo, J. A., Uryu, S., Wong, C. H. and Schultz, P. G. (2004). A new strategy for the synthesis of glycoproteins. *Science* **303**, 371–373.

105 van Swieten, P. F., Leeuwenburgh, M. A., Kessler, B. M. and Overkleeft, H. S. (2005). Bioorthogonal organic chemistry in living cells: novel strategies for labeling biomolecules. *Organic and Biomolecular Chemistry* **3**, 20–27.

106 Köhn, M. and Breinbauer, R. (2004). The Staudinger ligation – a gift to chemical biology. *Angewandte Chemie International Edition* **43**, 3106–3116.

107 Staudinger, H. and Meyer, J. (1919). Phosphinemethylene derivatives and phosphinimines. *Helvetica Chemica Acta* **2**, 635–646.

108 Bertozzi, C. R. and Saxon, E. (2000). Cell surface engineering by a modified Staudinger reaction. *Science* **287**, 2007–2010.

109 Saxon, E., Armstrong, J. I. and Bertozzi, C. R. (2000). A "traceless" Staudinger ligation for the chemoselective synthesis of amide bonds. *Organic Letters* **2**, 2141–2143.

110 Nilsson, B. L., Kiessling, L. L. and Raines, R. T. (2000). Staudinger ligation: a peptide from a thioester and azide. *Organic Letters* **2**, 1939–1941.

111 Kiick, K. L., Saxon, E., Tirrell, D. A. and Bertozzi, C. R. (2002). Incorporation of azides into recombinant proteins for chemoselective modification by the Staudinger ligation. *Proceedings of the National Academy of Sciences of the USA* **99**, 19–24.

112 Kolb, H. C., Finn, M. G. and Sharpless, B. K. (2001). Click chemistry: diverse chemical function from a few good reactions. *Angewandte Chemie International Edition* **40**, 2004–2021.

113 Tornoe, C. W., Christensen, C. and Meldal, M. (2002). Peptidotriazoles on solid phase: [1,2,3]-triazoles by regiospecific copper(I)-catalyzed 1,3-dipolar cycloadditions of terminal alkynes to azides. *Journal of Organic Chemistry* **67**, 3057–3064.

114 Link, A. J. and Tirrell, D. A. (2003). Cell surface labeling of *Escherichia coli* via copper(I)-catalyzed [3 + 2] cycloaddition. *Journal of the American Chemical Society* **125**, 11164–11165.

115 Link, A. J., Vink, M. K. S. and Tirrell, D. A. (2004). Presentation and detection of azide functionality in bacterial cell surface proteins. *Journal of the American Chemical Society* **126**, 10598–10602.

116 Deiters, A., Cropp, T. A., Mukherji, M., Chin, J. W., Anderson, J. C. and Schultz, P. G. (2003). Adding amino acids with novel reactivity to the genetic code of Saccharomyces cerevisiae. *Journal of the American Chemical Society* **125**, 11782–11783.

117 Deiters, A., Cropp, T. A., Summerer, D., Mukherji, M. and Schultz, P. G. (2004). Site-specific PEGylation of proteins containing unnatural amino acids. *Bioorganic and Medicinal Chemistry Letters* **14**, 5743–5745.

118 Brunner, J. (1993). New photolabeling and crosslinking

methods. *Annual Review of Biochemistry* **62**, 483–514.

119 Hohsaka, T. and Sisido, M. (2002). Incorporation of non-natural amino acids into proteins. *Current Opinion in Chemical Biology* **6**, 809–815.

120 Stromgaard, A., Jensen, A. A. and Stromgaard, K. (2004). Site-specific incorporation of unnatural amino acids into proteins. *ChemBioChem* **5**, 909–916.

121 Anthony-Cahill, J. S. and Magliery, T. J. (2002). Expanding the natural repertoire of protein structure and function. *Current Pharmaceutical Research* **3**, 299–315.

122 Chin, J. W., Martin, A. B., King, D. S., Wang, L. and Schultz, P. G. (2002). Addition of a photocross-linking amino acid to the genetic code of *Escherichia coli*. *Proceedings of the National Academy of Sciences of the USA* **99**, 11020–11024.

123 Chin, J. W. and Schultz, P. G. (2002). *In vivo* photocrosslinking with unnatural amino acid mutagenesis. *ChemBioChem* **3**, 1135–1137.

124 Pine, M. J. (1978). Comparative physiological effects of incorporated amino acid analogues in *Escherichia coli*. *Antimicrobial Agents and Chemotherapy* **13**, 676–685.

125 Hunt, S. (1991). Non-protein amino acids. In *Amino Acids, Proteins and Nucleic Acids* (Dey, P., Harborne, J. and Rogers, L., eds), vol. 5, pp. 55–137. Academic Press, New York.

126 Koe, B. K. and Weissman, A. (1967). Dependence of *m*-fluorophenylalanine toxicity on phenylalanine hydroxylase activity. *Journal of Pharmacology and Experimental Therapeutics* **157**, 565–573.

127 Pattison, F. L. M. (1953). Toxic fluorine compounds. *Nature* **172**, 1139–1141.

128 Minks, C., Alefelder, S., Moroder, L., Huber, R. and Budisa, N. (2000). Towards new protein engineering: *in vivo* building and folding of protein shuttles for drug delivery and targeting by the selective pressure incorporation (SPI) method. *Tetrahedron* **56**, 9431–9442.

129 Budisa, N., Minks, C., Medrano, F. J., Lutz, J., Huber, R. and Moroder, L. (1998). Residue-specific bioincorporation of non-natural, biologically active amino acids into proteins as possible drug carriers: structure and stability of the per-thiaproline mutant of annexin-V. *Proceedings of the National Academy of Sciences of the USA* **95**, 455–459.

130 Payne, J. W. (1986). Drug delivery systems: optimizing the structure of peptide carriers for synthetic anti-microbial drugs. *Drugs Experimental and Clinical Research* **12**, 585–594.

131 Evans, C. S. and Bell, E. A. (1980). Neuroactive plant amino acids and amines. *Trends in Neurosciences* **3**, 70–72.

132 Montenero, A. S. (1978). Toxicity and tolerance of tryptophan and its metabolites. *Acta Vitaminologica et Enzymologica* **32**, 188–194.

133 Koe, B. K. and Weissman, A. (1966). Marked depletion of brain serotonin by *p*-chlorophenylalanine. *Federation Proceedings* **25**, 452.

134 Stark, P. and Fuller, R. W. (1971). Behavioral and biochemical studies with *para* chlorophenylalanine pCPA 3-chlorotyrosine and 3-chlorotyramine. *Federation Proceedings* **30**, A504.

135 Nowak, M. W., Gallivan, J. P., Silverman, S. K., Labarca, C. G., Dougherty, D. A. and Lester, H. A. (1998). *In vivo* incorporation of unnatural amino acids into ion channels in *Xenopus* oocyte expression system. *Methods in Enzymology* **293**, 504–529.

136 Julius, D. (1991). Molecular biology of serotonin receptors. *Annual Review of Neuroscience* **14**, 335–360.

137 Copley, S. D. (2003). Enzymes with extra talents: moonlighting functions and catalytic promiscuity. *Current Opinion in Chemical Biology* **7**, 265–272.

138 Stryer, L. (2001). *Biochemistry*. Freeman, New York.

139 Dougherty, D. A. (1998). Is the brain ready for physical organic chemistry? *Journal of Physical Organic Chemistry* **11**, 334–340.

Epilogue

Over the past few decades, all of the important components of the protein synthesis apparatus have been more or less described, with the mechanistic structures of the ribosome crowning these efforts. These spectacular events have been gained at the height of predominantly mechanistic, reductive and dry descriptivist science that has dominated current biological and biochemical research. In parallel with these developments, a series of landmark experiments preformed over the last decade mainly by protein chemists (who were largely unnoticed by the public) yielded revolutionary possibilities for reprogramming the translation machinery and have subsequently opened a perspective to alter genetic code itself. This has marked the rebirth of a science dominated by conceptual intuition and inventive spirit, and one might hope that this will be dominant research trend in the coming decades.

The genetic code is a set of rules that govern the translation of the genetic information (saved in the DNA/RNA) into protein sequences. One notable property of the code is that all living beings perform their own protein synthesis with an invariant set of 20 canonical amino acids. However, many interesting amino acids containing fluorine, chlorine, bromine, iodine, boron, selenium, silicon or functional chemical groups (e.g. silyl, cyano, nitrozo, azido, nitro, etc.) are missing from this repertoire. It is therefore not difficult to imagine the possibilities protein synthesis might offer with such substances. This would include innovative technologies for design and production of novel classes of therapeutic and diagnostic proteins, non-invasive protein-based sensors, environmentally friendly or non-polluting biomaterials, etc.

Early experiments performed more than half century ago (with the intention of studying mainly cellular metabolism, physiology and mechanisms of protein synthesis) had shown that the number of amino acids that could be used as substrates for ribosome-mediated protein synthesis could be expanded far beyond the canonical 20. A recent renaissance of these experiments made it obvious that the interpretation of the genetic code can be changed (or its coding capacities expanded) in a controlled manner under defined experimental conditions. As a result of these efforts, engineering of the genetic code as a new research field emerged at the intersection between synthetic chemistry and molecular biology.

While the deciphering of the genetic code in 1960s was a result of intensive interactions between organic chemistry and genetics, the current marriage between

Engineering the Genetic Code. Nediljko Budisa

ISBN: 3-527-31243-9

these disciplines will yield its further evolution. Indeed, the present-day structure of the genetic code offers sufficient room for efficient repertoire expansion. However, such tailored (or target-engineered) genetic code will not be a result of environmentally driven natural selection, but rather of conscious human effort.

The conservation of the present-day code could considered to be the result of evolutionary historical developments that shaped its structure through a free interplay of chance and necessity. The current code, with its conserved structure, could be the "best possible" for all living beings on Earth. However, it is not the best possible for our technical and technological needs, requirements and demands. In this context, successful reprogramming of the translational apparatus through code engineering is the main prerequisite for biotechnology in the future. However, this should be not the only reason to challenge the conserved structure of the code. The evolution of living beings based on the conserved code brought us here not only to continue this process, but also to re-shape the genetic code and even living cells themselves, and to take their further evolution in our own hands.

Martinsried, June 2005 *Nedilijko Budisa*

Index

Engineering the Genetic Code. Nediljko Budisa

ISBN: 3-527-31243-9

c

g

o